The
Plant Disease Clinic
and
Field Diagnosis
of Abiotic Diseases

Malcolm C. Shurtleff
Professor Emeritus, University of Illinois
Urbana-Champaign

and

Charles W. Averre III
Professor Emeritus, North Carolina State University
Raleigh

APS PRESS
The American Phytopathological Society
St. Paul, Minnesota

Library of Congress Catalog Card Number: 96-85526
International Standard Book Number: 0-89054-217-1

Printed in the United States of America on acid-free paper

The American Phytopathological Society
3340 Pilot Knob Road
St. Paul, MN 55121-2097, USA

Preface

This book is written for plant professionals, including county agents, plant science specialists, plant diagnosticians, students in plant disease diagnosis classes, agribusiness personnel, agricultural inspectors, private consultants, and others, who diagnose plant health problems, disorders, and diseases in the field and in plant clinics. The person should have considerable training in botanical and related sciences and experience in growing plants. The book should help diagnosticians and field personnel improve their diagnostic skills and serve as a useful reference.

This book focuses on abiotic diseases or plant disorders that are commonly diagnosed by specialists in agronomy, forestry, and horticulture in the field and in clinics separate from those in which plant pathogens are isolated and identified. Its primary intent is to present basic field and laboratory methods for diagnosing plant diseases and to be a source of needed information. It contains illustrations and descriptions of many plant pathogens and other causal agents of plant diseases; suggestions for designing, equipping, and operating a modest plant disease clinic; keys for identifying many plant pathogens; references to literature on diagnosing plant problems; measurements and conversions for English and metric units; a glossary of terms; a comprehensive index; and tables on plant tolerances to air pollution, low temperatures, soil soluble salts, and pH. The book should be useful as a text for teaching diagnosis of abiotic plant diseases.

The material presented herein was obtained from original works of the authors, research articles, monographs, compendia, bulletins, books, and unpublished reports. Sources of borrowed material and contributions from colleagues are appreciated and acknowledged. All the methods described have been personally evaluated by one of the authors or in a few cases by respected colleagues; all have potential use for diagnosticians. Techniques for identifying plant diseases and their causal agents are constantly being improved, and new ones are introduced yearly; those described herein are not intended to be all inclusive.

Rapid advances are being made in methods of detecting some plant pathogens associated with host tissue, and these methods are useful in some plant disease-diagnosing situations. In general, they detect (within limits) specific pathogens if their concentrations and other factors are adequate. In some cases, they lack specificity, while in others the proper biotype of the pathogen must be present. Several of these methods are very useful and economical for routine assaying of large numbers of plant samples for a specific pathogen, and their use is suggested in quarantine, certification, and integrated pest management programs. Kits are available commercially for numerous pathogens, and the number of pathogens for which they are available is increasing rapidly. They are not now widely used in plant clinics for routine diagnoses because of the high cost and time input associated with single assays.

The authors are indebted to many staff and family members, friends, and organizations for their advice, encouragement, patience, and support in the preparation of this book. At the risk of thoughtless omissions, the authors specifically thank the administrative and support staffs of our universities, our cooperative extension services and departments, and our secretaries. At North Carolina State University, we especially thank the following administrators for support and encouragement: Robert Aycock, Chester D. Black, O. W. Barnett, and George Hyatt, Jr.; and for word processing and proofreading, Joanne B. Averre. Colleagues who provided helpful comments, ideas, and encouragement include Thomas E. Creswell, Alice M. Hisada, W. R. Jester, Ronald K. Jones, and numerous county agents, associates, and friends. We also thank Janet Shurtleff, a weed scientist with the North Carolina Department of Agriculture, for invaluable assistance in preparing the section on herbicide injury and proofreading this and other sections; A. Richard Bonanno, horticulturist at North Carolina State University, for his numerous suggestions in the herbicide injury section; and Daniel F. Marion, plant pathologist with the State University of New York, for critically reviewing the book. At the University of Illinois, we greatly appreciate the thoughtfulness of Richard E. Ford, H. Walker Kirby, Donald G. White, and former secretary Patricia C. Novak. We also appreciate the reviews of Gary W. Simone, extension plant pathologist at the University of Florida, and members of the APS Press review board. Special thanks go to Lenore Gray, who did the line drawings.

Any suggestions that readers may have for improving and making this volume more useful will always be welcome.

Malcolm C. Shurtleff
Charles W. Averre III

How to Use This Book for Diagnosing Plant Diseases

This book is designed for use with other books, bulletins, and references that describe the common diseases and disorders of specific crops and plants. These publications usually abound in extension offices and libraries of farmers, growers, agricultural suppliers, and plant professionals. Exceptionally valuable books include Flint's *Pests of the Garden and Small Farm*, Westcott's *Plant Disease Handbook*, *The Ortho Problem Solver*, Wellman's *Tropical American Plant Diseases*, and the American Phytopathological Society's compendia series. Many others are listed in Table 2.5 at the end of Chapter 2. This book provides basic information to diagnose or characterize a large proportion of plant problems that are routinely encountered in the field and in plant clinics. In cases in which the problem has not been diagnosed but has been characterized, more specialized texts such as the American Phytopathological Society's compendia may have to be used.

The usual and often best approach for diagnosing plant diseases is to ask, "Is this disease A or B caused by X and Y, respectively?" By referring to the index under causal agent X or the common name of the plant, the description of the causal agent (biotic or abiotic) and characteristics of the disease can be found. If they don't match those associated with the specimen, try disease B. Often with infectious diseases, this is done very rapidly by observing the pathogen.

Another approach for diagnosing plant diseases is to ask, "What is this disease, or what is this microorganism I see?" However, this approach is invariably time consuming and laborious and may involve a research project involving Koch's postulates to prove the causal agent. This is seldom necessary or done in plant clinics.

In both cases, the characteristics of the disease observed, such as symptoms, field distribution, and time of occurrence, must be determined to narrow the probable cause to a few possibilities. This is done by consulting tables, charts, keys, and figures in Chapter 4, Diagnosing in the Field. Chapters 4 and 5 give information for diagnoses of abiotic diseases such as those caused by air pollution, herbicides, and soil soluble salts. Chapter 5 also discusses methods for observing pathogens in host tissues. If a nematode, microorganism, or virus is associated with a disease, the diagnostic methods, keys, and illustrations found in the various texts listed in Chapter 2 can be used to identify pathogenic types.

A tree with 39 diseases. How many can you identify?

All would not occur on the same plant. However, each disease on a plant must be diagnosed in order to implement effective management steps.

 1 Root-lesion nematode injury
 2 Root-knot nematode injury
 3 Root-pruning nematode injury
 4 Stubby-root nematode injury
 5 Root rot
 6 Crown gall
 7 Fruit bodies of Armillaria root rot fungus
 8 Fruit body of Ganoderma wood and root rot fungus
 9 Fruit bodies of Phellinus (Fomes) wood rot fungi
10 Trunk canker
11 Cedar quince rust on hawthorn
12 Cedar hawthorn rust
13 Cedar apple rust
14 Mosaic
15 Downy mildew
16 Apple scab
17 Leaf spot
18 Powdery mildew
19 Black knot of plum
20 Wetwood (slime flux)
21 Fire blight
22 American mistletoe
23 2,4-D herbicide injury
24 Witches'-broom
25 Fruit rot of apple
26 Overwintering canker of fire blight
27 Wilt
28 Leaf curl or blister of peach, cherry, or plum
29 Leaf blister of oak
30 Sooty blotch and flyspeck of apple
31 Leaf blotch
32 Shot hole
33 Anthracnose
34 Ringspot
35 Sooty mold
36 Tar spot
37 Leaf scorch
38 Apple scab on fruit
39 Twig and branch canker

Contents

Color plates follow page 118

CHAPTER 1

Introduction to Plant Disease Diagnosis

Reasons for Diagnosing Plant Diseases

"A plant disease can be defined as a malfunctioning of host cells and tissues that results from their continuous irritation by a pathogenic agent or environmental factor and leads to the development of symptoms. Disease is a condition involving abnormal changes in the form, physiology, integrity or behavior of a plant. Such changes may result in partial impairment or death of the plant or its parts" (Agrios, 1988). A similar definition is given in the glossary. "Plant disorder" is a more restrictive term and can be defined as an abiotic disease, a harmful, nonpathogenic deviation from normal growth.

Plant diseases and disorders are usually diagnosed so that a grower can take corrective measures (disease management or control tactics) to reduce economic or aesthetic losses on existing plants or future plantings. The success of management measures usually is dependent on a prompt and accurate diagnosis and timely corrective actions. Good information on appropriate management measures is available in publications from land-grant universities, local county extension offices, federal and state departments of agriculture and natural resources, and plant disease and nematode societies and from sources in the private sector such as crop and landscape consultants, foresters, manufacturers of pesticides, and seed or plant suppliers. The diagnostician should obtain as many publications as possible.

The Diagnostician

The diagnosing of plant diseases is a scientific art that is enhanced with experience and constant study. Pertinent scientific technology and principles must be blended with practical knowledge of the host and grower practices along with recent growing conditions. Research experts in specific areas are constantly consulted because a plant disease diagnostician cannot be expert in all facets of crop husbandry or plant peculiarities, postharvest requirements, plant nutrition, soils, entomology, horticulture, forestry, weed science, air pollution, plant physiology, mycology, bacteriology, nematology, virology, microbiology, and molecular biology.

To interpret plant symptoms properly and relate them to causal agents, the diagnostician must understand basic plant anatomy, physiology, and nutrition; these are well described in elementary botany texts and are generalized in Figure 1.1. If any of the inputs (water, nutrients, oxygen, carbon dioxide, and light) are inadequate or in excess, the plant or its parts develop disease symptoms often called disorders. Plant functions (processes) occur in microscopic cells in specific areas of the plant tissues. Significant injury to cells by a pathogen or toxin, starvation, a toxic chemical in the air or soil, temperature extremes, lightning, fire, or mechanical force will impede one or more plant processes and cause microscopic and macroscopic symptoms (Plates 1–6). The injury may occur at a distant point. For example, symp-

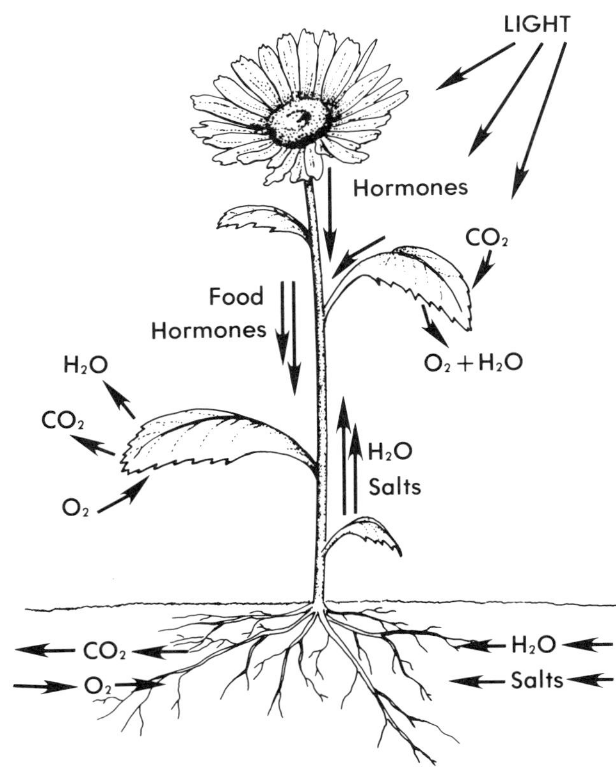

Fig. 1.1. Basic plant structures and functions that the diagnostician must understand to relate symptoms and cause.

toms on a leaf may be caused by injury to cells in the roots or stems. Table 4.6 lists symptoms and possible causes.

The diagnostician need not have expertise in plant disease management. After a plant disease has been properly diagnosed, information about appropriate management procedures is available in numerous state, federal, and industry publications and books. However, selection of an appropriate management procedure for a particular plant disease situation and suggestions to the grower should be deferred to a licensed or certified field expert such as a county agent, extension plant pathologist, or consultant. This is especially important if pesticides are involved. Ultimately, the grower must decide what course of action to take.

Diversity of Clinic Specimens

The plant species and problems received by plant clinics are diverse. For example, in 1984, the Plant Disease and Insect Clinic operated by North Carolina State University received 5,124 plants for disease diagnosis and 2,070 insects for identification. In 1993, the totals were 4,924 and 1,489, respectively. Most of the plants (32–50%) were ornamentals (Table 1.1). Of all plant problems diagnosed in 1984, the causes were as follows: 9% insects, 42% pathogens, and 28% abiotic agents; 21% were not diagnosed or were listed as "other plant problems" for which there was no definite answer. In 1993 (January through 18 November), the groupings were remarkably similar, with 41% of the specimens having a pathogen-induced disease (Table 1.2). Commonly diagnosed abiotic agents were air pollution (Plate 4), high soil soluble salts, unfavorable soil pH, nutritional problems (Plate 6), lack of or excess soil moisture, "winter kill," and pesticide injury, especially from herbicides (Plate 5). Major reasons for not diagnosing problems include inadequate specimens, deterioration of specimens during shipment, and lack of information concerning the specimen. Depending on the crop, the proportion of problems induced by pathogens can vary greatly (Table 1.2).

Table 1.1. Percentage of plant specimens from different crops received by the North Carolina State University Plant Disease and Insect Clinic during 3 years[a,b]

Crop	1981	1984	1993
Field crops	19	17	12
Forages	<1	<1	<1
Fruits and nuts	11	11	7
Ornamentals	32	33	50
Small grains	3	2	1
Turf	5	5	6
Vegetables	18	18	14
Trees	12	13	6
Miscellaneous	<1	1	4

[a] R. K. Jones, T. C. Creswell, and R. W. Kriz (*personal communication*).
[b] The total number received was 4,847 in 1981, 5,124 in 1984, and 4,924 in 1993. These numbers do not include insect samples received for identification by the clinic or nematode samples for assay by the North Carolina Department of Agriculture.

These figures frequently change drastically from year to year. For example, the high number of pathogen-induced diseases of peanuts in 1984 dropped in succeeding years as peanut growers became familiar with black root rot, a relatively new disease.

The numbers and types of specimens diagnosed will naturally vary from state to state and country to country, depending on the crops and plants grown, but will probably be similar to the percentages given for the North Carolina Plant Disease and Insect Clinic.

Plant Disease Characteristics

Plant Disease Definition

In the broadest sense and from a grower's point of view, a plant disease or disorder is any condition that prevents the normal growth and development of a plant and *usually* reduces its economic or aesthetic value. This definition has a large subjective facet relative to what is "normal." It stresses that disease is a continuing process involving time. It includes as a cause anything that can incite disease. A disease and the causal agent are not synonymous and should not be confused.

Two broad categories of causal agents of plant diseases are generally recognized: 1) "pathogens" or "biotic agents" encompass microorganisms, viruses, and other infectious agents and 2) "abiotic," "noninfectious," "nonparasitic," and "physiogenic" agents encompass a wide range of unfavorable growing conditions; an excess, deficiency, unavailability, or improper balance of temperature, light, air movement, air humidity, water, or essential soil elements; soil moisture-oxygen disturbances; injurious impurities in the air or soil; soil grade changes; girdling roots; mechanical and electrical agents; fire; lightning; and unfavorable preharvest or storage conditions (Plates 1–6).

It is important that the two broad categories of causal agents (pathogens and abiotic agents) be recognized because their diagnosis and management are different. Abiotic problems do not spread from diseased to healthy plants. They arise, often quite suddenly, at about the same time on a variety of different plant species growing in a given area or environment (see Chapter 4).

Plant Disease Signature

The concept of disease "signature" encompasses all factors and events, including plant symptoms, associated with a plant disease. It may include time of occurrence, field distribution, recent weather conditions, soil type, field topography, spectrum of plants involved, odors, and tastes plus external and internal plant symptoms, which include peculiarities in anatomy, cells, and physiology and the presence of toxic chemicals and microorganisms. In some cases, a single component of a disease signature is sufficient for an accurate diagnosis. In most cases, several disease signature components are needed to complete the diagnosis (Plates 1–6).

Number of Plant Diseases

The number of plant diseases is far too large to describe each in a text. Consider the combinations between plants and pathogens: there are about 300,000 known vascular plants and some 22,000 known plant pathogens, so the number of plant diseases would exceed 6.6 billion without counting those caused by abiotic agents!

Each cultivated plant species usually has 10 to 100 or more described diseases. Wild species would probably have similar numbers if extensively cultivated. For major crops such as maize (corn), tobacco, wheat, apple, and potato, the numbers of diseases reported are much higher, especially in the tropics (Table 1.3). The diseases of major crops have been studied extensively, and excellent books, monographs, bulletins, and compendia are available (see Table 2.5 at the end of Chapter 2). For most noncultivated plants and many "minor" crops and ornamentals, however, such books are not available. Fortunately, the descriptions and diagnostic procedures for well-described diseases are useful for diagnosing poorly described or undescribed diseases, regardless of the crop or plant species.

Concepts in Diagnosing Plant Diseases

It is useful to recognize two methods for diagnosing plant disease: method 1 confirms a preliminary diagnosis, and method 2 identifies an unknown or undescribed disease. The specific plant problem, expertise of the diagnostician, time constraints, resources available, and required level of diagnostic reliability (confidence) dictate which approach is appropriate.

Both methods require a grasp, even if incomplete, of the disease signature in the field and an understanding of plant

Table 1.2. Grouping of disease agents on major crops diagnosed by the North Carolina State University Plant Disease and Insect Clinic in 1984[a]

	Samples received		Diagnosis (%)								
Plant	Number	Ratio[b]	Ins[c]	Nem[d]	Fun[e]	Bac[f]	Vir[g]	Soil[h]	Pes[i]	Wea[j]	Other[k]
Alfalfa	32	1.3	7	0	69	0	0	0	0	0	24
Apple	125	1.6	15	0	56	1	0	0	0	0	27
Azalea	215	1.5	9	17	32	0	0	16	0	9	17
Bean	124	1.2	14	5	31	0	9	23	5	5	8
Boxwood	255	1.6	20	30	13	0	0	23	<1	1	12
Corn	118	1.2	6	15	8	<1	<1	45	19	0	5
Centipede grass	109	1.5	5	36	5	0	0	25	0	23	6
Dogwood	65	1.1	6	0	22	0	0	13	0	7	52
Fescue	73	1.4	7	0	47	0	0	30	0	0	16
Fir	105	1.0	4	0	31	0	0	31	2	2	30
Grape	65	1.1	13	0	36	3	3	6	0	7	32
Holly, Japanese	179	1.6	10	15	15	0	0	20	0	4	36
Juniper	76	1.1	22	0	15	0	0	15	0	4	44
Peanut	78	1.2	6	1	59	0	0	9	11	0	14
Pepper	81	1.0+	1	4	30	7	5	34	8	0	11
Peach	78	1.0+	12	4	31	6	0	13	10	4	20
Pine	111	1.1	14	<1	18	0	0	6	0	23	38
Rhododendron	107	1.0+	10	0	58	0	0	0	0	9	23
Rose	43	1.3	15	7	22	0	4	5	11	11	25
Soybean	141	1.2	2	23	39	0	0	24	6	0	6
Strawberry	80	1.2	6	13	44	0	1	13	4	7	12
Sweet potato	46	1.1	8	8	39	12	0	8	0	0	25
Tobacco	481	1.1	1	4	14	6	10	23	15	14	13
Tomato	215	1.0+	3	2	34	17	3	21	6	<1	13
Wheat	53	1.1	5	0	38	0	0	44	2	8	3
Average 1984	. . .	1.2	9	8	31	2	1	18	4	6	21
Average 1993	. . .	. . .	1	4	32	3	2	23	4	6	25

[a] The total number of specimens received was 5,124, excluding insect identifications by the clinic and nematode samples received by the North Carolina Department of Agriculture. In 1993, 4,924 plant specimens were submitted, and an approximate summary comparison for groups is given.
[b] Number of diagnoses to number of samples received (a sample may have more than one diagnosable problem).
[c] Insects, mites, and other animal pests associated with plant injury.
[d] Nematodes associated with plant injury.
[e] Fungi.
[f] Bacteria.
[g] Viruses and related pathogens.
[h] Soil pH, soluble salts, and in some cases nutritional disorders.
[i] Pesticide injury, especially herbicides.
[j] Adverse weather and air pollution, especially ozone.
[k] Inadequate samples, not diagnosed, and miscellaneous injuries and disorders.

symptoms and signs, general causes, and crop husbandry. For example, wilting of foliage may be caused by an infection of the roots or stems by microorganisms (bacteria, fungi, or nematodes) or by adverse conditions such as "wet feet"; drought; low or high temperatures; high levels of soluble salts in the soil; insect larvae feeding on the roots, stems, or leaves; or the presence of toxic substances in the air, soil, or growing medium. Obviously, the search for the specific cause of wilted foliage is focused on these areas. Many causes can be eliminated quickly by noting recent weather conditions, determining the soluble salts level in the soil, and noting the appearance of roots and internal tissues of the stem. In other cases, specific techniques are used to reject or confirm the cause. These techniques are described in later chapters.

Method 1: To Verify a Preliminary Diagnosis

This method is routinely used in plant clinics to diagnose *described* diseases by accepting or rejecting a preliminary diagnosis. A preliminary diagnosis is based on experience in recognizing certain diseases, its similarity to published descriptions and illustrations, or just hearsay and "hunch." The preliminary diagnosis is then verified or rejected by observing the presence of all, or as many as feasible, components of the disease signature, especially the causal agent. For example, in many diseases caused by fungi, the presence of plant symptoms (an indication of disease by reaction of the host) and close observation with a hand lens for visible signs of the pathogen or its parts (e.g., characteristic mycelium, fruiting structures, spores, and exudate) is sufficient for an accurate or useful diagnosis. In other cases, the use of a compound microscope is required. Consistency with other facets of the disease signature, such as time and place of occurrence, field distribution, recent weather conditions, and other hosts involved, increases the reliability of the diagnosis.

Method 2: To Identify an Unknown Disease

This method must be used when a disease has *not* been described (is unclear or not available) or when the sus-

Table 1.3. Numbers of pathogens causing disease on some commonly grown plants in the United States and tropical America[a]

Plant	Number	Plant	Number	Plant	Number
African violet	12	Fig	49	Pepper	64
Apple	112	Forsythia	14	Petunia	25
Arborvitae	29	Fuchsia	13	Philodendron	12
Ash	43	Gardenia	19	Pine	123
Asparagus	19	Geranium, florist's	32	Plum	59
Azalea	44	Grape	54	Poinsettia	20
Banana	(220)	Hackberry	26	Poplar	58
Barberry	17	Hawthorn	39	Potato	(175) 77
Beans	(280) 82	Hemlock	25	Privet	27
Beech	30	Hickory	44	Pyracantha	12
Begonia	20	Holly	57	Radish	30
Blueberry	67	Honey locust	24	Raspberry	52
Boxwood	28	Hydrangea	22	Rhododendron	46
Cabbage	53	Iris	15	Rhubarb	26
Cacao	(52)	Ivy	14	Rice	(600)
Caladium	9	Jasmine	13	Rose	83
Calendula	20	Juniper	63	Sage	24
Camellia	34	Kalanchoe	7	Snapdragon	30
Carnation	47	Lantana	7	Spinach	30
Carrot	42	Lettuce	35	Strawberry	66
Celery	40	Lilac	27	Sugarcane	(460)
Cherry	79	Lily	43	Sunflower	42
Citrus	(250)	Lupine	50	Sweet pea	37
Coconut	(35)	Magnolia	48	Sweet potato	(187) 48
Coffee	(400)	Maple	99	Tobacco	(221)
Corn (maize)	(130)	Marigold	22	Tomato	(278) 108
Crape-myrtle	9	Mimosa (*Albizia*)	9	Tulip	18
Crocus	5	Morning-glory	17	Tulip-tree	28
Cucumber	43	Muskmelon	45	Turnip	35
Cucurbits	(110)	Narcissus	28	Viburnum	34
Cypress	13	Nasturtium	14	Virginia creeper	12
Dahlia	33	Oak	100	Walnut	46
Daylily	10	Okra	21	Watermelon	38
Delphinium	47	Oleander	13	Willow	65
Dieffenbachia	9	Onion	40	Wisteria	10
Dogwood	37	Pea	50	Yew	19
Douglas fir	43	Peanut	54	Yucca	18
Eggplant	44	Pear	91	Zinnia	24
Elm	65	Pecan	36		

[a] Data from Horst, 1990, and in parentheses from Wellman, 1972. This listing does not include abiotic agents and weak or secondary organisms that invade injured tissues.

pected causal agent and disease are not certain. Method 2 is seldom used in plant clinics because it is difficult and time consuming. It usually involves a research project with specialized personnel and equipment to establish the causal relationship of the agent with the observed disease. The diagnostician must then attempt to prove the causal relationship or consult an appropriate specialist.

The use of keys for diagnosing common disorders of annual row crops (Table 4.7) and tables that relate disease symptoms, field distribution, and occurrence with biotic and abiotic causal agents in Chapter 4 may be helpful in suggesting some possible causes.

Levels of Reliability (Confidence) in Making Diagnoses

Experienced diagnosticians in well-established plant disease clinics satisfactorily diagnose 50–75% of the samples submitted. The low percentage of satisfactory diagnoses is the result of the poor quality of specimens submitted, the lack of background information on the specimens, and time constraints that do not permit a full investigation on "relatively unimportant specimens" such as a homeowner's potted tomato or African violet. Many times the diagnostician's report will read "cause undetermined" or "no evidence of infectious disease." In other cases, the clinician diagnoses a disease but realizes that the real problem is something else. For example, early blight of tomato is easily diagnosed but may not be "the problem" in a crop with nutritional difficulties. In the case of a report saying "cause undetermined," the grower should realize that the common diseases usually associated with the crop were absent. In both cases, useful information was provided to the grower. The grower may choose to pursue the matter by submitting additional samples or discussing the problem with the diagnostician or other specialists.

Five levels of reliability for diagnoses are recognized. They are ranked in descending order of complexity, time required, and cost. Each specific situation determines which level is possible or most appropriate.

1. Positive diagnosis (100% reliable) is rarely achieved because it involves essentially a research project in which the causal agent (biotic or abiotic) is known to be present, is described, and duplicates the disease when healthy plants are exposed to it. This principle of positive diagnosis was stated by Robert Koch in 1876 establishing the "germ theory" of disease, and the principle bears the name "Koch's postulates." This principle is universally accepted as final proof for the cause of a disease and is always followed when a disease is first described. However, Koch's postulates are seldom used in plant clinics because of the constraints of time, space, equipment, and personnel.

2. Accurate diagnosis (>99% reliable) is the goal of plant disease diagnosticians. The diagnosis is usually acceptable to plant pathologists, growers, and courts of law. For an accurate diagnosis, all facets of the disease signature must be present or their absence explained. For biotic diseases, the presence of the pathogen in the plant must be verified in appropriate tissues. For abiotic diseases, the presence of the causal agent at the appropriate time and place must be established. This information is often obtained from records of chemicals used by the grower, air pollution reports, evidence of construction activities, meteorological records, and reports from soil testing and plant assay laboratories.

3. Useful diagnosis (95–99% reliable) is the level at which most good diagnosticians operate. All important components of the disease signature are consistent with the disease. However, additional information and further research could implicate one or more remote possibilities. Growers and courts of law normally accept the diagnosis and act accordingly.

4. Indicative diagnosis (85–95% reliable) is generally the result of incomplete information. However, the information available and clinical observations are consistent with one or more causal agents. Several possible causes are usually listed in the clinic report. It is incumbent upon the grower to assess the merits of each and act accordingly.

5. Exclusion diagnosis (100% reliable) is often all the diagnostician can conclude. It is, however, extremely useful and is usually accepted by plant pathologists, growers, and courts of law. In this case, the cause of the problem (disease) cannot be diagnosed, but suspected diseases are eliminated because critical components of the disease signature are not present. If necessary, the grower and diagnostician can pursue the matter further. For example, bacterial leaf spot of tomato, caused by *Xanthomonas campestris* pv. *vesicatoria*, can be eliminated as a cause if bacteria cannot be observed or isolated from appropriate specimens. The grower can then confidently exclude a bactericide in the spray program and avoid unnecessary expense. If the leaf spotting is serious, the grower will likely pursue the matter.

Steps in Diagnosing Plant Diseases

1. Identify the plant. For commercial crops, it is desirable to know the cultivar and levels of resistance to different diseases. Information on plant resistance is available from seed and plant sources and published reports.

2. Develop a list of possible diseases.

   a. Refer to a host index, books, monographs, compendia, bulletins, and "fact sheets" that describe the common diseases and problems of the host plant. If the host is not listed, select a closely related species.

   b. Study the sections in Chapter 4 of this book on symptoms, field distribution, and time of occurrence for possible causes. Use diagnostic keys, an example of which is shown for annual row crops in Table 4.7. Diagnostic keys are often found in bulletins, leaflets, compendia, and books on plant diseases (see Table 2.5 at the end of Chapter 2).

3. Accept or reject the tentative diagnosis by comparing specific facets of the disease signature with those associated with the specimen or specimens.

   a. If the cause of the disease or disorder is suspected to be a pathogen, examine the specimen for presence of the microorganism or inclusion bodies (see Chapter 5) and verify its identity by using the illustrations and descriptions of plant pathogens in published descriptions.

   b. If the cause of the disease or disorder is suspected to be a nonpathogenic agent such as a toxic chemical, transitory insect or other animal feeders, or adverse weather conditions, read appropriate sections in Chapter 4 to determine the type of disorder each causes. Attempt to establish the presence of suspected disease agents from information submitted with the specimen, reports of pollution, and records of insect occurrence.

References

Agrios, G. N. 1988. Diagnosis of plant diseases. Pages 31-39 in: Plant Pathology. 3rd ed. Academic Press, San Diego, CA.

Aycock, R. 1976. The plant disease clinic—A thorn in the flesh or a challenging responsibility? Annu. Rev. Phytopathol. 14:165-175.

Ellis, D. E. 1976. Plant Pathology in North Carolina 1776–1976. School of Agriculture and Life Sciences, North Carolina State University, Raleigh.

Evans-Ruhl, G. 1982. Plant disease clinics—Past, present, and future. Plant Dis. 66:80-86.

Flint, M. L. 1990. Pests of the Garden and Small Farm. Division of Agriculture and Natural Resources, University of California, Oakland.

Grogan, R. G. 1981. The science and art of plant-disease diagnosis. Annu. Rev. Phytopathol. 19:333-351.

Horst, R. K. 1990. Westcott's Plant Disease Handbook. 5th ed. Van Nostrand Reinhold, New York.

Jones, R. K., Stephan, D. L., Hisada, A. M., and Dukes, D. P. 1985. The plant disease and insect clinic, N.C. State University at Raleigh. A partial summary—Diseases of major and key crops 1984. Plant Diagn. Q. 4(1):160-177.

Kucharek, T. 1986. The educational service performed by diagnoses of dysfunctions associated with samples of agronomic crops received by the Florida Plant Disease Clinic from 1970 to 1984. Proc. Soil Crop Sci. Soc. Fla. 45:109-115.

McIntyre, J. L., and Sands, D. C. 1977. How disease is diagnosed. Pages 35-53 in: Plant Disease—An Advanced Treatise, vol. 1. J. G. Horsfall and E. B. Cowling, eds. Academic Press, San Diego, CA.

Shurtleff, M. C. 1966. How to Control Plant Diseases in Home and Garden. 2nd ed. Iowa State University Press, Ames.

Streets, R. B., Sr. 1978. The Diagnosis of Plant Diseases. University of Arizona Press, Tucson. pp. 2.1-2.17.

Tisserat, N. 1993. The future of disease diagnosis. Grounds Maint. 4:59-61, 91.

Tuite, J. 1969. Diagnosing the cause of plant disease. Pages 117-118 in: Plant Pathology Methods. Burgess, Minneapolis.

Wellman, F. L. 1972. Tropical American Plant Diseases. Scarecrow Press, Metuchen, NJ.

Wellman, F. L. 1979. Dictionary of Tropical American Crops and Their Diseases. Scarecrow Press, Metuchen, NJ.

CHAPTER 2

The Plant Disease Clinic

The plant disease clinic can serve various functions in addition to its primary role of providing space and resources for the diagnosing of plant diseases. Plant disease laboratories are often used to develop diagnostic techniques, describe new diseases, and support field research projects and as centers for disseminating control information and assembling disease survey information. These secondary functions may detract from the primary role of the clinic.

The plant disease clinic (workshop or laboratory) can be very modest or quite sophisticated, depending on available resources (Figs. 2.1–2.13). The number and types of samples handled annually should dictate the type of equipment and size of the clinic. The plant disease laboratory de-

scribed in this chapter is designed to handle up to 5,000 samples annually in a midsize state in the United States. However, during the busy months of April to September or October, the facility is at times inadequate. It should be staffed by an experienced full-time diagnostician, a part-time secretary, a part-time plant pathologist, and in some clinics by a part-time entomologist. In addition, experts in recognizing herbicide injury and in horticultural and agronomic crops, soils, plant nutrition, plant pathology, mycology, virology, nematology, and entomology assist in diagnosing and making control recommendations.

In some manner, however, all plant disease clinics must provide space and facilities for receiving, storing, ex-

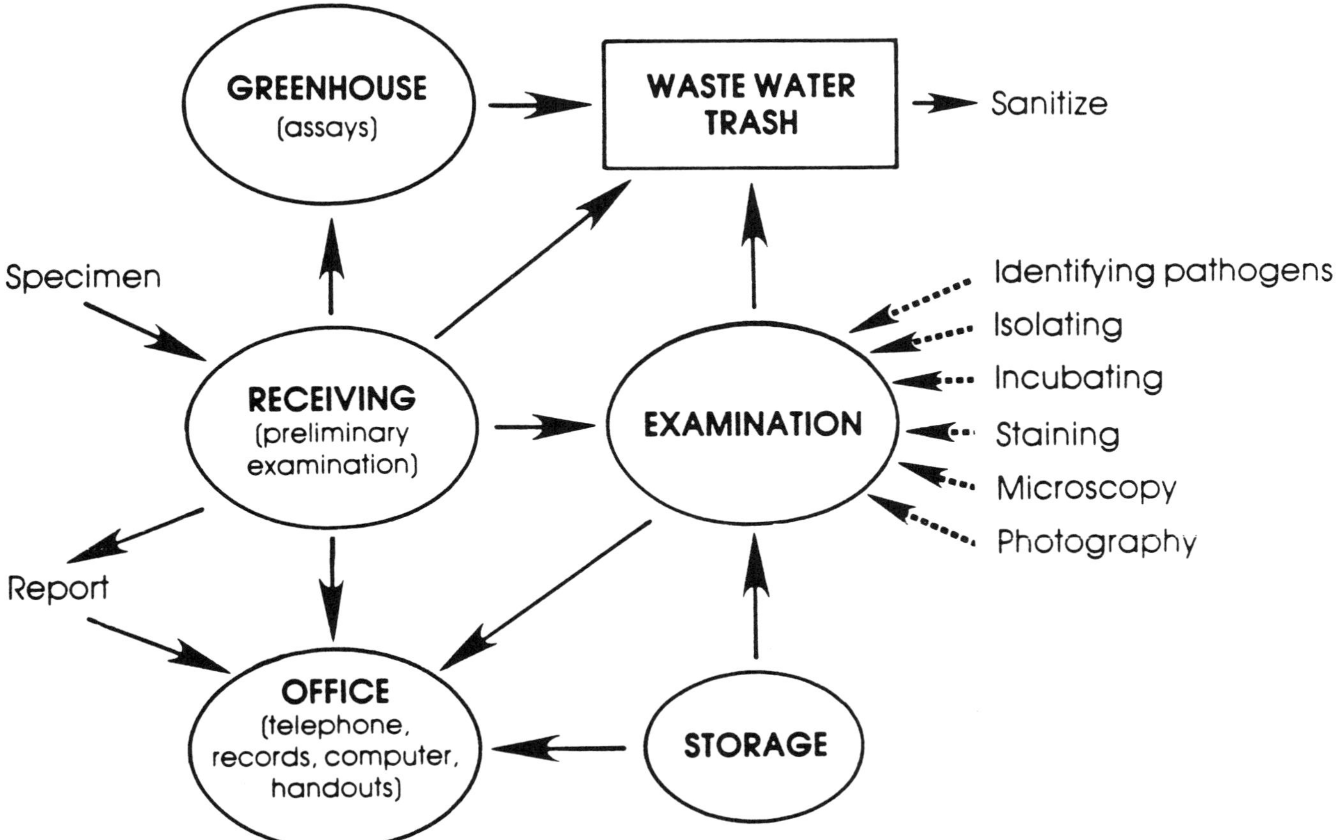

Fig. 2.1. Flow diagram of a plant disease clinic showing relationships between various areas and activities.

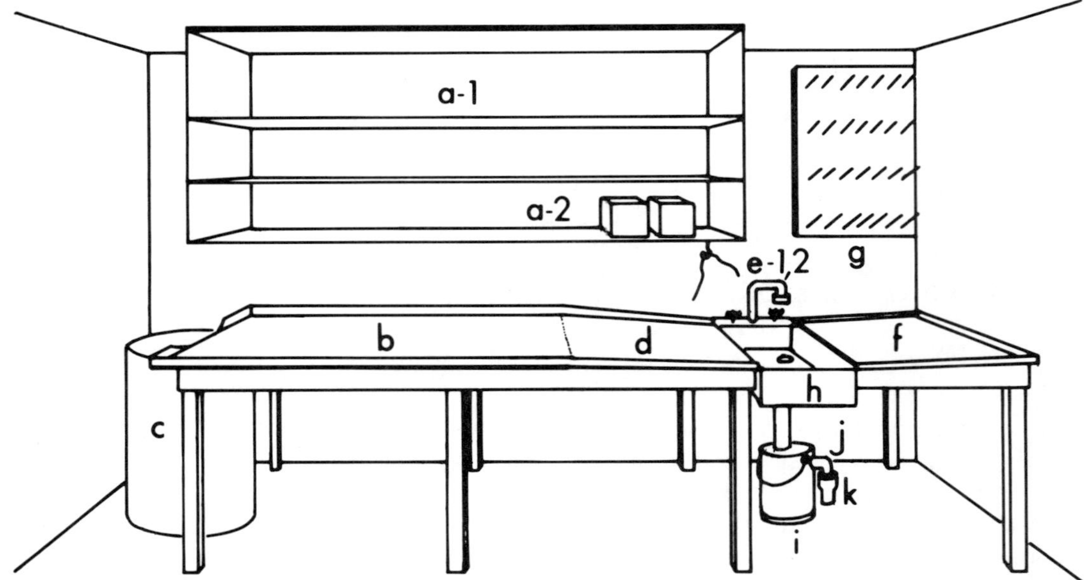

Fig. 2.2. Representative work area for efficiently opening packages and for preliminary examination of plant specimens. **a-1,** Open shelves for supplies, tools, and glassware; **a-2,** soil pH and soluble salts meters; **b,** stainless steel table with raised edges; **c,** heavy-duty plastic garbage can; **d,** 3-ft, slightly tilted surface (slope 1:40); **e-1,2,** distilled water spigot with aerator; **f,** stainless steel drainboard; **g,** pegs for drying glassware; **h,** sink with screen over drain; **i,** bucket soil trap; **j,** drain spout; and **k,** floor drain.

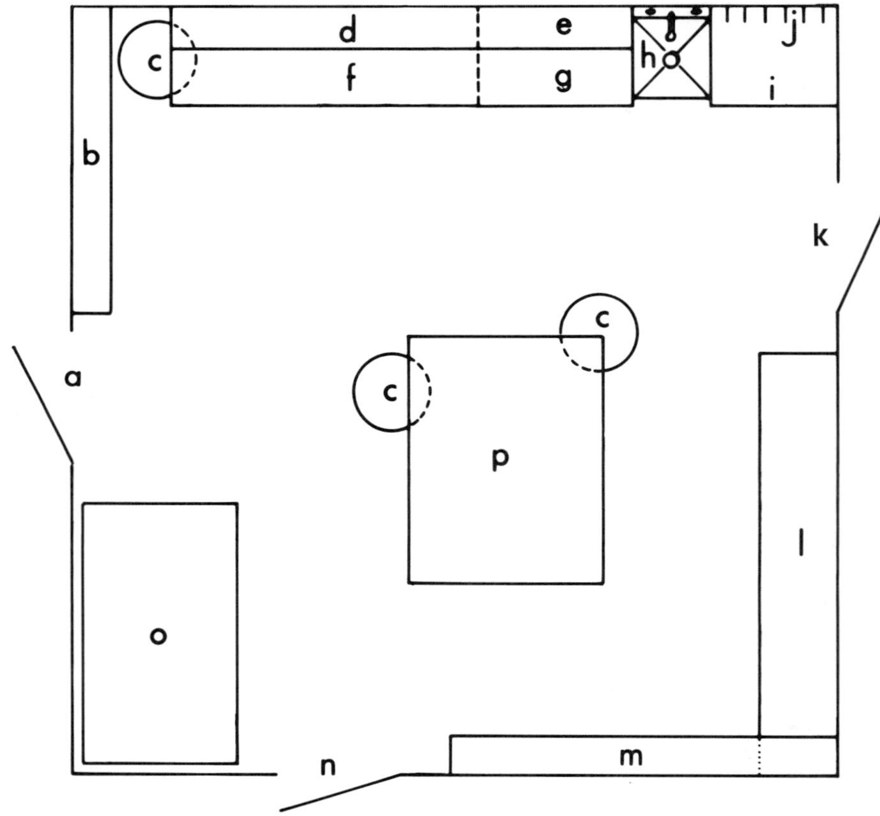

Fig. 2.3. Convenient floor plan of a receiving and preliminary examination room, 15 × 15 ft. **a,** Entrance with a Dutch door; **b,** wall shelves for moist chambers; **c,** heavy-duty plastic garbage cans; **d,** wall shelves; **e,** shelf area for pH and soluble salts meters; **f,** stainless steel countertop; **g,** sloped countertop; **h,** sink; **i,** drainboard; **j,** wall pegs; **k,** entrance to examination room; **l,** laboratory bench with dissecting and compound microscopes; **m,** wall bookcase; **n,** entrance to office with Dutch door; **o,** large refrigerator for specimens; and **p,** stainless steel table with raised edges. A wall telephone should be placed near **n.**

amining, recording, and reporting on the specimens received (Fig. 2.1). Large quantities of packaging material, plant debris, soil, and wash water must be disposed of efficiently, occasionally under conditions of quarantine. Expensive equipment must be protected from excessive amounts of dust, grit, and liquids of various types. Microbiological work areas should be protected from biological contamination. These considerations become increasingly important when more than five samples are received in a day.

Fig. 2.4. Soil pH and soluble salts meters. They are located near the sink and distilled water source in the receiving room.

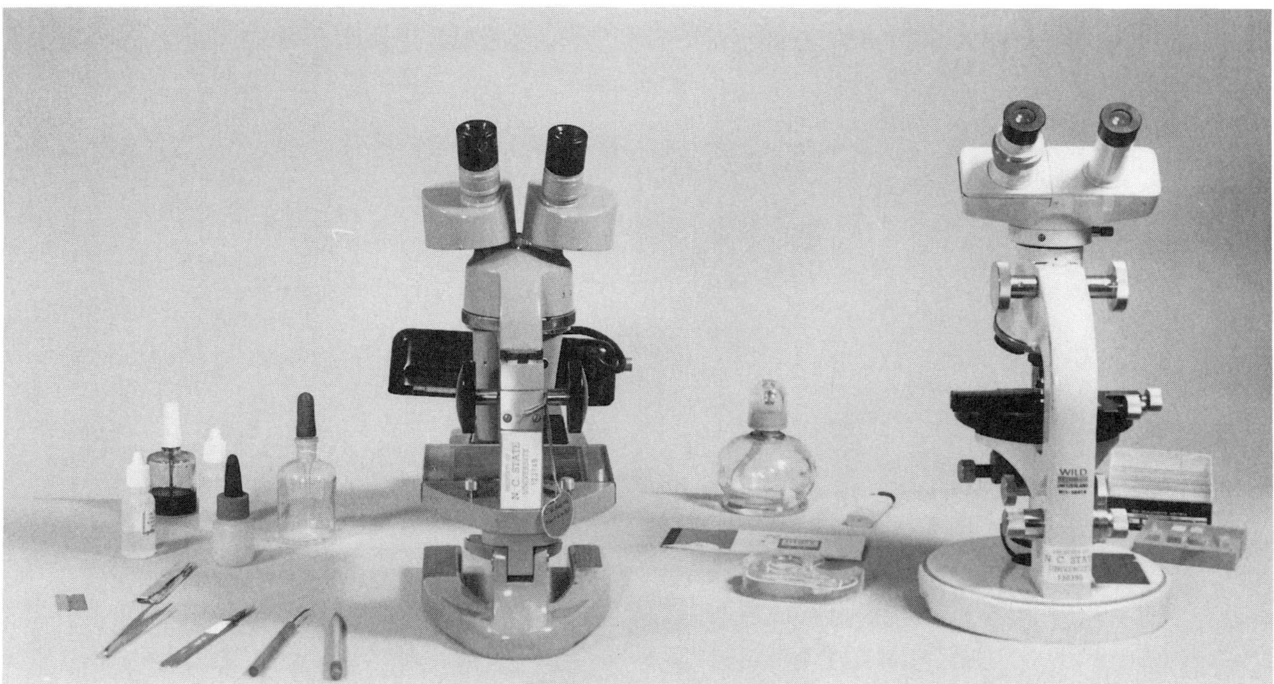

Fig. 2.5. Dissecting and compound microscope work area in the receiving room. Both microscopes should have built-in, variable-intensity light sources. The dissecting microscope has 10× eyepieces and 0.7×, 1.5×, and 2.5× objective lenses in the turret. The compound microscope is equipped with 10× eyepieces and 5×, 10×, and 44× objective lenses in the turret, a scaled mechanical stage, and an Abbey condenser with iris diaphragm. Slides, coverslips, tweezers, razor blades, scalpels, needles, and dropper bottles with water and various other solutions are nearby on the bench or table.

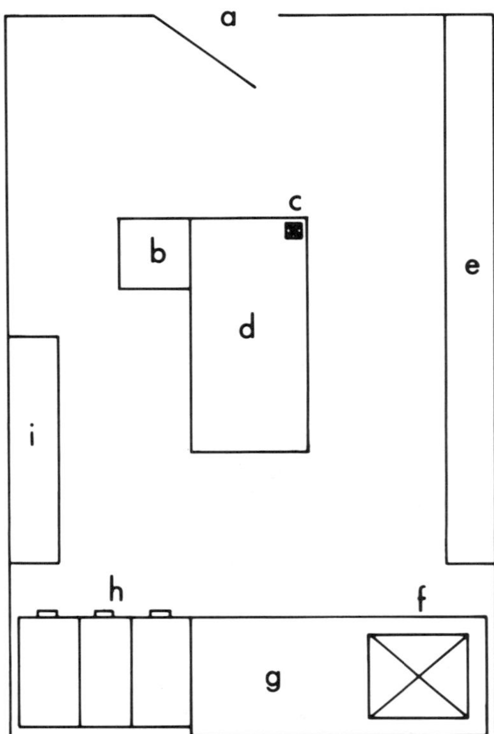

Fig. 2.6. Representative and efficient office floor plan for a plant diagnostic clinic. **a,** Entrance from receiving room; **b,** typewriter or word processor; **c,** telephone; **d,** secretary's desk; **e,** floor-to-ceiling shelves with pigeonholes for handouts; **f,** computer and printer; **g,** table; **h,** file cabinets; and **i,** bookcase on wall.

Receiving Area

Plant specimens are received by various types of mail and as "walk-in specimens" brought in by growers or their couriers. Specimens received by mail must be opened and examined promptly or placed in a cooler to prevent further deterioration for later examination. Usually, walk-in clients with specimens require an immediate, preliminary examination and, if possible, a diagnosis because the grower is present and anxious to know the cause of the problem and how to correct it.

The receiving area (Figs. 2.2 and 2.3) must have one or more large tables or countertops for preliminary examination of specimens. The surface must be suited for cutting and be able to withstand both water and soil. A stainless steel surface with raised edges is excellent. One end should be located and shaped so trash can easily be brushed directly into a large, watertight garbage can. At least 3 ft of the work surface on the other end should slope slightly (1:40) into a shallow sink. The sink must be provided with a large screen over the drain to catch things such as twigs and leaves. The drain must have a large, easily serviced soil trap, such as a bucket with a spout near the upper rim that overflows into a floor drain (Fig. 2.2i). Wall cabinets or shelves over work areas provide convenient space (Figs. 2.2 and 2.3) for a reference library, forms, supplies, and tools plus soil pH and soluble salts meters (Fig. 2.4) used in the preliminary examination of specimens. A source of distilled

Fig. 2.7. Large bookcase with pigeonholes containing specific plant disease information leaflets on symptoms and control. A file cabinet may also be used. This printed information may be included with the diagnostic report or is given to persons who bring in plant specimens. Distribution of these information sheets and other printed material on diseases and their control is a necessary function of a plant disease clinic. They should be prepared by plant pathologists, entomologists, or other specialists experienced with field control practices and legal ramifications concerning the use of pesticides.

water near the meters is necessary. Two microscope areas should be provided in the receiving room (Figs. 2.3 and 2.5). They should be somewhat removed from the preliminary examination table, the sink, and the soil pH and soluble salts meters. Each microscope area (Fig. 2.5) should have relatively inexpensive dissecting and compound microscopes and small storage spaces for slides, coverslips, lens paper, microscope brush, lens cleaning fluid, razor blades, various scalpels, dissecting needles, tweezers, and dropper bottles that contain tap water, 0.4% KOH solution, liquid household bleach (5.25% sodium hypochlorite), 0.05–0.10% cotton blue in lactophenol, and other stains.

Office and Records

The office (Fig. 2.6) should be adjacent to the receiving area so that the secretary can see and serve people coming into the clinic. Forms, records, and information handouts should be convenient to the receiving area. However, the office area must be sufficiently isolated so work can proceed in each area without interference from activities in other areas.

The office must have the standard secretarial furniture, equipment, and supplies. A large bookcase with pigeonholes should be included for holding the many handouts with plant disease information (Fig. 2.7). A large table (at least 2½ × 6 ft) is desirable for additional work such as sorting

clinic reports and for the computer. File cabinets are necessary for clinic reports and correspondence. Some books and reference materials not used routinely in diagnosing may be housed in the office. These would include catalogues for equipment and supplies, books on plant husbandry and culture, reprints, duplicate and obsolete books, and miscellaneous information on plant diseases and their control.

The plant disease clinic must maintain records on all specimens received and copies of the diagnoses issued or disposition of each sample. These records become important when follow-up specimens are received or when inquiries (occasionally of a legal nature) concerning a particular specimen are received. There are many different record-keeping systems in use by plant clinics.

Numerous clinics use computers with printers to print and file reports (Fig. 2.8). Other clinics use standard file cabinets to file typed or handwritten reports. The system adopted usually reflects the resources available, personal choice of the diagnostician, and perceived needs. The systems undergo continuous change, but all permit easy retrieval of diagnostic reports by date, plant species, county, grower, and causal agent.

The record system described here is based on that used at North Carolina State University and has evolved over more than 40 years.

1. The specimen is received with an 8- × 5-in. form that provides two color-coded carbon copies (Fig. 2.9). If the

Fig. 2.8. Computer and printer used to file diagnostic reports. The reports can be retrieved by the grower's name, county or other geographic area, date, crop, and causal agent. A computer is useful for sending and receiving electronic mail, preparing reports by verifying the spelling of causal agents, automatically printing standard control information for some diseases, and preparing annual summary reports.

information sheet has not been filled out, one is prepared, listing the specimen and what is known about it; or the specimen is rejected. The information sheet stays with the specimen during the entire diagnostic process. Observations made by the diagnostician are written directly in the space provided on the lower portion of the form or on supplementary sheets clipped to the form.

2. At the conclusion of the diagnosis, the secretary types in all relevant clinical observations, the diagnosis, and remedial options and adds enclosures, follow-up suggestions, the name of the diagnostician, and the date of mailing. The original copy (yellow) is mailed to the grower; the second copy (blue) is mailed to the county agent; and the third copy (green) is filed at the clinic.

Other states use one-page, carbonless forms and a letter is sent only to the client.

If the clinical observations, diagnosis, and suggestions are straightforward and legible, typing is omitted to save time. This is often the case with walk-in specimens. If a Plant Disease Information Sheet is given out, this is recorded on the form.

3. The clinic's copy is filed by crop and then by date. (In other states, the copy may be filed by county and then by date and plant species or by a sample number.)

After a diagnosis is completed, many plant disease clinics will give suggestions on remedial procedures. However, information on plant disease control is not the primary

---

COPY 2

Crop _Tomato_ — Variety _Manapal_ — When planted _4/12/85_ — Date collected _6/3/85_ — Clinic # ______

Grower _R. K. Smith_ — _Rt. 1, Box 123 Green, NC 27123 704/123-4567_
Name — No. & Street (Route & Box) — City — State — Zip — Telephone

Agent _S. R. James_ — County _Maple_ — Collected by _grower_ — in _same_

COPY 1

Crop _Tomato_ — Variety _Manapal_ — When planted _4/12/85_ — Date collected _6/3/85_ — Clinic # ______

Grower _R. K. Smith_ — _Rt. 1, Box 123 Green, NC 27123 704/123-4567_
Name — No. & Street (Route & Box) — City — State — Zip — Telephone

ORIGINAL

Crop _Tomato_ — Variety _Manapal_ — When planted _4/12/85_ — Date collected _6/3/85_ — Clinic # ______

Grower _R. K. Smith_ — _Rt. 1, Box 123 Green, NC 27123 704/123-4567_
Name — No. & Street (Route & Box) — City — State — Zip — Telephone

Agent _S. R. James_ — County _Maple_ — Collected by _grower_ — in _same_
City — County

Type planting: commercial ☒ yes ☐ no   Site: ☒ field ☐ greenhouse ☐ plant bed ☐ ______

Planting size (select one): ____ _5_ ____ acres ________ plants ________ plant bed; ____ _25_ ____ % affected

Crop history: Last year _tomato_ ____ 2 years ago _Tomato_ ____ 3 years ago _tobacco_ ____

SYMPTOMS:
☒ wilted ☒ yellowing
☒ stunted ☐ leaf spot
☐ dieback ☐ root rot
☒ death _stem rot_

DISTRIBUTION:
☒ localized
☒ scattered
☐ general

Fertilizers used:
_200 lb 10-10-10_
50 lb sidedress
_10-10-10_

Chemicals used:
maneb
sevin

Main concern/helpful information _What is the problem?_

CLINIC REPORT: (do not write in this space)

```
Southern blight caused by Sclerotium rolfsii.  This is a
soil-borne disease.  See inserts for information on control.
Leafspots are early blight caused by Alternaria solani.  The
poor leaf color may indicate a nutritional problem, and you
should send a tissue and soil sample to the Agronomic Division
for analyses.
```

C. H. Bradshed, Clinician

PLANT DISEASE AND INSECT CLINIC, BOX 7616, N. C. STATE UNIVERSITY, RALEIGH, N. C. 27695-7616

Fig. 2.9. Form used to submit a plant specimen. The grower fills in as many lined spaces as possible. The clinician writes in observations, diagnosis, and suggested remedial action (optional) in the space under Clinic Report. The grower receives the top yellow (original) sheet. The blue copy is mailed to the county agent, and the last copy (green) is retained and filed at the clinic. The Clinic # ______ entry at the top right is used in some clinics for logging in samples and filing.

function of a clinic and is best relegated to appropriate plant pathologists, entomologists, county agents, or farm advisors who specialize in pest and disease control. Where control recommendations are routinely stated on the diagnostic report, they are often standard replies typed on the report or added as inserts. For examples of a variety of forms used by land-grant university plant clinics, see the appendix at the end of this chapter.

Software for personal computers (PCs) is widely available that will organize, store, maintain, inquire, and issue reports on paper or by electronic mail. These systems, e.g., PClinic, have several advantages over traditional systems of filing and mailing reports on paper: 1) very little space is needed in the office; 2) standard answers can be quickly retrieved from a database and incorporated into the report; 3) reports can be sent directly by electronic mail or by printing a paper copy to be mailed or sent by facsimile; 4) reports can be retrieved quickly by one or more categories, e.g., host, pathogen, county, location, date, grower, or consultant; and 5) annual or summary reports can be quickly issued.

Examination Room

The examination room (Fig. 2.10) is designed for cleanliness and relative privacy for detailed diagnostic work. Clientele normally do not enter this area. Soil and plant specimens, or portions of them, are carried into this area only after being examined and sanitized in the receiving room. This minimizes trash, dust, and biological contamination.

The examination room is divided into areas for 1) preparing culture media, 2) aseptic culturing, 3) incubating cultures in petri dishes and specimens in moist chambers, 4) extracting nematodes from soil and plants, 5) preparing thin sections of plant tissues for microscopic examination, 6) identifying pathogens, nematodes, and insects microscopically (two areas), and 7) photographing interesting specimens.

Convenient to the examination area should be a storage room and small greenhouse. The storage room is used for supplies and equipment. The small greenhouse (20 × 10 ft) with two benches is used to grow differential plants for

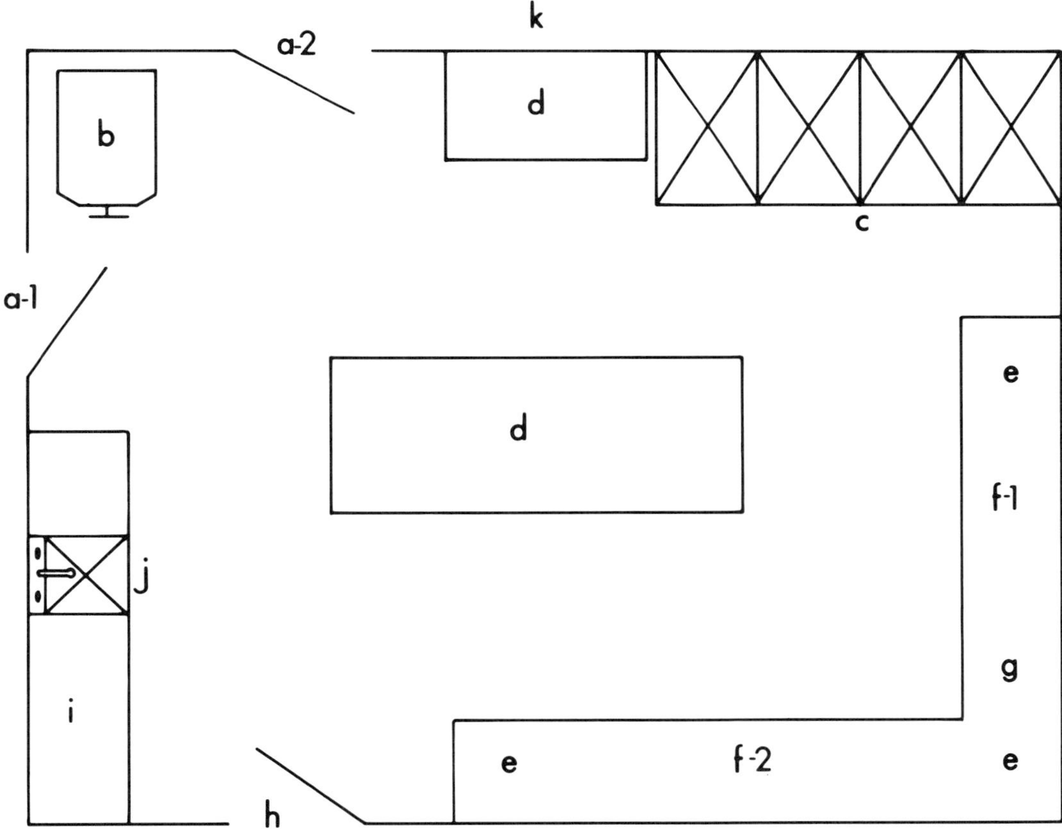

Fig. 2.10. Floor plan of an efficient examination room. **a-1,** Entrance from receiving room; **a-2,** entrance from greenhouse; **b,** autoclave; **c,** incubators (set at 15, 21, 27, and 33°C); **d,** table; **e,** laboratory bench with drawers and overhead open wall shelves; **f-1 and f-2,** microscope work areas, each with research-quality compound and dissecting microscopes; **g,** photographic area, for both macroscopic and microscopic work; **h,** entrance to storage room with floor-to-ceiling shelves; **i,** workbench with drawers, gas outlets, and culturing hood or chamber; **j,** stainless steel sink and drain surface for cleaning glassware, staining specimens, setting up nematode extractions, and preparing media; **k,** small greenhouse (20 × 10 ft) for identifying viruses on differential host plants, assaying soils and roots for nematodes, and establishing the pathogenicity of microorganisms.

identifying viruses, bioassaying nematodes and micro-organisms from appropriate soil and plant specimens, and establishing the pathogenicity of microorganisms.

Equipment and Supplies

Standard laboratory articles are available from numerous suppliers. A partial list of suppliers in the United States is given in Table 2.1, and the basic equipment and supplies used in the plant disease clinic are listed in Tables 2.2–2.4. Table 2.5 is a list of recommended books and reference materials. Laboratory equipment and supplies are expen-

Table 2.1. Some laboratory suppliers in the United States

Name and primary address	Comments
American Type Culture Collection 12301 Parklawn Drive Rockville, MD 20852-1776	Antisera and microorganisms
Capital Agricultural Service and Supply Company P.O. Box 550 Montgomery, AL 36101	Greenhouse and field items
Carolina Biological Supply Burlington, NC 27215	Biological models; prepared slide mounts; teaching supplies
Curtin Matheson Scientific 7381 Empire Drive Florence, KY 41042	General laboratory equipment and supplies
Difco Laboratories Detroit, MI 48201	Culture media
Fisher Scientific 50 Fadem Road Springfield, NJ 07081	General laboratory equipment and supplies
Forestry Suppliers, Inc. 205 West Ranking Street P.O. Box 8397 Jackson, MS 39204-9987	Field equipment; instruments; supplies; maps
E. C. Geiger Box 285, Rt. 63 Harleysville, PA 19438-0332	Florist and nursery supplies
A. H. Hummert Seed Co. 2746 Chouteau Ave. St. Louis, MO 63103	Horticultural supplies
LaMotte Chemical Products Co. Chestertown, MD 21620	Soil- and water-testing equipment; supplies
Sigma Chemical Company P.O. Box 14508 St. Louis, MO 63178-9974	Laboratory chemicals; equipment and supplies; pathogen-detection kits; books
Thomas Scientific Vine Street at Third P.O. Box 779 Philadelphia, PA 19105-0779	General laboratory equipment and supplies
Turtox/Cambosco MacMillan Science Co., Inc. 8200 South Hoyne Ave. Chicago, IL 60620	Biological models; teaching supplies; prepared mounts
Wards Natural Science Establishment, Inc. Rochester, NY 14692-9012	General laboratory supplies

Table 2.2. Field kit inventory

Article	Comments
*Metal tool box, 19 × 9 × 9 in.	To consolidate and protect items
First-aid kit	. . .
Soap and paper towels	. . .
Flashlight	. . .
Hand tools	. . .
String	To prepare samples for shipment
Tape (adhesive)	To prepare samples for shipment
Pruning saw	To remove branches
Hatchet	To cut into stumps, roots, and trunks
Chisel	To remove bark and samples from wood
Increment borer	To take samples from trees
Butcher knife	To cut large samples
Paring knife	To cut small samples
*Tweezers	To remove small samples and insects
Razor blades, single edge	To make close cuts
*Hand trowel	To dig up small plants; to perform root collar examinations on woody specimens
*Soil-sampling probe	To take soil samples; to assess soil profile
*Pocket knife	. . .
*Magnifying glass, 5×, 5×, 10× (total 20×)	To observe fruiting structures, small insects, mites, etc.
*Pencils, very soft lead (#1)	For use on damp paper
*Plastic bags	Gallon or larger and quart size
Tags with wire fasteners	To label samples
Vials with alcohol	For insect specimens
Rubber bands	35-mm film cartridge mailers make excellent containers for these and other small items
Paper clips	. . .
*Soil-testing boxes or bags	Available from appropriate service laboratories
*Nematode assay mailing boxes	Available from appropriate service laboratories
*Plant tissue analysis mailers	Available from appropriate service laboratories
*Forms for submitting samples to soil, nematode, plant tissue, and plant disease clinic laboratories	Available from appropriate service laboratories
File folders (bellows)	To organize forms, reference materials, and handouts
Agricultural chemicals manuals	Available from cooperative extension service in each state
Crop production manuals	Available from cooperative extension service in each state
Clipboard	Available from office suppliers
Receipt book	Some laboratories charge and growers need receipts
Camera: 35-mm, single-lens reflex with through-the-lens light meter	To record and document problems
Close-up lenses, nos. 2, 3, and 4	To photograph insects, leaf spots, etc.; a macrolens may be used but is more expensive
Polarizing filter	To reduce glare on foliage, waxy fruit, and background
Electronic flash, automatic	To highlight subject
Tripod	To steady camera
Shutter-release cable	To avoid camera movement
Black velvet (0.7 × 1.0 m)	To backlight light subjects; light blue for dark objects
Camel's-hair lens brush	To dust lenses
Lens paper	To clean lenses
Lens cleaning fluid	To clean lenses

*Essential item.

sive, but for most of the diagnoses done in the preliminary examination room, the least expensive articles from a scientific supplier are quite satisfactory. Secondhand equipment and supplies may be purchased from surplus sales conducted by hospitals and schools. The need and sources for high-quality equipment and supplies for certain diagnoses are stated when the procedure is described.

Field Kit

Diagnosticians in plant disease clinics may need to examine certain fields and other areas to gather appropriate samples and personally assess a disease situation. Because of uncertainties prior to going into the field, a prudent diagnostician will carry a field kit containing tools and supplies that may be needed (Fig. 2.11 and Table 2.2). The diagnostician should include items to support personal needs and interests. For example, a camera (Fig. 2.12), tape recorder, insect repellent, sun screen, and other items may be desirable.

Field Clinics

County agents, crop consultants, and other professionals can very adequately operate plant clinics to support local operations such as integrated pest management programs, farming and gardening operations, quarantine and quality-control stations, educational efforts, and plant-certification

inspections. Field clinics can be equipped in a general way with many items found in the receiving room of a formal clinic (Table 2.3 and Fig. 2.13). Field clinics are efficient and well received by growers, who find them convenient for rapid diagnoses of routine problems so that prompt

Table 2.3. Equipment and supplies for North Carolina county plant clinics

Description[a]	Possible source[b]
Clinic cabinet	. . .
Stereo microscope, 20×, 40×, with dual built-in illuminator	FS 19762
Replacement bulb	FS 19764
Dissecting forceps	FS 08-880
Dissecting scalpel, disposable (20)	FS 8-927B
Dissecting needle, straight (2)	FS 8-965A
Microscope, EHL-3; monocular, 40×, 100×, 400×, and 900× with mechanical stage	FS 19300
Replacement bulb	FS
Glass microscope slides, 3 × 1 in. (50)	FS 12-549
Microscope coverslips (plastic or glass), 18 × 18 mm (50)	FS 12-542A
Dropper bottle, 30 ml	FS 2-992A
Transparent tape, ¾ inch	. . .
pH meter	LCP 1906
Dissolved salts meter	LCP 1918
Thermometer, −10 to 110°C	FS 15-021-5B
Graduated cylinder, 50 ml	FS 8-549-17
Paring knife, 4-in. blade	. . .
Plastic freezer bags, quart (200)	. . .

[a] Approximate number needed is given in parentheses.
[b] There are many sources for equipment and supplies. Those listed describe a specific item and are given for the purpose of illustration. No discrimination is intended for similar items from other sources. FS = Fisher Scientific Catalog, P.O. Box 11666, Raleigh, NC 27604; LCP = LaMotte Chemical Products Co., Chestertown, MD 21620; other items are available from local home supply stores.

Table 2.4. Clinic equipment and supplies in addition to those suggested for county clinics

Number needed	Item
1	Refrigerator with freezer compartment
1	Electric stove with oven
1	Autoclave, electric (26 × 16 in.) or pressure cooker (11⅞ × 12⅝ in.)
2	Incubators (16 × 19 × 18 in.), 30–60°C
1	Incubator, refrigerated, 3–50°C (24 × 24½ × 34½ in.)
1	Research microscope, trinocular, Koehler illuminated, with 4×, 10×, 40×, and 100× objective lenses
1	Dark-field accessories
1	Instrument camera
1	Eyepiece micrometer
1	Stage micrometer, 5 and 0.05 mm
1	Polarization kit
1	Eyepiece pointer
1	Stereo microscope, 7×, 10×, 15×, 20×, and 25× with transilluminating base
1	Illuminator, dual, fiber optics
1	Computer with printer
1	Typewriter
1	Clean bench, 23¾ × 23¾ × 23¾ in. with transfer hood, for culturing
1	Alcohol lamp
1	Bunsen burner
2	Scissors
2	Inoculating loops
2	Dissecting needles with handle
1	Torsion balance, 120-g capacity, 2-mg readability
1	Set of gram weights, 1, 2, 5, 10, 25, 100, and 500
1	Dietary scale, 500-g capacity, 1-g readability
1	Fluorescent/incandescent desk lamp
1	Water purification still, 3.8 liters/h, or water deionizer, 19–38 liters/h, with six replacement cartridges
1	Nalgene rectangular carboy with spigot, 5 gal
10	Media bottles, 500 ml
10	Media bottles, 250 ml
10	Media bottles, 100 ml
6	Pyrex funnels, 127-mm diameter with 100-mm stems
6	Pyrex funnels, 75-mm diameter with 30-mm stems
1	Stainless steel funnel, 105-mm diameter, 111-mm length
1	Stainless steel funnel, 152-mm diameter, 168-mm length
2	Erlenmeyer flasks with screw caps, 2,000 ml
6	Erlenmeyer flasks with screw caps, 1,000 ml
6	Erlenmeyer flasks with screw caps, 500 ml
6	Erlenmeyer flasks with screw caps, 250 ml
6	Erlenmeyer flasks with screw caps, 125 ml
1	Polyethylene beaker with handle, 3,000 ml
1	Polyethylene beaker with handle, 1,000 ml
2	Stainless steel beakers, 1,200 ml
4	Stainless steel beakers, 250 ml
12	Pyrex tall beakers, 200 ml
12	Polypropylene beaker set, 50, 100, 250, 500, and 1,000 ml
12	Pyrex beakers, 10 ml
12	Pyrex beakers, 20 ml
12	Pyrex beakers, 50 ml
12	Pyrex beakers, 100 ml
12	Pyrex beakers, 250 ml
500	Culture (petri) dishes, plastic, disposable, sterile, 90 × 15 mm, or 48 Pyrex glass petri dishes, 90 × 15 mm

Table 2.5. Recommended plant clinic library[a]

Agrios, G. N. 1988. Plant Pathology, 3rd ed. Academic Press, San Diego, CA.

Ainsworth, G. C., Hawksworth, D. L., Kirk, P. M., Sutton, B. C., and Pegler, D. N. 1995. Ainsworth and Bisby's Dictionary of the Fungi, 8th ed. C.A.B. International, Wallingford, Oxon., United Kingdom.

Ainsworth, G. C., Sparrow, F. K., and Sussman, A. S., eds. 1973. The Fungi. An Advanced Treatise. Vol. IVA. A Taxonomic Review with Keys: Ascomycetes and Fungi Imperfecti. Vol. IVB. A Taxonomic Review with Keys: Basidiomycetes and Lower Fungi. Academic Press, San Diego, CA.

Alexopoulos, C. J., and Mims, C. W. 1979. Introductory Mycology, 3rd ed. John Wiley & Sons, New York.

Alfieri, S. A., Jr., Langdon, K. R., Wehlburg, C., and Kimbrough, J. W. 1984. Index of Plant Diseases in Florida. Bull. 11. Division of Plant Industry, P. O. Box 1269, Gainesville, FL.

Anonymous. 1994. PClinic Plant Diagnostic Laboratory Software. Plant Diagnostics Quarterly 15(3):148-152. Department of Botany and Plant Pathology, Purdue University, West Lafayette, IN.

Anonymous. 1981. Weeds of the North Central States. Bull. 772. College of Agriculture, University of Illinois, Urbana.

Anonymous. 1970. Virus Diseases of Small Fruits and Grapevines (A Handbook). University of California, Division of Agricultural Science, Berkeley.

American Phytopathological Society, Disease Compendium Series, The Society, 3340 Pilot Knob Road, St. Paul, MN 55121-2097:

Alfalfa Diseases, 2nd ed. 1990. Stuteville, D. L., and Erwin, D. C., eds.

Apple and Pear Diseases. 1990. Jones, A. L., and Aldwinckle, H. S., eds.

Barley Diseases. 1982. Mathre, D. E., ed.

Bean Diseases. 1991. Hall, R., ed.

Beet Diseases and Insects. 1986. Whitney, E. D., and Duffus, J. E., eds.

Blueberry and Cranberry Diseases. 1995. Caruso, F. L., and Ramsdell, D. C., eds.

Citrus Diseases. 1988. Whiteside, J. O., Garnsey, S. M., and Timmer, L. W., eds.

Corn Diseases, 2nd ed. 1980. Shurtleff, M. C., ed.

Cotton Diseases. 1981. Watkins, G. M., ed.

Cucurbit Diseases. 1996. Zitter, T. A., Hopkins, D. L., and Thomas, C. E., eds.

Elm Diseases. 1981. Stipes, R. J., and Campana, R. J., eds.

Flowering Potted Plant Diseases. 1995. Daughtrey, M. L., Wick, R. L., and Peterson, J. L.

Grape Diseases. 1988. Pearson, R. C., and Goheen, A. C., eds.

Onion and Garlic Diseases. 1995. Schwartz, H. F., and Mohan, S. K., eds.

Ornamental Foliage Plant Diseases. 1987. Chase, A. R.

Pea Diseases. 1984. Hagedorn, D. J., ed.

Peanut Diseases. 1984. Porter, D. M., Smith, D. H., and Rodríquez-Kábana, R., eds.

Potato Diseases. 1981. Hooker, W. J., ed.

Raspberry and Blackberry Diseases and Insects. 1991. Ellis, M. A., Converse, R. H., Williams, R. N., and Williamson, B., eds.

Rhododendron and Azalea Diseases. 1986. Coyier, D. L., and Roane, M. K., eds.

Rice Diseases. 1992. Webster, R. K., and Gunnell, P. S., eds.

Rose Diseases. 1983. Horst, R. K.

Sorghum Diseases. 1986. Frederiksen, R. A., ed.

Soybean Diseases, 3rd ed. 1989. Sinclair, J. B., and Backman, P. H., eds.

Stone Fruit Diseases. 1995. Ogawa, J. M., Zehr, E. I., Bird, G. W., Ritchie, D. F., Uriu, K., and Uyemoto, J. K., eds.

Strawberry Diseases. 1984. Maas, J. L., ed.

Sweet Potato Diseases. 1988. Clark, C. A., and Moyer, J. W.

Tobacco Diseases. 1991. Shew, H. D., and Lucas, G. B., eds.

Tomato Diseases. 1991. Jones, J. B., Jones, J. P., Stall, R. E., and Zitter, T. A., eds.

Tropical Fruit Diseases. 1994. Ploetz, R. C., Zentmyer, G. A., Nishijima, W. T., Rohrbach, K. G., and Ohr, H. D., eds.

Turfgrass Diseases, 2nd ed. 1992. Smiley, R. W., Dernoeden, P. H., and Clarke, B. B., eds.

Wheat Diseases, 2nd ed. 1987. Wiese, M. V.

Ayoul, S. M. 1980. Plant Nematology, An Agricultural Training Aid, rev. ed. NemaAid, P.O. Box 23058, Sacramento, CA.

Baker, K. F., ed. 1972. The UC System for Producing Healthy Container-Grown Plants Through the Use of Clean Soil, Clean Stock, and Sanitation. Manual 23. Division of Agricultural Sciences, Agricultural Experiment Station Extension Service, University of California, Berkeley.

Bakshi, B. K. 1971. Indian Polyporaceae. Indian Council of Agricultural Research, New Delhi.

Barnett, H. L., and Hunter, B. B. 1987. Illustrated Genera of Imperfect Fungi, 4th ed. Macmillan, New York.

Barr, M. E. 1972. Preliminary studies on the Dothideales in temperate North America. Contrib. Univ. Mich. Herb. 9:523-638.

Barr, M. E. 1978. The Diaporthales in North America, with Emphasis on *Gnomonia* and its Segregates. Mycol. Mem. 7:1-232.

Barr, M. E. 1979. A classification of Loculoascomycetes. Mycologia 71: 935-957.

Barron, G. L. 1968. The Genera of Hyphomycetes from Soil. Williams and Wilkins, Baltimore, MD.

Beckman, C. H. 1987. The Nature of Wilt Diseases of Plants. American Phytopathological Society, St. Paul, MN.

Bennett, W. F., ed. 1993. Nutrient Deficiencies and Toxicities in Crop Plants. American Phytopathological Society, St. Paul, MN.

Booth, C., ed. 1971. Methods in Microbiology. Vol. 4. Academic Press, San Diego, CA.

Brako, L., Rossman, A. Y., and Farr, D. F. 1995. Scientific and Common Names of 7,000 Vascular Plants in the United States. American Phytopathological Society, St. Paul, MN.

Bradbury, J. F. 1986. Guide to Plant Pathogenic Bacteria. C.A.B. International, Wallingford, Oxon., United Kingdom.

Brandenburger, W. 1985. Parasitische Pilze an Gafaerrspflanzen in Europe. Gustav Fischer Verlag, New York (in German).

Breitenbach, J., and Kränzlin, F., eds. 1984, 1986. Fungi of Switzerland. A contribution to the knowledge in the fungal flora of Switzerland. Vol. 1, Ascomycetes; Vol. 2, Nongilled Fungi, Heterobasidiomycetes, Aphyllophorales, Gasteromycetes. Verlag Mykologia, Lucerne, Switzerland.

Brunt, A., Crabtree, K., and Gibbs, A. 1990. Viruses of Tropical Plants. Descriptions and Lists from the VIDE Database. C.A.B. International, Wallingford, Oxon., United Kingdom.

Brunt, A., Crabtree, K., Gibbs, A., and Watson, L. 1994. Viruses of Plants. C.A.B. International, Wallingford, Oxon., United Kingdom.

Burnett, H. C. 1965. Orchid Diseases. State of Florida Department of Agriculture, Division of Plant Industry, vol. 1, no. 3, Gainesville.

Bushnell, W. R., and Roelfs, A. P., eds. 1984. The Cereal Rusts. Vol. 1, Origins, Specificity, Structure, and Physiology. Academic Press, San Diego, CA.

C.A.B. International Mycological Papers, monographs on individual genera of fungi. C.A.B. International, Wallingford, Oxon., United Kingdom.

Cannon, P. F., Hawksworth, D. L., and Sherwood-Pike, M. A. 1985. The British Ascomycotina: An Annotated Checklist. C.A.B. International, Wallingford, Oxon., United Kingdom.

Carmichael, J. W., Kendrick, W. B., Conners, I. L., and Sigler, L. 1980. Genera of Hyphomycetes. University of Alberta Press, Edmonton, Alberta, Canada.

(continued on next page)

---

[a] Unfortunately, no clinic library could possibly afford or use all these publications. Additional references are listed throughout the text when special topics are discussed. Many excellent bulletins on diseases of specific crops are available from land-grant universities and county cooperative extension offices, state departments of agriculture, federal departments or ministries of agriculture, corporations, and associations. They are too numerous to list here, and new ones are constantly being published.

Table 2.5. (*continued*)

Carter, J. C. 1979. Diseases of Midwest Trees. Spec. Publ. 35, College of Agriculture, University of Illinois at Urbana-Champaign.

Chase, A. R., and Broschat, T. K., eds. 1991. Diseases and Disorders of Ornamental Palms. American Phytopathological Society, St. Paul, MN.

Chu, H. F. 1949. How to Know the Immature Insects. Wm. C. Brown, Dubuque, IA.

Clarke, B. B., and Gould, A. B. 1993. Turfgrass Patch Diseases Caused by Ectotrophic Root-Infecting Fungi. American Phytopathological Society, St. Paul, MN.

Commonwealth Mycological Institute. 1964–present. Descriptions of Pathogenic Fungi and Bacteria.

Cook, A. A. 1975. Diseases of Tropical and Subtropical Fruits and Nuts. Hafner Press, New York.

Cook, A. A. 1978. Diseases of Tropical and Subtropical Vegetables and Other Plants. Hafner Press, New York.

Cook, A. A. 1981. Diseases of Tropical and Subtropical Field, Fiber and Oil Plants. Macmillan, New York.

Cook, R. J., and Veseth, R. J. 1991. Wheat Health Management. American Phytopathological Society, St. Paul, MN.

Cooper, J. I. 1993. Virus Diseases of Trees and Shrubs. Chapman & Hall, London.

Couch, H. B. 1995. Diseases of Turfgrasses, 3rd ed. Krieger, Melbourne, FL.

Cummins, G. B. 1962. Supplement to Arthur's Manual of the Rusts in the United States and Canada. Hafner, New York.

Cummins, G. B. 1971. The Rust Fungi of Cereals, Grasses and Bamboos. Springer-Verlag, New York.

Cummins, G. B. 1978. The Rust Fungi of Composites and Legumes in North America. University of Arizona Press, Tucson.

Cummins, G. B., and Hiratsuka, Y. 1983. Illustrated Genera of Rust Fungi, rev. ed. American Phytopathological Society, St. Paul, MN.

Curran, W. S., Gray, M. E., and Shurtleff, M. C. 1991. Field Crop Scouting Manual. A Guide to Identifying & Diagnosing Pest Problems. Cooperative Extension Service, University of Illinois at Urbana-Champaign.

Davidson, R. H., and Lyon, W. F. 1979. Insect Pests of Farm, Garden, and Orchard. John Wiley & Sons, New York.

Dennis, R. W. G. 1981. British Ascomycetes. J. Cramer, Vaduz, Liechtenstein.

Dhingra, O. D., and Sinclair, J. B. 1995. Basic Plant Pathology Methods, 2nd ed. CRC Lewis, Boca Raton, FL.

Domanski, S. 1965. Fungi II. Polyporaceae I (resupinatae) Mucronoporaceae I (resupinatae). Translated from Polish 1972. U.S. Dep. Commer., Natl. Tech. Information Serv., Springfield, VA.

Domanski, S., Orlós, H., and Skirgiello, A. 1967. Fungi III. Polyporaceae II (pileatae) Mucronoporaceae II (pileatae), Ganodermataceae, Bondarzewiaceae, Boletopsidaceae, Fistulinaceae. Translated from Polish 1973. U.S. Dep. Commer., Natl. Tech. Information Serv., Springfield, VA.

Domsch, K. H., Gams, W., and Anderson, T-H. 1980. Compendium of Soil Fungi. Vols. 1 (text) and 2 (bibliography). Academic Press, San Diego, CA.

Donk, M. A. 1974. Checklist of European Polypores. Verhand. Koninklijke Nederlandse Akad. van Wetenschappen, afd. Natuurkunde 62:1-469.

Dropkin, V. H. 1989. Introduction to Plant Nematology, 2nd ed. John Wiley & Sons, New York.

Durán, R. 1987. Ustilaginales of Mexico. Taxonomy, Symptomatology, Spore Germination, and Basidial Cytology. Washington State University, Pullman.

Durán, R., and Fischer, G. W. 1961. The Genus *Tilletia*. Washington State University, Pullman.

Ellis, M. B. 1971. Dematiaceous Hyphomycetes. C.A.B. International, Wallingford, Oxon., United Kingdom.

Ellis, M. B. 1976. More Dematiaceous Hyphomycetes. C.A.B. International, Wallingford, Oxon., United Kingdom.

Ellis, M. B., and Ellis, J. P. 1985. Microfungi on Land Plants. An Identification Handbook. Macmilllan, New York.

Engelhard, A. W. 1989. Soilborne Plant Pathogens: Management of Diseases with Macro- and Microelements. American Phytopathological Society, St. Paul, MN.

Erwin, D. C., Bartnicki-Garcia, S., and Tsao, P. H. 1983. *Phytophthora*: Its Biology, Taxonomy, Ecology, and Pathology. American Phytopathological Society, St. Paul, MN.

Evans, E., ed. 1993. Plant Parasitic Nematodes in Temperate Agriculture. C.A.B. International, Wallingford, Oxon., United Kingdom.

Fahy, P. C., and Persley, G. J., eds. 1983. Plant Bacterial Diseases. A Diagnostic Guide. Academic Press, San Diego, CA.

Farr, D. F., Bills, G. F., Chamuris, G. P., and Rossman, A. Y. 1989. Fungi on Plants and Plant Products in the United States. American Phytopathological Society, St. Paul, MN.

Fichter, G. S., and Zim, H. S. 1966. Insect Pests. Golden Press, New York.

Flannigan, B., ed. 1986. Spoilage and Mycotoxins of Cereals and Other Stored Products. C.A.B. International, Wallingford, Oxon., United Kingdom.

Flint, M. L. 1990. Pests of the Garden and Small Farm. Publ. 3332. Division of Agriculture and Natural Resources, University of California, Oakland.

Forsberg, J. L. 1979. Diseases of Ornamental Plants. Spec. Publ. 3, College of Agriculture, University of Illinois at Urbana-Champaign.

Fox, R. T. V. 1993. Principles of Diagnostic Techniques in Plant Pathology. C.A.B. International, Wallingford, Oxon., United Kingdom.

Francois, L. E. 1980. Salt Injury to Ornamental Shrubs and Ground Covers. USDA Home and Garden Bull. 231. U.S. Government Printing Office, Washington, DC.

Funk, A. 1981. Parasitic Microfungi of Western Trees. Canadian Forest Service, Pacific Forest Research Centre, Victoria, British Columbia.

Funk, A. 1985. Foliar Fungi of Western Trees. Canadian Forest Service, Pacific Forest Research Centre, Victoria, British Columbia.

Gallegos, H. L., and Cummins, G. B. 1981. Uredinales (Royas) de Mexico. Vols. 1 and 2. Secretaria de Agricultura y Recursos Hidraulicos, Sinaloa, Mexico. (in Spanish).

Gilbertson, R. L., and Ryvarden, L. 1986, 1987. North American Polypores. Vol. 1, *Abortiporus–Lindtneria*; Vol. 2, *Megasporoporia–Wrightoporia*. Fungiflora, Oslo.

Ginns, J., and Lefebvre, M. N. 1993. Lignicolous Corticioid Fungi (Basidiomycota) of North America: Systematics, Distribution, and Ecology. American Phytopathological Society, St. Paul, MN.

Goodman, R. N., and Novacky, A. J. 1994. The Hypersensitive Reaction in Plants to Pathogens: A Resistance Phenomenon. American Phytopathological Society, St. Paul, MN.

Goto, M. 1992. Fundamentals of Bacterial Plant Pathology. Academic Press, San Diego, CA.

Grand, L. F. 1985. North Carolina Plant Disease Index. Agric. Res. Serv. Tech. Bull. 240, rev., North Carolina State University, Raleigh.

Hammerschlag, F. A., and Litz, R. E., eds. 1992. Biotechnology of Perennial Fruit Crops. C.A.B. International, Wallingford, Oxon., United Kingdom.

Hampton, R., Ball, E., and DeBoer, S., eds. 1990. Serological Methods for Detection and Identification of Viral and Bacterial Plant Pathogens: A Laboratory Manual. American Phytopathological Society, St. Paul, MN.

Hanlin, R. T. 1990. Illustrated Genera of Ascomycetes. American Phytopathological Society, St. Paul, MN.

Harrington, T. C., and Cobb, F. W., Jr., eds. 1988. Leptographium Root Diseases on Conifers. American Phytopathological Society, St. Paul, MN.

Hawksworth, D. L. 1974. Mycologist's Handbook. An Introduction to the Principles of Taxonomy and Nomenclature in the Fungi and Lichens. C.A.B. International, Wallingford, Oxon., United Kingdom.

Hayward, A. C., and Hartman, G. L., eds. 1994. Bacterial Wilt: The Disease and Its Causative Agent, *Pseudomonas/Solanacearum*. C.A.B. International, Wallingford, Oxon., United Kingdom.

Hepting, G. H. 1971. Diseases of Forest and Shade Trees of the United States. USDA Forest Service Handbook 386. U.S. Government Printing Office, Washington, DC.

(*continued on next page*)

Table 2.5. (*continued*)

Hill, D. S., and Waller, J. M. 1982. Pests and Diseases of Tropical Crops. Vol. 1, Principles and Methods of Control. Longman, London.

Hill, S. A., 1984. Methods in Plant Virology. Methods in Plant Pathology, Vol. 1. Blackwell Scientific, London.

Hillocks, R. J., ed. 1992. Cotton Diseases. C.A.B. International, Wallingford, Oxon., United Kingdom.

Hirschhorn, E. 1986. Las Ustilaginales de la Flora Argentina. Publicacion Especial, Comision de Investigaciones Cientificas, La Plata, Argentina. (in Portuguese).

Hoffard, W. H., Anderson, R. L., and Sites, W. H. 1980. How to Collect and Prepare Forest Insects and Disease Organisms and Plant Specimens for Identification. SA-GR 13. USDA Forest Service, Atlanta, GA.

Holliday, P. 1980. Fungus Diseases of Tropical Crops. Cambridge University Press, Cambridge, United Kingdom.

Holliday, P. 1989. A Dictionary of Plant Pathology. Cambridge University Press, Cambridge, United Kingdom.

Horst, R. K. 1990. Westcott's Plant Disease Handbook, 5th ed. Van Nostrand Reinhold, New York.

Hughes, S. J. 1958. Revisiones hyphomycetum aliquot cum appendice nominibus rejiciendis. Can. J. Bot. 36:727-836.

Hunt, D. J. 1993. Aphelenchida, Longidoridae and Trichodoridae: Their systematics and bionomics. C.A.B. International, Wallingford, Oxon., United Kingdom.

Hunt, N. T., and Baker, J. R., eds. 1982. Insect and Related Pests of Field Crops. North Carolina Agricultural Extension Service AG-271. North Carolina State University, Raleigh.

Jarvis, W. R. 1992. Managing Diseases in Greenhouse Crops. American Phytopathological Society, St. Paul, MN.

Jepson, S. B. 1987. Identification of Root-Knot Nematodes. C.A.B. International, Wallingford, Oxon., United Kingdom.

Johnston, A., and Booth, C., eds. 1983. Plant Pathologist's Pocketbook, 2nd ed. C.A.B. International, Wallingford, Oxon., United Kingdom.

Jones, R. K., and Lambe, R. C., eds. 1982. Diseases of Woody Ornamental Plants and Their Control in Nurseries. North Carolina Agricultural Extension Service AG-286. North Carolina State University, Raleigh.

Juelich, W. 1984. Die Nichtblaetterpilze, Gallertpilze und Bauchpilze. Aphyllophorales, Heterobasidiomycetes, Gasteromycetes. Gustav Fischer Verlag, New York. (in German).

Juelich, W., and Stalpers, J. A. 1980. The Resupinate Non-poroid Aphyllophorales of the Temperate Northern Hemisphere. North-Holland, Amsterdam.

Junell, L. 1967. Erysiphaceae of Sweden. Symb. Bot. Ups. 19:1-117.

Kader, A. D. 1992. Postharvest Technology of Horticultural Crops. Publ. 3311. Division of Agriculture and Natural Resources, University of California, Oakland.

Kakishima, M. 1982. A taxonomic study on the Ustilaginales in Japan. Mem. Inst. Agric. For., Univ. Tsukuba 1:1-124. (in Japanese with English summary).

Keifer, H. H., Baker, E. W., Kono, T., Delfinado, M., and Styer, W. E. 1982. An Illustrated Guide to Plant Abnormalities Caused by Eriophyid Mites in North America. USDA Agric. Handbook 573, U.S. Government Printing Office, Washington, DC.

Kobayashi, T. 1970. Taxonomic Studies of Japanese Diaporthaceae with Special Reference to their Life-Histories. Bull. Gov. For. Exp. Stn. (Japan) 226:1-242.

Korf, R. P. 1972. Synoptic key to the genera of the Pezizales. Mycologia 64:937-994.

Kranz, J., Schmutterer, H., and Koch, W., eds. 1977. Diseases, Pests and Weeds in Tropical Crops. Verlag Paul and Perrey, Berlin.

Lacasse, N. L., and Treshow, M., eds. 1976. Diagnosing Vegetation Injury Caused by Air Pollution. Appl. Sci. Assoc. Inc. for U.S.-EPA Library, EPA Air Pollution Institute, Research Triangle Park, NC.

Lelliott, R. A., and Stead, D. E. 1987. Methods for the Diagnosis of Bacterial Diseases of Plants. Methods in Plant Pathology, Vol. 2. Blackwell Scientific, London.

Lenné, J. M., ed. 1994. Diseases of Tropical Pasture Plants. C.A.B. International, Wallingford, Oxon., United Kingdom.

Lindsey, J. P., and Gilbertson, R. L. 1978. Basidiomycetes That Decay Aspen in North America. J. Cramer, Vaduz, Liechtenstein.

Lorenz, O. A., and Maynard, D. N. 1988. Knott's Handbook for Vegetable Growers, 3rd ed. John Wiley & Sons, New York.

Luc, M., Sikora, R. H., and Bridge, J., eds. 1990. Plant Parasitic Nematodes in Subtropical and Tropical Agriculture. C.A.B. International, Wallingford, Oxon., United Kingdom.

Lucas, G. B. 1975. Diseases of Tobacco, 3rd ed. Biological Consulting Association, P.O. Box 572-C, Raleigh, NC.

Lucas, G. B., Campbell, C. L., and Lucas, L. T. 1988. Introduction to Plant Diseases. Identification and Management. 2nd ed. Van Nostrand Reinhold, New York.

MacNab, A. A., Sherf, A. F., and Springer, J. K. 1983. Identifying Diseases of Vegetables. The Pennsylvania State University, University Park.

Mai, W. F., and Lyon, H. H. 1975. A Pictorial Key to Genera of Plant-Parasitic Nematodes, 4th ed. Comstock Publishing, a Division of Cornell University Press, Ithaca, NY.

Malloch, D. 1981. Moulds—Their Isolation, Cultivation, and Identification. University of Toronto Press, Toronto.

Manion, P. D. 1991. Tree Disease Concepts, 2nd ed. Prentice-Hall, Englewood Cliffs, NJ.

Manion, P. D., and Lachance, D. 1992. Forest Decline Concepts. American Phytopathological Society, St. Paul, MN.

Maramorosch, K., and Raychaudhuri, S. P., eds. 1988. Mycoplasma Diseases of Crops. Basic and Applied Aspects. Springer-Verlag, New York.

Martens, J. W., Seaman, W. L., and Atkinson, T. G., eds. 1984. Diseases of Field Crops in Canada. The Canadian Phytopathological Society, Harrow, Ontario.

Matthews, G. A., and Hislop, E. C. 1993. Application Technology for Crop Protection. C.A.B. International, Wallingford, Oxon., United Kingdom.

Matthews, R. E. F. 1991. Plant Virology, 3rd ed. Academic Press, San Diego, CA.

Matthews, R. E. F., ed. 1993. Diagnosis of Plant Virus Diseases. CRC Press, Boca Raton, FL.

Maude, R. B. 1995. Seedborne Diseases and their Control. C.A.B. International, Wallingford, Oxon., United Kingdom.

McGee, D. C. 1988. Maize Diseases: A Reference Source for Seed Technologists. American Phytopathological Society, St. Paul, MN.

McGee, D. C. 1992. Soybean Diseases: A Reference Source for Seed Technologists. American Phytopathological Society, St. Paul, MN.

Melouk, H. A., and Shokes, F. M., eds. 1995. Peanut Health Management. American Phytopathological Society, St. Paul, MN.

Metcalf, R. L., and Metcalf, R. A. 1993. Destructive and Useful Insects: Their Habits and Control, 5th ed. McGraw-Hill, New York.

Miller, J. F. 1972. Weeds of the Southern States. Cooperative Extension Service, University of Georgia, Athens.

Miller, O. K., Jr. Mushrooms of North America. E. P. Dutton, New York.

Mitchell, R. T., and Zim, H. S. 1964. Butterflies and Moths. Golden Press, New York.

Mordue, J. E. M., and Ainsworth, G. C. 1984. The Ustilaginales of the British Isles. Mycol. Pap. 154:1-96.

Morrill, W. L. 1995. Insect Pests of Small Grains. American Phytopathological Society, St. Paul, MN.

Nickle, W. R., ed. 1991. Manual of Agricultural Nematology. Marcel Dekker, New York.

Nyvall, R. F. 1989. Field Crop Diseases Handbook, 2nd ed. Van Nostrand Reinhold, New York.

O'Brien, M. J., and Rich, A. E. 1976. Potato Diseases. USDA Agricultural Handbook 474. U.S. Government. Printing Office, Washington, DC.

O'Donnell, K. L. 1979. Zygomycetes in Culture. University of Georgia Press, Athens.

Ogawa, J. M., and English, H. 1991. Diseases of Temperate Zone Tree Fruit and Nut Crops. Publ. 3345. Division of Agriculture and Natural Resources, University of California, Oakland.

(*continued on next page*)

Table 2.5. (*continued*)

Ou, S. H. 1985. Rice Diseases, 2nd ed. C.A.B. International, Wallingford, Oxon., United Kingdom.

Parker, C. A., Rovira, A. D., Moore, K. J., Wong, P. T. W., and Kollmorgen, J. F., eds. 1985. Ecology and Management of Soilborne Plant Pathogens. American Phytopathological Society, St. Paul, MN.

Persley, D., ed. 1993. Diseases of Fruit Crops. Department of Primary Industries, Brisbane, Queensland, Australia.

Persley. D., ed. 1994. Diseases of Vegetable Crops. Department of Primary Industries, Brisbane, Queensland, Australia.

Persley, G. J., ed. 1995. Biotechnology and Integrated Pest Management. C.A.B. International, Wallingford, Oxon., United Kingdom.

Pfleger, F. L., and Linderman, R. G. 1994. Mycorrhizae and Plant Health. American Phytopathological Society, St. Paul, MN.

Pirone, P. P. 1970. Diseases and Pests of Ornamental Plants, 5th ed. John Wiley & Sons, New York.

Pirone, P. P., Hartman, J. R., Sall, M. S., and Pirone, T. P. 1988. Tree Maintenance, 6th ed. Oxford University Press, New York.

Powell, C. C., and Rossetti, R. 1992. The Healthy Indoor Plant. Roswell, Columbus, OH.

Riffle, J. W., and Peterson, G. W. 1986. Diseases of Trees in the Great Plains. USDA Forest Service, General Technical Report RM-129, Fort Collins, CO.

Rossman, A. Y., Palm, M. E., and Spielman, L. J. 1987. A Literature Guide for the Identification of Plant Pathogenic Fungi. American Phytopathological Society, St. Paul, MN.

Rotem, J. 1994. The Genus *Alternaria*: Biology, Epidemiology, and Pathogenicity. American Phytopathological Society, St. Paul, MN.

Rowe, R. C., ed. 1993. Potato Health Management. American Phytopathological Society St. Paul, MN.

Ryvarden, L. 1976. The Polyporaceae of North Europe. Vol. 1, *Albatrellus–Incrustoporia*. Fungiflora, Oslo.

Ryvarden, L. 1978. The Polyporaceae of North Europe. Vol. 2, *Inonotus–Tyromyces*. Fungiflora, Oslo.

Ryvarden, L., and Johansen, I. 1980. A Preliminary Polypore Flora of East Africa. Fungiflora, Oslo.

Saettler, A. W., Schaad, N. W., and Roth, D. A., eds. 1989. Detection of Bacteria in Seed and Other Planting Material. American Phytopathological Society, St. Paul, MN.

Savulescu, T. 1957. Ustilaginalele din Republica Populara Romina. Vols. 1 and 2. Editura Academiei Republicii Populare Romine. (in Rumanian).

Schaad, N. W., ed. 1988. Laboratory Guide for Identification of Plant Pathogenic Bacteria, 2nd ed. American Phytopathological Society, St. Paul, MN.

Schenck, N. C., ed. 1982. Methods and Principles of Mycorrhizal Research. American Phytopathological Society, St. Paul, MN.

Schots, A., ed. 1994. Modern Assays for Plant Pathogenic Fungi: Identification, Detection and Quantification. C.A.B. International, Wallingford, Oxon., United Kingdom.

Semancik, J. S., ed. 1987. Viroids and Viroid-like Pathogens. CRC Press, Boca Raton, FL.

Sherf, A. F., and MacNab, A. A. 1986. Vegetable Diseases and Their Control, 2nd ed. John Wiley & Sons, New York.

Shukla, D. D., and Ward, C. W. 1994. The Potyviruses. C.A.B. International, Wallingford, Oxon., United Kingdom.

Shurtleff, M. C. 1966. How to Control Plant Diseases in Home and Garden, 2nd ed. Iowa State University Press, Ames.

Shurtleff, M. C., Fermanian, T. W., and Randall, R. 1987. Controlling Turfgrass Pests. Prentice-Hall, Englewood Cliffs, NJ.

Siddiqi, M. R. 1985. Tylenchida. Parasites of Plants and Insects. C.A.B. International, Wallingford, Oxon., United Kingdom.

Sigee, D. C. 1993. Bacterial Plant Pathology. Cellular and Molecular Aspects. Cambridge University Press, Cambridge, United Kingdom.

Sinclair, W. A., Lyon, H. H., and Johnson, W. T. 1987. Diseases of Trees and Shrubs. Comstock Publishing, a division of Cornell University Press, Ithaca, NY.

Singleton, L. L., Mihail, J. D., and Rush, C. M., eds. 1992. Methods for Research on Soilborne Phytopathogenic Fungi. American Phytopathological Society, St. Paul, MN.

Sivanesan, A. 1984. The Bitunicate Ascomycetes and Their Anamorphs. J. Cramer, Vaduz, Liechtenstein.

Skelly, J. M, Davis, D. D., Merrill, W., Cameron, E. A., Brown, H. D., Drummond, D. B., and Dochinger, L. S., eds. 1987. Diagnosing Injury to Eastern Forest Trees. Agricultural Information Service, The Pennsylvania State University, University Park.

Skerritt, J. H., and Appels, R., eds. 1995. New Diagnostics in Crop Sciences. C.A.B. International, Wallingford, Oxon., United Kingdom.

Smith, A. H., Smith, H. V., and Weber, N. S. 1981. How to Know the Non-gilled Mushrooms. Wm. C. Brown, Dubuque, IA.

Smith, A. H., and Weber, N. S. 1980. The Mushroom Hunter's Field Guide. University of Michigan Press, Ann Arbor.

Smith, D., and Onions, H. S. 1995. The Preservation and Maintenance of Living Fungi, 2nd ed. C.A.B. International, Wallingford, Oxon., United Kingdom.

Smith, D. M., ed. 1984. The Ortho Problem Solver, 2nd ed. Ortho Information Service, Chevron Chemical Co., San Francisco, CA.

Smith, I. M., Dunez, J., Lelliott, R. A., Phillips, D. H., and Archer, S. A., eds. 1988. European Handbook of Plant Diseases. Blackwell Scientific, London.

Smith, J. D., Jackson, N., and Woolhouse, A. R. 1989. Fungal Diseases of Amenity Turf Grasses, 3rd ed. E. & F. N. Spon, London.

Sneh, B., Burpee, L., and Ogoshi, A. 1991. Identification of *Rhizoctonia* Species. American Phytopathological Society, St. Paul, MN.

Snell, W. H., and Dick, E. A. 1971. A Glossary of Mycology. Harvard University Press, Cambridge, MA.

Snowdon, A. L. 1990. A Color Atlas of Post-Harvest Diseases and Disorders of Fruits and Vegetables. Vol. 1, General Introduction and Fruits. CRC Press, Boca Raton, FL.

Snowdon, A. L. 1992. A Color Atlas of Post-Harvest Diseases and Disorders of Fruits and Vegetables. Vol. 2, Vegetables. CRC Press, Boca Raton, FL.

Sorenson, K. A., and Baker, J. R. 1983. Insect and Related Pests of Vegetables. Agricultural Extension Service AG-295. North Carolina State University, Raleigh, NC.

Spencer, D. M., ed. 1978. The Powdery Mildews. Academic Press, San Diego, CA.

Spencer, D. M., ed. 1981. The Downy Mildews. Academic Press, San Diego, CA.

Sprague, H. B. 1964. Hunger Signs in Crops. A Symposium. David McKay, New York.

Stalpers, J. A. 1978. Identification of wood-inhabiting Aphyllophorales in pure culture. Stud. Mycol. 16, Centralb. Schimmel-kult., Baarn, Switzerland.

Stevens, R. B., ed. 1981. Mycology Guidebook. University of Washington Press, Seattle.

Stover, R. H. 1972. Banana, Plantain and Abaca Diseases. C.A.B. International, Wallingford, Oxon., United Kingdom.

Streets, R. B., Sr. 1978. The Diagnosis of Plant Diseases. University of Arizona Press, Tucson.

Strider, D. L., ed. 1985. Diseases of Floral Crops. Vols. 1 and 2. Praeger, New York.

Stuckey, J. M., Monaco, T. J., and Worsham, A. D. 1980. Identifying Seedling and Mature Weeds Common in the Southeastern United States. Agricultural Extension Series. North Carolina State University, Raleigh.

Sutton, B. C. 1980. The Coelomycetes. Fungi Imperfecti with Pycnidia, Acervuli and Stromata. C.A.B. International, Wallingford, Oxon., United Kingdom.

Swain, R. W., ed. 1985. Plant Physiological Disorders. Reference Book 223. Ministry of Agriculture, Fisheries and Food, Bristol, United Kingdom.

Tattar, T. A. 1989. Diseases of Shade Trees, rev. ed. Academic Press, New York.

Teng, P. S., ed. 1987. Crop Loss Assessment and Pest Management. American Phytopathological Society, St. Paul, MN.

Thorne, G. 1961. Principles of Nematology. McGraw-Hill, New York.

(*continued on next page*)

Table 2.5. (*continued*)

Thurston, H. D. 1984. Tropical Plant Diseases. American Phytopathological Society, St. Paul, MN.

Tuite, J. 1969. Plant Pathology Methods. Fungi and Bacteria. Burgess, Minneapolis, MN.

Tuite, J. 1988. Plant Pathology Methods. Laboratory Exercises. Purdue University, West Lafayette, IN.

Tuite, J., and Cantone, F. 1990. Plant Pathology Methods. Laboratory Exercises. Teacher's Supplement. Purdue University, West Lafayette, IN.

Vanky, K. 1985. Carpathian Ustilaginales. Symb. Bot. Ups. 24:2:1-309.

Van Regenmortel, M. H. V. 1982. Serology and Immunochemistry of Plant Viruses. Academic Press, San Diego, CA.

von Arx, J. A. 1981. The Genera of Fungi Sporulating in Pure Culture, 3rd ed. J. Cramer, Vaduz.

von Arx, J. A., and Mueller, E. 1975. A re-evaluation of the bitunicate Ascomycetes with keys to families and genera. Stud. Mycol. 9:1-159.

Walkey, D. G. A. 1991. Applied Plant Virology, 2nd ed. Chapman and Hall, London.

Walla, J. M., Holderness, M., and Ritchie, B. J. 1994. Plant Clinic Handbook. C.A.B. International, Wallingford, Oxon., United Kingdom.

Watschke, T. L., Dernoeden, P. H., and Shetlar, D. J. 1995. Managing Turfgrass Pests. CRC Press, Boca Raton, FL.

Weber, G. F. 1973. Bacterial and Fungal Diseases of Plants in the Tropics. University of Florida Press, Gainesville.

Webster, J. 1980. Introduction to Fungi, 2nd ed. Cambridge University Press, Cambridge, United Kingdom.

Wellman, F. L. 1972. Tropical American Plant Diseases. (Neotropical Phytopathological Problems). Scarecrow Press, Metuchen, NJ.

Westcott, C. 1964. The Gardener's Bug Book. Doubleday, New York.

Wilson, M., and Henderson, D. M. 1966. British Rust Fungi. Cambridge University Press, Cambridge, United Kingdom.

Wilson, E. E., and Ogawa, J. M. 1979. Fungal, Bacterial, and Certain Nonparasitic Diseases of Fruit and Nut Crops in California. University of California, Berkeley.

Wingfield, M. J., Seifert, K. A., and Webber, J. F., eds. 1993. *Ceratocystis* and *Ophiostoma*: Taxonomy, Ecology, and Pathogenicity. American Phytopathological Society, St. Paul, MN.

Wyllie, T. D., and Scott, D. H. 1988. Soybean Diseases of the North Central Region. American Phytopathological Society, St. Paul, MN.

Zeigler, R. S., Leong, S. A., and Teng, P. S., eds. 1994. Rice Blast Disease. C.A.B. International, Wallingford, Oxon., United Kingdom.

Zheng, R.-Y. 1985. Genera of the Erysiphaceae. Mycotaxon 22:209-263.

Ziller, W. G. 1974. The Tree Rusts of Western Canada. Canadian Forest Service, Department of the Environment, Victoria, British Columbia.

Zillinsky, F. J. 1983. Common Diseases of Small Grain Cereals. A Guide to Identification. CIMMT, Mexico, DF, Mexico (available from Agribookstore, IADS Operations, 1611 N. Kent St., Arlington, VA).

Zuckerman, B. M., Mai, W. F., and Krusberg, L. R. 1990. Plant Nematology Laboratory Manual, rev. ed. University of Massachusetts Agricultural Experiment Station, Amherst.

Zycha, H., Siepmann, R., and Linnemann, G. 1969. Mucorales, eine Beschreibung aller Gattungen und arten dieser Pilzgruppe mit einem Beitrag zur Gattung *Mortierella*. J. Cramer, Lehrte, Germany. (in German).

Fig. 2.11. Field kit suggested for county agents in North Carolina. **a,** Metal tool box; **b,** file folder (bellows) with compartments for forms and information sheets; **c,** Agricultural Chemicals Manual; **d,** vials with alcohol; **e,** pruning saw; **f,** butcher knife; **g,** hand trowel; **h,** pruning clippers; **i,** hand tools; **j,** 50-foot tape; **k,** soil-sampling probe; **l,** tape and string; **m,** tags with wire fasteners; **n,** first aid kit; **o,** plastic bags; **p,** 35-mm film canisters for paper clips, rubber bands, specimens, etc., and **q,** soft-leaded pencil. Not pictured and stored in the tool box: paper mailing canister, stakes, hatchet, flashlight, soap, paper towels, chisel, increment borer, paring knife, tweezers, razor blades, magnifying glass, thermometer, and pocket knife. For other items, see Table 2.2.

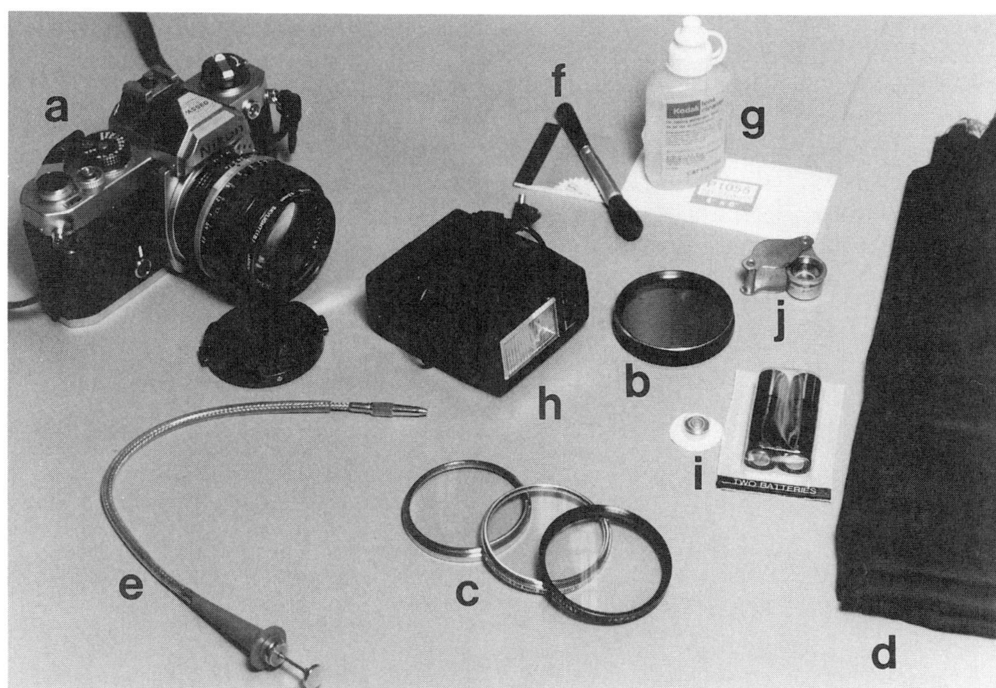

Fig. 2.12. Camera equipment for field use. **a,** Camera, 35-mm, single-lens reflex with through-the-lens light metering and standard 55-mm lens and equipped with a delayed exposure lever; **b,** polarizing filter to reduce sheen from foliage and waxy fruit; **c,** series of screw-on close-up lenses; **d,** black velvet background cloth; **e,** remote shutter-release cable to avoid camera movement with time exposures; **f,** camel's-hair lens brush; **g,** lens cleaning fluid and paper; **h,** automatic electronic flash for highlighting subject and underexposing background; **i,** extra batteries; and **j,** a 20× magnifying glass. Not illustrated: carrying case to protect equipment from dust, rain, and shock (a white exterior is preferred to reduce heat buildup in the sun); 35- to 80-mm zoom lens for framing distant views; macrolens for close-up photography; automatic film advance for rapid photography; and tripod.

Fig. 2.13. Small field clinic. **a,** Cabinet on coasters for storing and moving equipment, supplies, and reference materials; **b,** secondhand dissecting and compound microscopes; **c,** soil pH and soluble salts meters; **d,** wash bottle; **e,** area for dissecting needles, tweezers, forceps, paring knife, scalpel, razor blades, dropper bottles, glass microscope slides, coverslips, pencils, transparent tape, etc.; **f,** packaging material; **g,** plastic freezer bags; **h,** forms and laboratory specimen mailers; **i,** clinic forms; and **j,** bottles. For other items not pictured, see Table 2.3.

remedial action can be taken. These clinics must have access to formal clinics and laboratories (e.g., state and land-grant universities) as backups to diagnose samples they cannot handle. A growing number of consultants and practitioners also have fair to excellent private diagnostic clinics.

Field clinics can be established for the diagnosis of specific diseases, which may involve sophisticated procedures. However, these clinics generally lack breadth and are often unable to handle routine diagnoses of other diseases. For example, a clinic may have equipment and a few inexpensive supplies to serologically detect the presence of potato virus Y (PVY) in pepper plants intended for shipment and transplanting. This clinic would be efficient in diagnosing PVY in pepper but would be unable to diagnose many diseases caused by other viruses, pathogens, and agents that require different procedures, equipment, and supplies.

Grower's Name___________________________   File No.___________________

TREE AND SHRUB DISEASE IDENTIFICATION FORM

Information to Supplement the Plant Disease Identification Form

Each year we receive a number of twig and branch samples accompanied by the question "What's wrong with my tree?" or "What killed my shrub?" Many times the cause of the problem is not actually located on the specimen. Such problems are often best diagnosed only after careful inspection of the tree or shrub and its growing site. However, because we cannot visit the growing site, your observations and information can be of tremendous help to us. The more information you are able to supply, the better we are able to narrow down the cause of the problem with your plant.

Please complete the following information in addition to the Plant Disease Identification Form.

Symptoms

1. What is the distribution of affected branches? (Indicate location of damage on the diagram below.)

☐ Scattered branches all over tree/shrub

☐ One portion of tree/shrub only

☐ Several portions of tree/shrub

2. What is the progression of symptoms throughout the tree/shrub (e.g. from lower branches to top; from branch tips toward trunk, etc.)?

____

____

____

3. What is the order of symptom development on the leaves or needles? _______________________________________

____

4. Has the problem occurred: slowly_____ or rather suddenly_____? How rapidly? _______________________

5. Are there any cankers or sunken areas in the main limbs or trunk? _______________________________

_______________________________________ Any injuries? ________

Where? _______________________________________

6. If there are splits or cracks in the trunk, please indicate extent of damage and the side of the tree/shrub with damage:

____

7. Does the tree (especially maples) have a girdling root up near the trunk, at soil level or below soil level? (Look for a large root forming a "noose" around the trunk within 18 inches below the soil surface. Mature trees (15 years or more) lack normal flare on buttress roots.) ____________

8. Are other plants of this type showing symptoms? ________ Are there healthy plants of the same type near by? ________

(continued on reverse side)

UNIVERSITY OF KENTUCKY • COLLEGE OF AGRICULTURE • COOPERATIVE EXTENSION SERVICE
Agriculture • Home Economics • 4-H • Development

Growing Site

1. Is the tree/shrub surrounded by pavement or buildings? ________

2. Is the tree/shrub in: full sun _____ partial shade _____ full shade _____?

3. Is the soil around the tree/shrub compacted? ________

4. Have the roots been disturbed in the past 5 years (e.g. by construction, etc.)? ________ Has the grade around the tree/shrub been changed in the past 5 years (i.e. soil added or removed from root area)? ________

5. Is the tree/shrub in a site where the topsoil was removed (e.g. in a new home development)? ________

6. Is the tree/shrub located in a low, wet area or at the base of a downspout? ________ Does water stand on the soil after a heavy rain? ________

Cultural Practices

1. What were the results of the last soil test taken for the site?

pH__________________   P__________________   K__________________   Ca-Mg__________________

2. When was the last time the tree/shrub or surrounding lawn was fertilized? __________________________
(Be sure to indicate type of fertilizer and rates on the Plant Disease Identification Form.)

3. How was the fertilizer applied (e.g. cores dug, broadcast, injection, etc.)? __________________________

____

4. Have herbicides been used to control weeds? __________________________

5. Is the tree/shrub watered thoroughly during dry periods (i.e. the equivalent of 1 inch of rain per week)? __________

6. Is there mulch around tree/shrub base? ________

7. Has the tree/shrub been bumped by lawn mowers, string trimmers or other equipment? ________

8. If plant was transplanted in the last 2 to 3 years, describe method __________________________

____

Was plant: balled and burlapped ________ container grown ________ bare rooted ________?

Additional Information

5M—7-83

PLANT DISEASE IDENTIFICATION FORM

File No. _______________________

**Send all copies to: Plant Disease Diagnostic Laboratory
Department of Plant Pathology, University of Kentucky, Lexington, KY 40546; OR P. O. Box 469, Princeton, KY 42445**

Grower _________________________________   Name of Plant _________________________________

Address _________________________________   Variety _________________________________

_________________________ Zip_______   Date Collected ____________ LAB USE ___________
(date rec'd.)

Parts Diseased: □ roots □ stems □ leaves/needles □ twigs/branches □ trunk □ buds □ fruit □ flowers

Symptoms: □ spot □ mottling □ distortion □ yellowing □ stunting □ wilt □ canker □ shot hole □ dieback
□ root rot □ stem rot □ burn or scorch □ galls or swellings □ fruit decay □ other ____________

Location of Plant: □ field □ plant bed □ garden □ yard □ orchard □ greenhouse □ nursery □ indoors

Pattern of Diseased Plants: □ single plant □ scattered plants □ group of plants □ large area □ every plant
Can you associate pattern with: □ low, wet area   □ high, dry area   □ slopes   □ shaded areas   □ sunny area

% Affected   Date problem   Planting date   Soil type ____________
Plants____________   first noticed__________   or age of plant _________   Soil draining __________

Previous Crop(s) (include year): ____

Recent weather (temperature, rainfall, watering, etc.): ____

Chemicals Applied to This Crop: (include name, rate and date of application)
Fertilizer____
Herbicides____
Fungicides____
Insecticides____

Unusual disturbances: □ lightning □ hail □ construction □ pruning □ injuries □ soil compaction
□ transplanting

Additional information: ____
____
____

Signed_________________________________   County _________________________________
County Extension Agent or Specialist

Disease Diagnosis and Control

____
____
____
____
____
____
____
____
____
____
____

Date__________________   ____
Plant Pathologist

Distribution of copies: 1—Grower; 2—County Extension Office; 3—Plant Diagnostic Laboratory

FLORIDA COOPERATIVE EXTENSION SERVICE
UNIVERSITY OF FLORIDA
INSTITUTE OF FOOD AND AGRICULTURAL SCIENCES

| COOPERATIVE EXTENSION SERVICE | AGRICULTURAL EXPERIMENT STATIONS |
| SCHOOL OF FOREST RESOURCES AND CONSERVATION | COLLEGE OF AGRICULTURE |

white: Pest Management
yellow: Grower
pink: County Agent

PLANT PEST IDENTIFICATION

Grower: _____________________ Address: _____________________ Date: _____________

Location of Field With Problem: _____________________

Crop Variety: _____________________

Area of Injury (Circle): Leaves Stem Branch Fruit Roots Flowers Others: _____________

Nature of Injury (Circle): Leaf Spot Root Rot Fruit Rot Mosaic Blossom Blight

   Wilt Stem Canker or Spot Gall or Swelling Feeding Injury Other: _____________

Symptoms Appeared in Past: _____ Days _____ Weeks _____ Months

Prevalence of Injury (Circle): Entire Planting Localized Area Scattered Plants

Crop Rotation (Last 3 Crops on Field): _____________________

Weather Conditions Prior to Symptom Development:

   Temperature Range _____________ Rainfall _____________ High Winds _____________

Remarks: (Include any other information you think might help identify the trouble)

Forward To:
Pest Management (IPM)
Agricultural Research and Education Center
University of Florida
Institute of Food and Agricultural Sciences
18905 S. W. 280 Street
Homestead, Florida 33031
County

Diagnosis and Recommendation

Ken Pohronezny
Associate Professor
(Pest Management Specialist)

Collecting Plant Specimens
for Disease Diagnosis

1) **Examine** the entire plant(s) and growing site very carefully.

2) **Collect several plant specimens.** Submit a number of samples showing various stages of disease development. One sample is frequently not enough.

3) **Collect the entire plant,** whenever possible. Plants should be dug (not pulled) so the root system remains intact. When the whole plant cannot be submitted, collect as large a portion of the diseased plant as possible. Branch specimens should be at least 8 to 12 inches long.

4) **Select samples that are still alive** and showing symptoms. Completely dead plants are of no value in plant disease diagnosis. We cannot perform autopsies.

5) **Fill out this form completely.** Provide **all** the information requested. Woody specimens must also be accompanied by a Tree and Shrub Disease Identification Form.

Packaging Specimens for Mailing

1) **Use plastic bags.** Plant specimens often dry out enroute and therefore all samples should be enclosed in a plastic bag. (Exceptions are fleshy fruits, vegetables or tubers in the early stage of decay which should be wrapped individually in newspaper. This also applies to mushrooms submitted for identification.)

2) **Do not add water or wet paper towels.** Moisture hastens decay and the growth of saprophytic organisms, obscuring the real problem.

3) **Identifying labels and diagnostic forms should never be packaged** where they will come **in contact with any soil.**

4) **Tie off the root ball.** Place the root system (with any accompanying soil) in a plastic bag and tie it securely to the lower stem of the plant. Place a second bag over the foliage portion of the plant in order to reduce desiccation.

5) **Package delicate material in a sturdy box or mailing tube.** Padded envelopes are good for flat specimens that are not likely to be easily crushed.

6) **Mail all packages FIRST CLASS.**

7) **Mail early in the week** to avoid weekend delays.

UNIVERSITY OF ILLINOIS
PLANT CLINIC SPECIMEN DATA FORM

Plant Clinic
1401 W. St. Mary's Road
Urbana, IL 61801 (217)333-0519

Submitter_______________________________________

Grower____

    Commercial__________   Home Grower________

County____

Office Use Only	
Plant Clinic #	
Date Received	
County	
Charge	
Date Paid	Ck.#

Send response to: Name____

              Address______________________________________

              City________________________ State________ Zip__________

              Phone # (________)__________________

Crop or Plant Name________________________ Variety____________________________

ACCURATE DIAGNOSIS IS DEPENDENT ON COMPLETE INFORMATION

Describe Problems or Symptoms/Sketch Distribution:

Symptoms Appeared in Past: Days________ Weeks________ Months__________
Describe Conditions Prior To Symptom Development:
    Temperature__________ Rainfall__________ Other__________________________

Planting History: Crop Two Years Ago__________________ Crop One Year Ago________________
Soil Type: _____________________ pH________ % Organic Matter________________________
Soil Test Information: ____
Type of Nitrogen Application: ____
Chemicals Applied This Year:
    Fertilizer_______________________ Type of Application ________________________
    Herbicide(s)____
         Rates____
         Type of Application____
Chemicals Applied Last Year____

Ornamentals:
Approximate Age and Size:____
Condition of Nearby Species:____

NEMATODE SOIL SAMPLE FORM

Clinic No. (s)
N-

University of Illinois at Urbana-Champaign
PLANT CLINIC, St. Mary's Road, Urbana, IL 61801
(217) 333-0519

Submitted by________________________ Date of Sampling________________

Address____________________________ Date Submitted__________________

County__________________ Phone__________ Date Received________________

Clinic Numbers	Sample Number	Soil Type	Present Crop (Summer)		Previous Crop (Summer)		Crop 2 yrs ago (Summer)		Next Crop To Be Grown	Past Nematicide/ Insecticide Treatments	
			Crop	Variety	Crop	Variety	Crop	Variety		Year	Chemical
N-											
N-											
N-											
N-											
N-											
N-											
N-											
N-											
N-											
N-											

Circle Appropriate Information

Distribution of Symptoms: Scattered   Clustered in Spots   Uniform

Association with Terrain or Soil Type: Yes   No

Weather Conditions Prior to Symptom Development —
　Rainfall: Low Medium High　　Temperature: Low Medium High

Soil Test Information or Fertilizer Application — 2 yrs past________ 1 yr past________ Current yr________

Herbicides Applied This Year________________________________

COMMENTS:

For Lab Use	Soil	Roots	Juveniles	Cysts	Comments:
Date Processed					
Date Read					

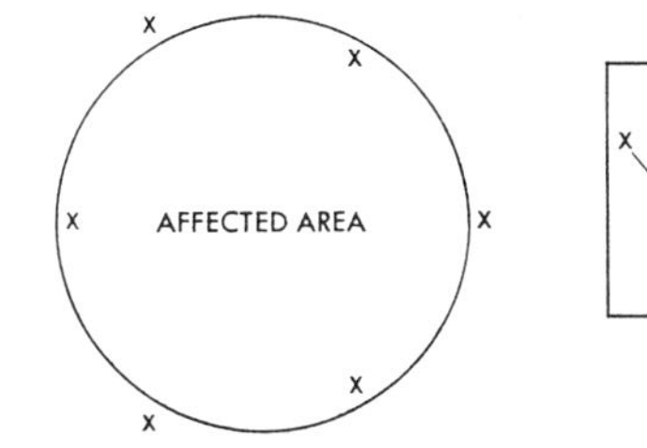

Key Points for Submitting Nematode Samples

Collecting Samples

1. Take the sample from the margin of the affected areas to a depth of 8-10". When sampling from sandy soils, go to a depth of 10-12".
2. Dig several samples from the affected area, mix; and package 1 qt. of the mixed soil.
3. Include a handful of roots, including feeder roots if possible.

Packing and Shipping Samples

1. Use a nematode soil sample bag or a thick, sturdy plastic bag.
2. Tie the bags closed with twist ties, light wires, or use zip-lock bags.
3. Label each bag clearly and simply. Write directly on the bag or attach the label securely.
4. Place the bags in a strong container and pack them with newspaper or another insulating material.
5. Complete the reverse side of this form.
6. Ship the samples immediately, so they remain as fresh as possible. Send them early in the week to avoid drying out in weekend storage or mailing.
7. Keep the samples cool; don't allow them to dry out, but don't add water.

SAMPLING PATTERNS

PLANT DISEASE IDENTIFICATION FORM

Submit specimens to: EXTENSION PLANT PATHOLOGY
DEPARTMENT OF PLANT PATHOLOGY
UNIVERSITY OF ARIZONA   Date collected______________
TUCSON, ARIZONA 85721   P.C. #______________

☐ CA   ☐ C   ☐ GA
☐ N   ☐ HG   ☐

How to Collect and Ship Plant Specimens for Disease Diagnosis

The accurate diagnosis of a plant disease depends on receiving a fresh sample. All specimens should be fresh when collected, placed in a plastic sack and shipped immediately in a mailing tube or strong cardboard carton to prevent crushing enroute. When specimens arrive unidentified, wilted, crushed or in advanced stages of decay, diagnosis is often impossible. If the sample is in good condition, the disease can be diagnosed more rapidly. Most specimens will be diagnosed and acknowledged within a few days. Proper completion of this form will aid in prompt, accurate service.

GENERAL INFORMATION: Telephone Number______________

Owner or grower______________Address______________
Collector______________Address______________Zip Code______________
Plant (or crop) attacked______________Variety______________Acreage______________
   Plant part injured: Roots______; stem or branch______; leaves______; flower______; fruit______
   General appearance: Wilted______; dead areas______; yellowed______; stunted______;
                abnormal growth______; leaf spot or blight______; leaf mottle______;
                other ______________
   Distribution of diseases: Scattered plants______; groups of plants______; most of field______;
                on slopes______; low areas______; upland areas______
   When were symptoms first noticed?______________
   Weather conditions of previous week:______________
   Cropping history:______________
   Chemicals applied, rates and dates:______________
      Fertilizer
      Fungicide
      Herbicide
      Insecticide
      Other
   Location in relation to buildings:______________

ADDITIONAL INFORMATION FOR SPECIAL CROPS AND ORNAMENTALS:

	Year established______________ Approximate age and size______________
Trees	Location: street or terrace______________; landscape______________; flower bed______________
and	Depth to caliche layer:______________
Shrubs	Shaded area______________; relation of injury to wind and sun______________
	What is the watering schedule?______________

DIAGNOSIS AND PREVENTION:

Date reply:______________   L:______   OC:______   T:______   By______________

Extension
Cooperative Extension Service
Clemson University

CLEMSON UNIVERSITY
PLANT PROBLEM CLINIC

Mail two copies to: Plant Problem Clinic, Cherry Road, Clemson University,
           Clemson, SC 29634-0395   Phone # (803) 656-3125

Please fill out form completely.   Username: PPCLINIC

GROWER________________________________   COUNTY________________________________
ADDRESS_______________________________   DATE COLLECTED________________________
CITY________________________ ZIP______   DATE SUBMITTED________________________
FORM FILLED OUT BY:___________________________   TITLE__________________________
SENDER____   TITLE__________________________

A. FOR PLANT IDENTIFICATION: 1. Origin: ( )cultivated ( )native ( )uncertain
   2. Type plant: ( )tree ( )shrub ( )herb ( )vine ( )grass ( )other__________

B. FOR INSECT OR WEED IDENTIFICATION: 1. Where found__________________________
   2. Degree of infestation: ( )slight ( )moderate ( )severe
   3. Type insect damage___________________________ 4. Control request: ( )yes ( )no
   5. Previous insecticides or herbicides____
   ____

C. FOR PLANT PROBLEM IDENTIFICATION:
   1. Name of plant (crop) and variety____
   2. Nematode sample enclosed: ( )Yes ( )No
   3. Soil sample enclosed: ( )Yes ( )No
   4. Tissue analysis sample enclosed: ( )Yes ( )No
   5. Planting date or age of plant_____________ 6. Plant height______________
   7. Acreage or number of plants__________ 8. Previous crop_________________
   9. Location of planting: ( )home grounds ( )garden ( )greenhouse ( )nursery
      ( ) field ( )orchard ( )indoors ( )other______________________________
   10. Plant part affected: ( )roots ( )stems ( )leaves/needles ( )twigs/branches
      ( )trunk ( )buds ( )fruit ( )flowers ( )pods ( )seed ( )entire plant
   11. Symptoms: ( )spot ( )rot ( )blight ( )wilt ( )dieback ( )abnormal growth
      ( )canker ( )stunting ( )galls/swellings ( )burn/scorch ( )yellowing
      ( )defoliation ( )mosaic ( )other____
   12. Degree of injury to individual plants: ( )slight ( )moderate ( )severe
   13. Development of problem: ( )sudden ( )gradual
   14. Distribution of problem: ( )single plant ( )scattered plants
      ( )localized area ( )large area ( )every plant
   15. Lighting: ( )full shade ( )morning shade ( )afternoon shade ( )full sun
   16. Soil texture: ( )artificial mix ( )clay ( )sand ( )loam
   17. Moisture situation: Rainfall- ( )below normal ( )normal ( )above normal
      Irrigation- ( )No ( )Yes   Drainage- ( )poor ( )fair ( )good ( )extensive
   18. Tillage: ( )no till ( )minimal till ( )normal
   19. Chemicals applied to this crop:
      Fertilizer :What_____________ When________________ Rate__________
      Herbicide :What_____________ When________________ Rate__________
      Fungicide :What_____________ When________________ Rate__________
      Insecticide:What_____________ When________________ Rate__________
      Nematicide :What_____________ When________________ Rate__________
   20. Suspected diagnosis____
   21. Additional comments____
   ____
   ____

Cooperative Extension Service, Purdue University
Department of Botany and Plant Pathology

PLANT DISEASE IDENTIFICATION FORM

To: Plant Diagnostic Clinic, Department of Botany and Plant Pathology,
Lilly Hall of Sciences, Purdue University, West Lafayette, Indiana 47907

PLEASE COMPLETE as much as possible of the information below. This will help in identifying the problem and will result in better and faster service.

FROM: _______________________________
Agent's name   *Agent's address*   County

GROWER: ☐ Commercial   _______________________________
☐ Home
Grower's name and address

Crop or plant name: __________ Variety/Species: __________ Date sent: __________

Size of planting: __________ Acres affected: __________ Date first noticed: __________

Previous cropping history (last 2 yrs.): __________

Rainfall: ☐ High ☐ Average ☐ Low   Temperature: ☐ High ☐ Average ☐ Low

Soil type: ☐ Sandy ☐ Clay ☐ Silt ☐ Loam ☐ Organic Other: __________

PLANT LOCATION		AREA AFFECTED	NATURE OF INJURY	
☐ Field	☐ Orchard	☐ Leaves	☐ Wilt	☐ Stem Canker
☐ Yard/Garden	☐ Golf Course	☐ Stem/Branch	☐ Leaf Spot	☐ Top Dieback
☐ Nursery	☐ Plant Bed	☐ Roots	☐ Fruit Spot	☐ Yellowing
☐ Greenhouse	☐ Other ____	☐ Flowers	☐ Blossom Blight	☐ Fruit Rot
		☐ Fruit		

PREVALENCE OF INJURY	DEGREE OF INJURY	DISTRIBUTION OF INJURY	SOIL DRAINAGE
☐ Entire Planting	☐ Light	☐ Low areas	☐ Good
☐ Localized Area	☐ Moderate	☐ Upland	☐ Moderate
☐ Scattered Plants	☐ Severe	☐ Slopes	☐ Poor
		☐ No Trend	

Chemicals applied (Name, Date, Rate): __________

TREES, SHRUBS & TURF

Approximate age and size: __________ ☐ Sunny ☐ Shady ☐ Exposed ☐ Sheltered
Relation to nearest house/construction (ft): __________ Relation to road/sidewalk: __________
Maintenance program (sprays, mulch, etc.): __________
Other helpful information: __________

DIAGNOSIS AND RECOMMENDATION

Date __________ By __________
Cooperator's copy, white; Extension office copy, yellow; Diagnostic Clinic copy, pink; Forward all copies to the Diagnostic Clinic.

---

cooperative EXTENSION SERVICE ● STATE OF INDIANA

PURDUE UNIVERSITY
and the
U. S. DEPARTMENT of
AGRICULTURE cooperating
West Lafayette, Indiana 47907

Date __________

Address correspondence to:

Plant Diagnostic Clinic
Department of Botany & Plant Pathology
Lilly Hall of Life Sciences
Purdue University
West Lafayette, IN 47907

Dear __________:

We received your sample and our diagnosis is given below. If you have further questions, please call the Plant Diagnostic Clinic at (317) 494-4641 or write to the clinic at the above address.

Sincerely,

Extension Plant Pathologist

Diagnosed by: __________

Specimen: __________

Diagnosis: __________

It is the policy of the Cooperative Extension Service of Purdue University that all persons shall have equal opportunity and access to its programs and facilities without regard to race, religion, color, sex or national origin.

COOPERATIVE EXTENSION WORK. U. S. DEPARTMENT OF AGRICULTURE AND PURDUE UNIVERSITY COOPERATING
PLANT DIAGNOSTIC CLINIC

SAMPLE_________________________ REFERENCE NO.__________ DATE__________

Dear___________________________,

____ Sample must be cultured. Results will be sent in approximately a weeks time.

____ Sample was deteriorated upon arrival. Please submit additional sample. Specimen should be spread and pressed between paper, or wrapped in paper, and sent immediately. Do not use saran wrap or aluminum foil. Plastic bags may be placed around roots and soil.

____ Sample is inadequate for diagnosis. Please submit additional specimen. Include roots___, leaves___, stem___, soil___, whole plant___.

____ Turf samples should include the transition area between diseased and healthy portion of grass. Plugs should include soil & roots.

THANK YOU FOR YOUR COOPERATION. Sincerely,

Botany & Plant Pathology Dept.   Extension Plant Pathologist

Plant ________________________________ Sample ____________________ Date __________

Diagnosis ____

Owner ____________________________   Address & Phone

Rec'd. from ____________________   ________________________________

________________________________   ________________________________

Field__ Nursery__ Greenhouse__ Dooryard__ Exp. plot__ Size of Planting ________

Per cent affected   Distribution
0-10   ⟨50   50⟩   90-100   general__   spotty__   local__

Age of crop __________ First symptoms ____________________ Soil type __________

Rotation & Weather ____

Pesticides ____

Fertilization ____

Symptoms ____

____

____

(BEFORE FILLING OUT THIS FORM, PLEASE FOLLOW INSTRUCTIONS ON REVERSE SIDE)

INFORMATION SHEET TO ACCOMPANY SPECIMENS SUBMITTED FOR DIAGNOSIS TO: THE PLANT DISEASE AND INSECT CLINIC, P.O. BOX 5397, N.C. STATE UNIVERSITY, RALEIGH, N.C.   27607

CROP_____________________ VARIETY____________________ DATE PLANTED________________

GROWER & ADDRESS ____

AGENT__________________ COUNTY__________________ COLLECTED BY_____________________

DATE COLLECTED________________   TOWN & COUNTY WHERE COLLECTED ________________________
                                  IF OTHER THAN GROWER'S ADDRESS

DESCRIBE SYMPTOMS AND HOW LONG OBSERVED____

____

PLANTS AFFECTED______% CROP LAST YEAR________________ CROP YEAR BEFORE LAST____________

LOCATION	SIZE OF PLANTING (TOTAL)	DISTRIBUTION OF SYMPTOMS
☐ field		(several may apply)
☐ garden	Total number of acres ______	☐ general
☐ home grounds	OR	☐ scattered
☐ nursery	Total number of plants______	☐ in spots
☐ greenhouse	OR	☐ in rows
☐ golf course	Approximate dimensions______	☐ certain variety
☐ orchard	OR	☐ in low areas
☐ plant bed	Other description	☐ in high areas
☐ _______(other)	_______________	☐ ___________(other)

CHEMICALS AND RATES USED: (e.g. fertilizers, top or side dressings, herbicides, insecticides, fungicides, nematicides, fumigants, etc.; comment on recent applications)

____

____

SOIL: TYPE   TERRAIN   DRAINAGE   HAS SOIL BEEN SENT IN FOR NUTRIENT
        ☐ sandy   ☐ sloped   ☐ good   ANALYSIS OR NEMATODE ASSAY TO:
        ☐ clay   ☐ level   ☐ moderate   Yes No
        ☐ loam   ☐ low   ☐ poor   ☐   ☐   Soil Testing Division
        ☐ organic   ☐   ☐   Nematode Assay Laboratory
        ☐ potting mixture
          composed of:___________   Significant results:___________

        ___________________   ___________________

TREES &   LOCATED NEAR: Describe any disturbance of roots or sod; also any
SHRUBS   ☐ building   injuries observed on trunk, branches, or roots
             ☐ downspout   (e.g. digging or construction injury, lawnmower
             ☐ road, highway   injury, animal injury, borer injury, cold injury)
             ☐ driveway
             ☐ or in wooded area   ____
             ☐ _________(other)
                                        ____

COMMENTS: (significant weather, greenhouse conditions, cultural practices or other factors which may relate to problem)

SUSPECTED CAUSE:

Clinic # _____________

Crop _____________ Variety _____________ When planted _____________ Date collected _____________

Grower ____
Name   No. & Street (Route & Box)   City   State   Zip   Telephone

Agent _____________ County _____________ Collected by _____________ in _____________
City   County

Type planting: _commercial, _field, _greenhouse, _plant bed, _other_____________
Planting size : acres____ plants____ plant bed____ other_____________. % affected____
Crop history: Last year _____________ 2 years ago _____________ 3 years ago _____________

SYMPTOMS ON:_________	DISTRIBUTION:	APPEARANCE:	Chemicals/fertilizers:
☐ wilted ☐ yellowing ☐ stunted ☐ leaf spot ☐ dieback ☐ root rot ☐ death _________	☐ localized ☐ scattered ☐ general	_ sudden. _gradual. On: _new, _maturing or _old growth.	_________ _________ _________ _________

Main concern/helpful information _______________________________________
____
____
____

Field distribution, characteristics, areas walked:

Relevant weather:

____

(Indicate:
North, roads,
buildings, low
areas, hard
pans, etc.)

Sources of planting material: _______________________________________

Plant Species Showing Symptoms:
 In the field: _______________________________________
 Borders: _______________________________________
 Woods: _______________________________________
 Other Fields: _______________________________________
 Other Growers: _______________________________________
_____________ : _______________________________________

Dates	Item	Comments/Findings
_______	Grower first noticed	_______________________
_______	First visit	_______________________
_______	Plant Disease & Insect Clinic	_______________________
_______	Soil analysis lab.	_______________________
_______	Tissue analysis lab.	_______________________
_______	Nematode assay lab.	_______________________
_______	Specialist visit	_______________________
_______	Other visits	_______________________
_______	Photograph	_______________________

Conclusion: _______________________________________

Crop ________________________________ Variety ______________ #

Grower ______________________________ ______ % of ______ Ac.

Grower's Address ____

Agent ____________________ County ______________________

REPORT:

PLANT DISEASE AND INSECT CLINIC, N. C. STATE UNIVERSITY AT RALEIGH

PLANT DISEASE AND INSECT CLINIC

Clinic # ______________

Crop ______________ Variety ______________ When planted ____________ Date collected ____________

Grower ____

Name   No. & Street (Route & Box)   City   State   Zip   Telephone

Agent ______________ County ______________ Collected by ______________ in ______________

City   County

Type planting: _commercial, _field, _greenhouse, _plant bed, _other ______________

Planting size : acres____ plants____ plant bed____ other________________. % affected____

Crop history: Last year ______________ 2 years ago ______________ 3 years ago ______________

SYMPTOMS:		DISTRIBUTION:	APPEARANCE:	Chemicals used:
☐ wilted	☐ yellowing	☐ localized	_ sudden. _gradual.	________
☐ stunted	☐ leaf spot	☐ scattered	On: _new, _maturing	________
☐ dieback	☐ root rot	☐ general	or _old growth.	________
☐ death				________

Main concern helpful information ______________________________

CLINIC REPORT: (do not write in this space)

DIAGNOSTIC PROCEDURE:

Clinic # ___________

Crop ________________ Variety ________________ When planted ________ Date collected ________

Grower ____
        (name)   (address)   (telephone)

Agent ________________ County ________________ Collected by __________ in _________

(Check or Enter Correct Choices)

Main concern: ________________________________ Parts affected ________________

Crop grown: Last year ________________ 2 years ago ________________ 3 years ago ________________

Planting size ________________ ________________ % Affected ______ Other plants affected ________________

Planting: ☐commercial ☐field ☐greenhouse ☐plant bed ☐potted ☐other ________________

Symptom Distribution: ☐localized ☐scattered ☐general ☐other ________________

Symptom Onset: ☐sudden ☐gradual ☐other ________________ , on: ☐new ☐mature ☐old ☐all growth

SYMPTOMS:
☐wilt   ☐mosaic   ☐distortion
☐stunt   ☐spot   ☐scorch
☐death   ☐yellow   ☐dieback
☐rot   ☐other ________

Chemicals/Fertilizers: ________________
____
____
____

Helpful Information: ____

CLINIC REPORT (do not write in this space)

COUNTY PLANT CLINIC

AGENT'S RECORDS AND NOTES

SKETCH field characteristics, disease
distribution, areas walked, etc.
RELEVANT WEATHER: __________
____
SOURCE OF PLANTING MATERIAL:

(Indicate: north,
roads, buildings,
low areas, hard
pans, etc.)

OTHER SPECIES SHOWING SYMPTOMS IN:

  Field ________________   Other fields ________________
  Borders ________________   Other growers ________________
  Woods ________________   Other areas ________________

DATE	ITEM	COMMENTS/FINDINGS/CONCLUSIONS
________	Grower first noticed.	________________
________	First visit.	________________
________	P.D. & I. Clinic.	________________
________	Soil Anal. Lab.	________________
________	Tissue Anal. Lab.	________________
________	Nema. Assay Lab.	________________
________	Specialist contact.	________________
________	Other contacts.	________________
________	Photographs.	________________
________	________	________________

DIAGNOSIS:
☐accur. ☐usef.
☐indic. ☐excl.

CONCLUSION: ________________
____
____

I understand the University of Tennessee has entered an agreement with ICI Americas, Inc., whereby ICI will subsidize the cost of analyzing fescue at the Plant and Pest Diagnostic Center in Nashville. I have enclosed an original grass-analysis coupon and agree to having the Plant and Pest Diagnostic Center mail a copy of this Fescue Endophyte Assay Report to ICI Americas, Inc. The participation of the University of Tennessee in this program is solely for the purpose of providing educational information to the public and does not imply approval of ICI Americas, Inc., products to the exclusion of others which may be of similar, suitable composition.

CODE: _________________

COUNTY: _________________

AGENT: _________________

DATE: _________________

____
Grower

FESCUE ENDOPHYTE SPECIMEN FORM

Mail to: Plant and Pest Diagnostic Center, Agricultural Extension Service, University of Tennessee, P. O. Box 110019, Nashville, TN 37222-0019

Name:____

Address: ____

City: _________________ State: _________________ Zip: _________________

Sample Type: ☐ Plant   ☐ Seed

Test Desired: ☐ Stain   ☐ Grow-Out

Fescue Variety: _________________ Size of Field:_________________

Name of Field or Seed Lot No.: _________________________________

Field Composition: _________ % Fescue   _________ % Clover   _________ % Other

Age of Seed Lot or Year Pasture Established: _________________

Type of Livestock: _________________
____

Animal Performance (Good, Fair, Poor, Unknown)
____
____

———————————— **LAB USE ONLY** ————————————

Sample No.: _________________ Date In: _________________ Date Out: _________________

Condition of Sample: _________________
____

_________________ % Infected Plants   _________________ % Infected Seed

Date Reply: _________________

By: _________________

White — Grower Copy   Pink — Extension Office Copy   Yellow — Diagnostic Center Copy

Forward all copies to Plant and Pest Diagnostic Center

CHAPTER 3

Gathering Information

Diagnosing plant diseases requires a grasp of the disease signature, e.g., symptoms, distribution of affected plants, previous weather, and grower practices (see Chapter 1). This information is best obtained by the county agent, consultant, agricultural specialist, or diagnostician by direct observation of the field problem and consultation with the grower. Usually, however, this information is indirectly obtained from the grower or from the form submitted with the specimen. Unfortunately, indirect sources of information are often inadequate for diagnosticians to formulate a disease signature, and in these cases, they are handicapped in their ability to make a quality diagnosis and assess the problem or problems. If the grower is present in the laboratory, the diagnostician can ask relevant questions and request more information and perhaps additional samples. Although many plant diagnosticians do not visit problem sites, they should be familiar with the protocol suggested for gathering the field components of the disease signature, which is presented in this chapter, in order to ferret out significant information from the clinic form.

Visiting a field with a perplexed grower can be an unsettling experience. The diagnostician often feels inadequate to satisfy the immediate needs of the grower for a diagnosis and fears embarrassment. The field diagnostician's biggest temptation is to succumb to the desire for a rapid diagnosis. Even experienced diagnosticians observing a familiar disease take time to ask questions and gather all the components of the disease signature and, if necessary, collect specimens and samples for later laboratory examination. The best way to be embarrassed and unprofessional is to give a hasty diagnosis after a cursory examination only to find later that the diagnosis was wrong or that the diagnosis was correct

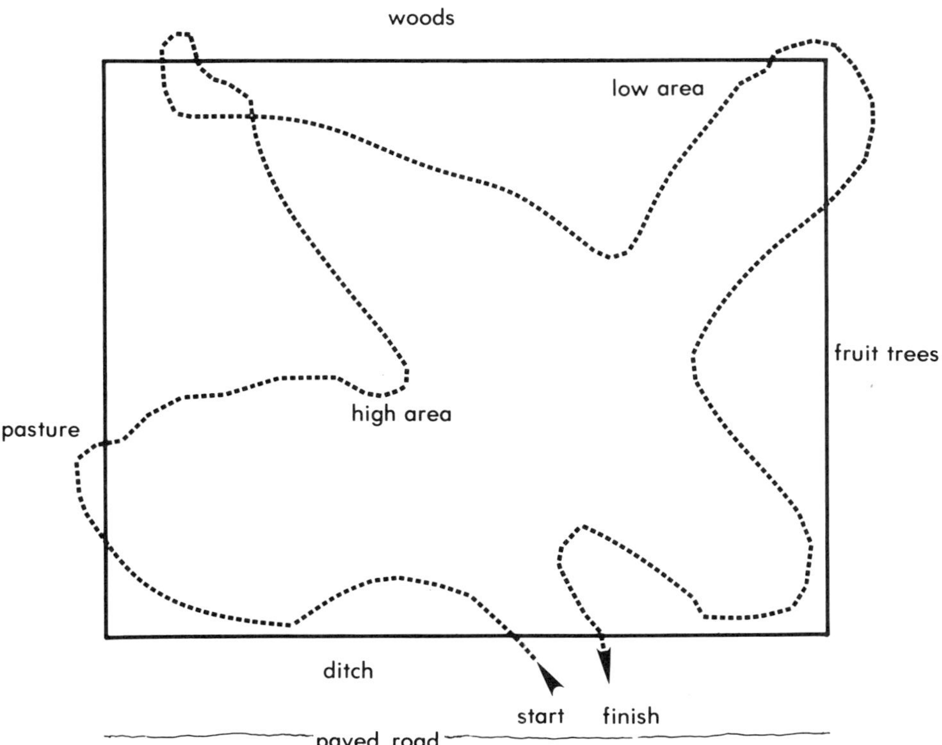

Fig. 3.1. Actual pattern traversed in the examination of a plant problem in a field based on a systematic pattern.

but other important diseases or problems were missed. For example, a diagnostician may proclaim that the leaf spot affecting a cucumber planting is anthracnose and miss the fact that other threats to the crop are the minor presence of powdery mildew on a few plants and the extensive occurrence of root-knot nematodes. The diagnostician should feel good about having done a professional job, even if the problem was not diagnosed. A negative diagnosis, if well done, has considerable value, as discussed in Chapter 2.

Examining Field Problems

Obtain from the grower a description of the problem, including a statement of problems with the previous crop or crops; the recent history of the field and practices (e.g., fertilizers, herbicides, insecticides, methods of application, and flooding); and the grower's ideas on the probable cause.

Inspect the entire planting:

1. Look at the front part of the field, traverse the field diagonally and across the back side, and then return to the starting point. The pattern is flexible to permit inspection of unusual (e.g., low and high) areas. The ends of rows and turn rows are especially interesting and should always be carefully examined. Also inspect fence rows and adjacent fields (Fig. 3.1).

2. Note the distribution of the problem in the field as it relates to things such as field characteristics and drainage patterns, and also observe the crop and conditions in neighboring fields. Note whether the problem is associated with certain rows, low or high areas, soil type, or edges of the field. Observe other plant species for symptoms (Plate 1A–D).

3. Examine individual plants in detail, and notice the position of symptoms on the plant such as on old leaves, stems, exposure (e.g., west), and new growth. Look for insects and insect damage. Cut stems to check for vascular discoloration. Hold leaves up to the light to check for mosaic, other viral symptoms, or the presence of webbing and mites. Look for leaf abnormalities in color, size, shape, and texture (Plates 2–6).

4. Carefully dig roots and examine them. Check for galls, rots, abnormal root color, and feeder root condition, and assess growth.

5. Dig into the root zone. Notice odors, insects, fertilizer placement, and depth of planting.

6. Take soil samples from the root zone in "good" and

Crop __Tomato__   Variety __Manapal__ (Mountain Pride)   When planted __4/12/85__   Date collected __6/3/85__   Clinic # ________

Grower __R. K. Smith__   __Rt.1, Box 123__   __Green, NC__   __27123__   __704/123-4567__
     Name   No. & Street (Route & Box)   City   State   Zip   Telephone

Agent __S. R. James__   County __Maple__   Collected by __grower__   in __same__
     City   County

Type planting: commercial ☒ yes ☐ no   Site: ☒ field ☐ greenhouse ☐ plant bed ☐ ________

Planting size (select one): __5__ acres ________ plants ________ plant bed; __25__ % affected

Crop history: Last year __tomato__   2 years ago __tomato__   3 years ago __tobacco__

SYMPTOMS:

☒ wilted   ☒ yellowing

☒ stunted   ☐ leaf spot

☐ dieback   ☐ root rot

☒ death   __stem rot__

DISTRIBUTION:

☒ localized

☒ scattered

☐ general

Fertilizers used: __200 lb. 10-10-10__   __50 lb. sidedress__   __10-10-10__

Chemicals used: __maneb__   __sevin__

Main concern/helpful information __What is the problem?__

CLINIC REPORT: (do not write in this space)

PLANT DISEASE AND INSECT CLINIC, BOX 7616, N. C. STATE UNIVERSITY, RALEIGH, N. C. 27695-7616

Fig. 3.2. Plant disease clinic form submitted with a specimen.

"bad" areas for nematode assay and chemical analyses. While probing the soil, note soil compaction, the presence and depth of hardpans, and soil structure, texture, and organic matter.

After completing the field examination, listen to the grower's ideas on the problem. The views of the grower on the cause of the problem may be modified after the field examination. After questioning, the grower often recalls additional information. Record the field information (Fig. 3.2).

If further examination is necessary, take soil and plant samples and mail them, along with properly completed forms, to the appropriate laboratories. Be sure to follow all directions on the forms provided by the laboratory or clinic. Consider using out-of-state university and private laboratories to supplement the diagnosis; often there is a charge (some are nominal) for these assays and diagnoses.

Recording Field Information

Experienced diagnosticians can often diagnose the major problem after examining plants in the field and point out other problems as well. As previously stated, the diagnosis must be reported in writing to the grower. If a diagnosis is not made in the field, the field information must be recorded and attached to the specimen.

Recording field information is best done on standard clinic forms to ensure that all relevant information is recorded (Fig. 3.2). There are many forms in use (see Appendix to Chapter 2), and many more have been tried and discarded as unsuitable. There is a paradox in that the diagnostician wants detailed information that entails long and complicated forms. At a certain point, however, a significant number of growers, agricultural specialists, home owners, and consultants will not fill them out properly, and simple but critical information is lost. The form then becomes useless. We have seen clinic forms consisting of only a 3- × 5-in. index card requesting the grower's name, address, name of crop, and complaint. On the other hand, we have seen forms involving two and one-half pages of questions. Obviously, diagnosticians must develop forms suitable for their needs and acceptable to growers or their advisers. In situations involving legal operations or a high risk of litigation, a special form may be used to record more details, diagnostic steps taken, and progress of the diagnosis, which may involve other laboratories.

A successful form is one that pleases the grower by being easy to fill out and gently coaches him to recall information useful for the diagnosis. In developing a clinic form, much consideration is given to the format to achieve these ends. The information form has five major components.

1. Clerical: address and telephone number of the grower, clinic or specimen number, dates, crop, and disposition of the final report

2. Site: crop, size of planting, type of planting, type of plant, cropping practices, 3-year cropping history, and source of planting stock (data would be somewhat different for nonagricultural crops; information on trees, for example, might include age, diameter at breast height [dbh], spread, height, and recent construction work)
3. Disease description: symptoms, field distribution, percentage of plants affected, and first appearance
4. Grower comments and tentative diagnosis
5. Diagnostic report: usually a part of the same form; however, it can be a separate form or written in a letter. In either case, the following is given:

a. Diagnosis. Give the approved common name of the disease as listed in *Common Names for Plant Diseases* (compiled by the American Phytopathological Society) and the causal agent. Avoid using local common names. The level of reliability of the diagnosis should be indicated. If not diagnosed, follow-up suggestions are stated.

b. Management. Information on management is a function separate from diagnosing and should be prepared by an experienced farm adviser or county agent, an extension specialist (e.g., a plant pathologist), or a licensed, experienced consultant. However, these functions are often combined.

c. List of inserts.

d. Name of diagnostician and date of report. The form should have information (perhaps coded) on the quality of the sample received and reliability of the diagnosis (see Chapter 1).

Collecting, Preparing, and Shipping Specimens

The development of a plant problem (disease) is a process involving time. It starts with exposure of the plant to the causal agent followed by, sequentially, the development of primary symptoms, development of secondary symptoms, and finally colonization of plant parts by secondary organisms. If the problem is caused by a pathogen, the microorganism will be going through various life stages, including the development of various morphological structures such as spores, which are extremely useful in diagnoses. Collecting specimens provides a small "time window" in the entire process. Often the window is inappropriate for diagnostic purposes. By carefully selecting specimens, the window often can be expanded to increase the chances of obtaining specimens suitable for diagnostic purposes. This is done by collecting specimens showing different intensities of symptoms. Often, plants showing mild symptoms have been exposed to the causal agent for a short period of time, while those showing advanced symptoms have been exposed for a long time. Usually, plants that have been dead for some time are poor specimens for diagnosing.

Fig. 3.3. Method of wrapping a container-grown plant or a plant with soil on the roots before it is placed in a mailing box. Left, plastic bag and wire fastener. Center, plant sample with a ball of soil and roots. Right, the entire sample is placed in the plastic bag, which is then tied above the soil at the base of the stem and again at the top. Holes are punched in the plastic covering the foliage to avoid rot.

Once the specimens have been collected, most deteriorate quite rapidly, especially if succulent. The following procedures minimize deterioration during transit and expedite the diagnosis:

1. Give complete information requested on the clinic form (Fig. 3.2).
2. Place specimens in plastic bags at the time of collection, not hours or days later (see below for specific instructions on various crops). Do *not* seal the plastic bags. Do *not* enclose moist paper towels. Do *not* wet plants, because free water causes rapid decomposition. Plants wet from dew, rain, or irrigation should be dried in the office. Do *not* overstuff plastic bags with plant material, because plant tissues will "cook" from their own heat and lack of air. Perforated plastic bags are suggested for succulent plant samples. Wrap the roots or soil ball securely to prevent soil from mixing with foliage during transit (Fig. 3.3).
3. Send specimens immediately after collecting. Mail early in the week to avoid weekend delays. If the specimen cannot be mailed immediately, refrigerate it until it can be shipped. Dead plants or materials that are dry or decomposed on arrival are difficult if not impossible to diagnose. (But dry is better than rotted and wet!)

Quarantines

When specimens are submitted from areas under state or federal quarantine, the package must be tight to contain the sample and accompanying soil and debris. Soil should be sealed in plastic bags. Usually, shipping permits must be obtained from the state department of agriculture and attached to the package. For more details, see Appendix K in the general appendixes.

Woody Plants with Symptoms of Decline or Low Vigor

See Plates 4P and 5F.

1. Collect a generous handful of feeder roots and a quart of soil from representative plants that have slight to moderate symptoms. Place the roots and soil together in a plastic bag and seal. For plants such as junipers, cedars, pines, or fruit trees, include pieces of large roots (if rotted or showing an internal decay) in addition to the feeder roots. An early stage of root rot is needed for an accurate diagnosis.
2. In a separate plastic bag, place several branches or large twigs with foliage showing a range of symptoms.
3. Be sure to mention on the form whether there are known injuries, cankers, or discoloration on the roots, large branches, or trunk.

Seedlings and Rooted Cuttings

See Plates 1B and F, 2S, 5I and J, and 6B.

1. Collect a dozen or more seedlings or rooted cuttings with at least 0.5 liter (1 pint) of soil or growing medium. If the seedlings are in a greenhouse tray, send the entire flat or representative portions. Seedlings should be carefully dug with a trowel or spade (not pulled) with soil left intact on the roots.
2. Select specimens with symptoms in an early to intermediate stage of development.
3. Package the sample to keep soil intact on the roots and to prevent soil from getting on leaves. Enclose the entire sample in a plastic bag with holes punched in the top portion (Fig. 3.3).

Container-Grown Floriculture Plants and Nursery Stock

Note. The following does not apply to seedlings.

1. Select two to four entire plants showing a range of symptoms.
2. Place the entire plant in a sturdy container in a plastic bag. Tie off the plastic bag above the root system to keep soil off the leaves, bring the rest of the bag up and over the top of plant, and close by tying. Holes should be punched in the plastic covering the leaves to avoid decay before the specimen reaches the clinic (Fig. 3.3).
3. Mail the specimen in a sturdy cardboard box with protective padding material (e.g., old newspapers or Styrofoam granules) around the plant to prevent movement in the box during transit. A small floriculture plant, such as an African violet, may be sent in a large, padded envelope, but a cardboard box, with padding material around the specimen, is preferred.

Large Potted Plants

1. Collect 250 ml (1 cup) of representative soil from the root zone and about 15 ml (1 tablespoon) of roots about 5 cm (2 in.) long. Place roots and soil together in a plastic bag and seal.
2. In a separate plastic bag, place two or three samples of the foliage. When possible, select foliage with a range of symptoms. If insects are observed on stems and leaves, do not remove them. Leave them on the plant material enclosed in a plastic bag.
3. Do *not* add water to any of these specimens.

Ground Covers (Grasses, Forage Crops, and Small Grains)

See Plates 1J and 5H and K.

1. Dig one or more representative samples in the area where diseased plants meet healthy plants, leaving soil and plants intact. Each sample should be approximately 8–10 cm (3–4 in.) deep and 15–20 cm (6–8 in.) in diameter.
2. Enclose the entire specimen in a plastic bag in which holes have been punched. Use a separate bag for each specimen if more then one area is sampled. Do not let soil come in contact with the foliage. If the plant is more than 15 cm (6 in.) tall, tie off the bag above the roots and soil to keep the foliage soil free, and then pull the rest of the plastic bag up and over the plant top and seal again. If the bag is not large enough to cover the entire specimen (Fig. 3.3), a second plastic bag can be used to enclose the top portion.

Herbaceous Plants

The following applies to plants beyond the seedling stage:

1. Select six to 12 entire plants representing a range of symptoms.
2. Carefully dig up the plants with their roots and 1 liter (or quart) of soil.
3. Place the specimen or specimens in a large plastic bag to cover the entire sample, tie off the bag above the roots and soil, bring the rest of the bag up over the tops of the plants, punch holes in the bag, and seal (Fig. 3.3).
4. Package the sample securely in a large, padded envelope or cardboard box with padding material around the specimen.

Annual Crops (Field Crops, Vegetables, and Flower Bedding Plants)

1. Dig (do not pull) two or more entire plants with their root systems and about 1 liter (or quart) of attached soil.

If the plant is very large, it may be cut into sections suitable for packaging. Enclose tops in a large plastic bag. Cover roots and soil with the plastic bag, and tie the bag shut around the base of the stem (Fig. 3.3).
2. If plants have succulent leaves, such as cauliflower, or the amount of foliage is large, the plastic bag should be perforated or left open.
3. Package the specimen in a secure container for mailing.

Small Trees (Ornamentals, Christmas Trees, and Fruit Trees)

1. Dig three to six trees carefully. Include the roots and 1 liter (or quart) of attached soil. Never remove soil from roots because they dry out quickly, even in a plastic bag.
2. When possible, place each plant in a perforated plastic bag large enough to easily cover the entire tree. Tie off the root and soil portion, and then bring the rest of the bag up and over the entire tree and tie again. Large specimens may be cut in half if necessary.
3. If trees are quite small, several may be placed in one plastic bag. Be sure all samples are put in plastic bags at the time of collection. Identify the location and conditions under which the tree or shrub was growing.

Strawberries

1. Select plants with a range of symptoms, including plants with live (white) roots, because these are needed to diagnose red stele and root knot diseases. Do *not* send dead plants.
2. Dig four to eight entire plants, leaving at least 1 liter (or quart) of soil intact on the roots. Package the specimens securely to prevent soil from spreading over the leaves during transit. Enclose the entire sample in a plastic bag in which holes have been punched (Fig. 3.3).

Trees with Vascular Diseases

1. Select a part of the tree that shows wilting, yellowing, or dieback but still has live wood. Dead parts are generally useless for diagnosis.
2. Using a small knife or increment borer, look for dark brown, black, tan, or greenish streaks just under bark, i.e., in vascular tissues. This is often difficult with Verticillium wilt of maple and many oak species affected with oak wilt. Wood just under the bark that is not discolored usually does not contain the causal organism.

   Note. Verticillium wilt on green ash seldom causes vascular streaking in the stems; a leaf petiole assay is needed.
3. Collect six to 12 branch sections, 1 cm (½ in.) or more in diameter and 15–20 cm (6–8 in.) long, from portions of a tree meeting the criteria described above, and place the samples immediately in a plastic bag.

Large Samples with Localized Symptoms

1. Select and collect several plant portions representing a range of symptoms, and place them in a plastic bag.
2. For cankers, make cuts several centimeters (about an inch) beyond each end of the affected area to include some live wood. Place samples in a plastic bag.
3. For vegetables, fruits, tubers, bulbs, or corms, wrap specimens individually in newspaper, and package them in a cardboard box to prevent crushing. Do not package succulent tissues that have advanced soft rots.
4. If the plant or tree has symptoms of low vigor or dieback in addition to localized rot, include a liter (or quart) of soil with a generous sample of roots in a plastic bag.

Mushrooms

1. Select several specimens showing different stages of growth or maturity. If the mushroom is growing in the soil, remove it by digging it out rather than by pulling it to ensure that all of the diagnostically important base is included. No soil is needed.
2. Wrap each mushroom gently in waxed paper, place it in a paper cup or similar container, and close the cup by crimping. Do *not* put mushrooms in plastic bags because they deteriorate rapidly.
3. Package the cups in a crush-proof mailing box with a lightweight filler material to prevent them from moving in the box during transit.

Insects

1. Most insects and similar animals (e.g., roaches, termites, bugs, beetles, flies, wasps, ants, maggots, and spiders) are sent in leak-proof vials containing 70% ethyl or isopropyl alcohol.
2. For mites, scales, aphids, and thrips, send the living insects on affected plant parts. Handle as a plant specimen.
3. Send butterflies and moths in jars with a small amount of ethyl acetate to kill them. Package lightly with tissue paper in a crush-proof box.
4. Send caterpillars alive with host plant tissue in a plastic bag packaged in a mailing box. Diagnosticians generally prefer mailing in a crush-proof container.
5. Send living grubs in 0.5–1.0 liter (1–2 pints) of soil enclosed in a plastic bag and packaged in a mailing box.

Soil Samples

NEMATODE ASSAY SAMPLES

Plant-parasitic nematodes are obligate parasites, and their populations in the soil decrease rapidly in the absence of a living host. Consequently, the reliability of soil and tissue nematode assays rapidly declines after death of the roots and plant. The sample should be taken from the 5- to 20-cm (2- to 8-in.) soil horizon in the root zone from plants that are showing decline but that are not dead. Nematode assays usually depend on recovering living or recognizable nematodes. Because of their small size and delicate nature, nematodes are quickly killed by freezing, heating, or desiccation and decompose. Soil samples for nematode assays must be handled very carefully because excessive and vigorous mixing of soil samples will crush many nematodes. The presence of feeder roots in the soil sample is highly desirable. If endoparasitic nematodes (those within plant tissue) are involved, root samples must be gathered.

Plant tissue samples for nematode assay should be collected and handled as previously described.

Soil samples for determining the presence of potentially damaging nematodes prior to planting should be collected according to the directions of the testing laboratory. General directions include the following:

1. Take the samples from areas of a field with a common crop history.
2. If the soil in the area to be sampled is fairly uniform and the area is 1.6 hectares (ha) (about 4 acres) or less in size, one composite sample will often suffice. If the field is larger than 1.6 ha but less than 3 ha (8 acres), divide the field into two blocks of approximately equal size, and sample each block separately. When fields are larger than 3 ha and the previous crop grew uniformly, select two blocks not larger than 1.6 ha (4 acres) that are representative of the field, and take a sample from each with a minimum of 12 borings taken on a regular grid over the area (Fig. 3.4).
3. If the soil in the area to be sampled is variable, such as a heavy clay soil in one portion and a sandy soil in the remainder, take one sample from each soil type.

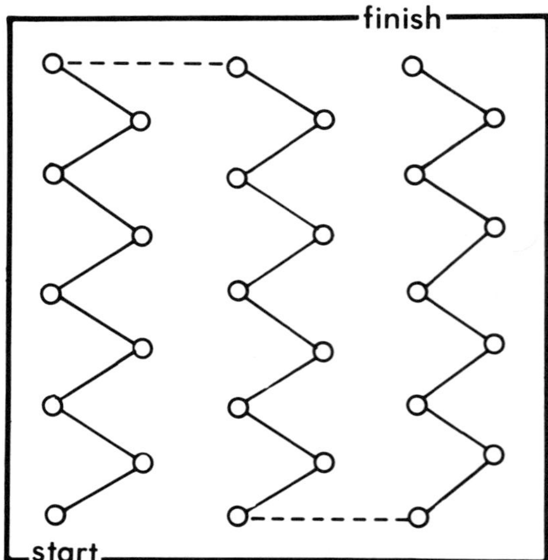

Fig. 3.4. Pattern for taking soil core samples on a 23-m (76-ft) systematic grid in a 4-acre field (maximum sampling unit size) to determine the presence of potentially damaging nematodes before planting. The sampling unit should be homogenous in topography, soil type, and cropping history.

Fig. 3.5. Depth of soil sampling for nutrient assay and pH determination. For nematode assay, the surface soil (2.5–5.0 cm or 1–2 in.) is discarded.

4. Take samples only when the soil is in good working condition, not too wet or dry or frozen.
5. The sample should consist of core borings taken on a 23-m (76-ft) grid pattern (27–30 cores per 1.6 ha or 4 acres). Avoid taking cores from the same rows. As each core is taken, discard the top 5 cm (2 in.) (Fig. 3.5), and place it in a clean bucket, preferably plastic. After the cores have been collected, gently mix the soil in the bucket, transfer 1 liter (or 1 quart) of soil to a sturdy plastic bag, and close the bag by tying. The excess soil may be used for fertility and pH analyses (Fig. 3.6).

   For annual or perennial plants where a soil nematode problem is suspected, soil samples are taken in the area where plants show mild to moderate symptoms. Usually, nematode problems in the field are randomly distributed in oval to elongated areas 15–30 m (50–100 ft) in diameter with a gradual transition from normal plants at the periphery to plants with the most severe symptoms in the center. At least 20 soil samples should be taken in the root zone of plants in the transition zone. In addition to the soil samples, two or three root systems should be carefully dug with soil attached and submitted for assay. It is always good to submit companion samples from areas where plants appear normal. In many cases, these samples can be split and a sample sent for soil nutrient and pH analyses.
6. Place each sample, properly labeled, in a plastic bag, seal the bag tightly to keep the soil moist, and put it into a liter (or quart) soil-sample box (Fig. 3.6). Write the appropriate field number in the space provided on each

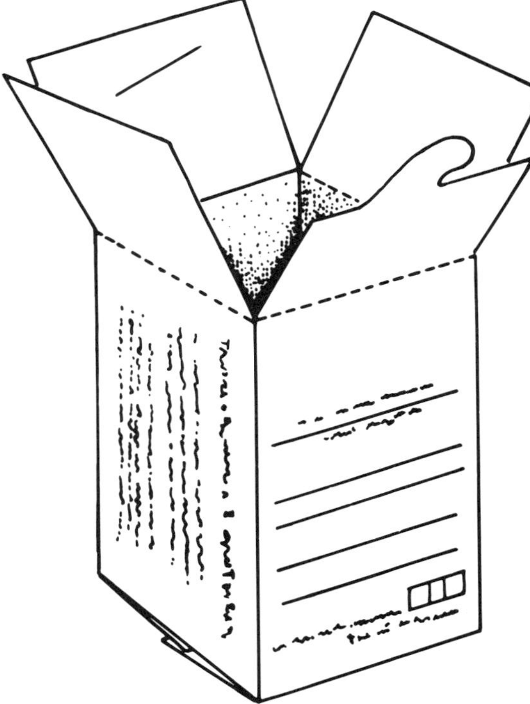

Fig. 3.6. Typical ½-pint (250-cm³) soil-sample box provided by a soil analysis laboratory. Space is provided on the box to write the grower's name and address and a three-digit field number. Nematode soil-sample boxes are similar, 1 pint (500 cm³) or larger. The soil is sealed in a plastic bag before it is placed in the box. Nematode assay samples are short lived (1 week) and must be kept refrigerated. Overheating (>90°F or 32°C) and freezing will ruin the nematode assay sample.

Table 3.1. Appropriate samples for tissue analysis

Crop Growth stage	Part to sample	Number of plants to sample
Field crops		
Alfalfa		
Prior to or at one-tenth bloom stage	Mature leaf blades taken about one-third the way down the plant	40–50
Clover and other legumes		
Prior to bloom	Mature leaf blades taken about one-third the way down from top of the plant	50–75
Corn (maize)		
Seedling, <12 in.	All aboveground portion	15–20
Prior to tasselling	Entire leaf fully developed below the whorl	10–15
From tasselling and shooting to silking	Entire leaf at the ear node or leaf just below or above it	10–15
Sampling after silking not recommended		
Cotton		
Prior to or at first bloom or when first squares appear	Youngest fully mature leaves on main stem	30–40
Forage grasses, hay or pasture		
Prior to seed head emergence or at optimum stage for best quality forage	Four uppermost leaf blades	50–75
Peanuts (groundnuts)		
Prior to or at the bloom stage	Mature leaves from both the main stem and a cotyledon lateral branch	40–50
Small grains (wheat, barley, oats, rye)		
Seedling, <12 in.	All aboveground portion	60
Prior to heading	Four uppermost leaves	60
Sampling after heading not recommended		
Sorghum (milo)		
Prior to or at heading	Second leaf from top of plant	15–25
Soybean and other beans		
Seedling, <12 in.	All aboveground portion	20–30
Prior to or during initial flowering	Two or three fully developed leaves at top of plant	25
Sampling after pods begin to set not recommended		
Tobacco		
Before bloom	Uppermost fully developed leaf; fourth leaf from top of plant	8–12
Vegetable crops		
Beans		
Seedling, <12 in.	All aboveground portion	20–30
Prior to or during initial flowering	Two or three fully developed leaves at top of plant	20–30
Celery		
Midgrowth (12–15 in. tall)	Petiole of youngest mature leaf	15–30
Head crops (e.g., cabbage)		
Prior to heading	First mature leaves from center of whorl	10–20
Leaf crops (e.g., lettuce, spinach)		
Midgrowth	Youngest mature leaf	35–55
Cucurbits (e.g., cucumber, muskmelon, squash, watermelon)		
Early stages of growth prior to fruit set	Mature leaves near basal part of the main stem	20–30
Peas		
Prior to or during initial flowering	Leaves from the third node down from top of plant	30–60
Peppers		
Prior to bloom	Most recently mature leaf	25

(continued on next page)

box. These numbers are the only means by which the samples can be accurately identified in the laboratory, and they should correspond to the numbers in the column "Your Field Number" on the Nematode Assay Information Sheet. Send the samples to the assay laboratory promptly.

7. Keep samples out of direct sunlight to avoid overheating. Samples may also be damaged by heat if they are kept for several hours in a car or truck. A Styrofoam cooler should be kept in the vehicle. Record all information requested on the information sheet provided by the laboratory.

For established perennial crops (e.g., in a landscape), collect soil samples from around each plant species showing decline. Do not sample directly around dead plants. Turfgrass samples should come from the edges of damaged areas and be 10 cm (4 in.) deep. Soil samples should be collected and mixed as for annual crops.

NUTRIENT AND pH ANALYSES SAMPLES

Soil testing every 3 to 4 years or more frequently is an accepted practice to avoid plant nutritional problems and to maintain production by proper liming and fertilizing. For various reasons, poor growth of plants is often the result of

Table 3.1. (*continued*)

Crop Growth stage	Part to sample	Number of plants to sample
Vegetable crops (*continued*)		
Potato		
Prior to or during early bloom	Third to sixth leaf from the growing tip	20–30
Root crops (e.g., beets, carrots, onions)		
Prior to root or bulb enlargement	Center mature leaves	20–30
Sweet corn		
Prior to tasselling	Entire fully mature leaf below the whorl	20–30
At tasselling	Entire leaf at the ear node	20–30
Tomato (field)		
Prior to or during early bloom stage	Third or fourth leaf from the growing tip	20–25
Tomato (greenhouse)		
Prior to or during fruit set	Young plants: leaves next to second or third cluster	20–25
	Older plants: leaves from fourth to sixth clusters	20–25
Fruits and nuts		
Almond, apple, apricot, cherry, peach, pear, plum		
Midseason	Leaves near base of current year's growth or from spurs	50–100
Grapes		
From end of bloom period through August	Petioles and leaves adjacent to fruit clusters	60–100
Pecan		
6–8 weeks after bloom	Leaves from terminal shoots (select leaflet pairs from the middle of the leaves)	30–45
Raspberry		
Midseason	Youngest mature leaves on laterals or "primo canes"	20–40
Strawberry		
Midseason	Youngest, fully expanded mature leaves	50–75
Walnut		
6–8 weeks after bloom	Mature leaflet pairs from mature shoots	30–35
Ornamentals		
Azaleas		
Prior to flowering	Most recently matured leaves	50–75
Begonias, Rieger elatior		
Prior to abundant flower formation	First leaf from top that is 2 in. or more wide	20
Carnations		
Unpinched plants	Fourth or fifth leaf pairs from base of plant	20–30
Pinched plants	Fifth and sixth leaf pairs from top of primary laterals	20–30
Chrysanthemums		
Prior to or at flowering	Most recently matured leaf from top of plant	20–30
Ornamental shrubs		
Current year's growth	Fully developed leaves	30–100
Ornamental trees		
Current year's growth	Fully developed leaves	30–100
Poinsettias		
Prior to flowering	Most recently mature, fully expanded leaves	15–20
Roses		
During flower production	Five-leaflet leaves below bud	20–30
Turf		
During normal growing season	Leaf blades (clip by hand to avoid contamination with soil)	½ pint of material

an unsatisfactory soil pH, nutrient levels, or excess soluble salts. Unlike soil samples for nematode assay, soil samples for pH and nutrient analyses need not be protected from freezing, high temperatures, or drying out. In fact, dry soil samples may be desirable to reduce shipping weight.

1. Collect about 500 ml (1 pint) of soil to represent a whole field. The field may vary in size from 1.2 to 6 ha (3 to 15 acres) or more but should be uniform. The sample should be a mixture of cores taken from several spots throughout an entire field or sampling area. Obtain 15–20 cores from the whole field or from an area that ap-

pears quite uniform. Bear in mind that a sample consisting of only a few cores can lead to false information about lime and fertilizer needs.

Various types of spades may be used to obtain a small core or plug of soil at the desired depth, but a soil sampling tube is recommended (Fig. 2.11). Use a clean plastic bucket or container to collect the individual cores that will make up the total soil sample, especially if micronutrients are to be determined. Metal containers, such as galvanized buckets, may contaminate the soil sample with mineral elements such as zinc and copper and lead to misleading results. Never expose the soil sample to tools, containers, or floor surfaces that may contaminate it with chemicals or fertilizers.

Take each soil core to the specific depth that the field is to be worked, generally about 20 cm (8 in.), because this is the depth of the previous mixture of lime and fertilizer and the depth at which most of the roots will grow (Fig. 3.5). It is easier to judge the depth of the probe and to remove the core by first stepping on the spot before inserting the tube.

For established perennial crops, such as fescue, alfalfa, or turf, soil samples are taken to a depth of 10 cm (4 in.). For shrub and tree crops, samples are taken to a depth of 30–50 cm (12–20 in.).

2. Because the 15–20 cores collected from each sampling area will, no doubt, be more than the soil-sample box will hold, pulverize each core so the entire collection in the bucket can be blended or mixed.
3. Fill the soil-sample box about two-thirds full, and mark it according to the identifying number on the farm or other map (Fig. 3.6).

Use only standard soil-sample boxes (Fig. 3.6), usually provided by the soil analysis laboratory. Samples sent in bags or other containers will not fit into the "assembly line" system used in the laboratory and may be rejected. Unlike soil samples for nematode assay, do not put plastic bags in the sample box. Seal the shipping box if soil is from a quarantined area.

In plant-problem areas, samples are collected somewhat differently from those collected for routine testing. Subsoil samples should be taken at the time that topsoil samples are taken, but the two should be kept separate. However, the same general principles of collecting a good representative sample should be followed, even though the "bad area" of the field may be relatively small. At the same time, a sample representative of a "good area" of the same field should be taken.

Tissue Samples

Plant problems are often caused by a nutrient deficiency or an excess of some element. The availability of plant nutrients can be determined by assaying the soil. However, certain soil or environmental conditions can impede the uptake of adequate amounts of nutrients by the plant from the soil, growing medium, or hydroponic solution, and symptoms develop. Tissue analysis reports are extremely valuable, but these analyses are beyond the scope of most plant clinics. These samples must be very carefully selected, usually from designated foliar parts of the plant. When these samples are taken, a companion soil (or growing medium) sample also should be collected and sent to the soil analysis laboratory. The directions from the specific plant tissue analysis laboratory should be followed. A general outline for taking appropriate samples follows.

At the outset, remember that plant analyses and evaluations are compromised unless the proper plant parts are selected for testing (Table 3.1).

1. Do *not* sample
   a. dead or diseased plants;
   b. insect-damaged or mechanically injured plants;
   c. stressed plants (e.g., those exposed to temperature or moisture extremes);
   d. plants with soil-covered leaves;
   e. plants in advanced stages of fruiting;
   f. plants that have had no rainfall after a foliar application of insecticides, fungicides, or nutrient elements;
   g. plants that have shown deficiency symptoms for a prolonged period of time;
   h. early in the morning or on extremely cloudy days because nitrates accumulate under these conditions.
2. Place plant tissue samples inside the large envelope usually provided with the mailing kit. Do *not* wrap or enclose tissue samples in plastic bags or other impermeable containers. Succulent or wet tissue samples should be air dried for at least 1 day before they are mailed to the tissue analysis laboratory.
3. A companion sample from a normal plant should be taken to compare with the suspected nutrient-deficient plant. An individual mailing container must be used for each sample.

When sampling instructions are not given for the crop you want to check, a good rule of thumb is to sample the most recently matured leaves, but it is best to consult the laboratory. Instructions for obtaining appropriate samples are subject to change as tissue analysis laboratories develop improved techniques.

CHAPTER 4

Diagnosing in the Field

After the plant symptoms, field patterns, and a disease history have been determined, the cause of a plant problem can usually be diagnosed, or at least categorized, in a general way. Many possible causes can be excluded, or at least relegated to "highly improbable," and need not confuse the investigation further. Determination of the exact cause or confirmation of the tentative field diagnosis must often be deferred until appropriate samples can be examined in the laboratory with special equipment. Some diagnosticians, county agents, and crop consultants routinely carry to the field a 20× hand lens, dissecting and compound microscopes, soil pH and soluble salts kits or meters, and references to complete as many diagnoses in the field as possible (Chapter 2).

Although a field examination and tentative diagnosis are probably the most important steps in diagnosing diseases of plants, most specimens received by plant clinics must be investigated without them. The clinic diagnostician must rely on the usually inadequate information accompanying the specimen to formulate a diagnosis. Many growers minimize the importance of field information and the field diagnosis on the basis of the erroneous assumption that a detailed examination of symptomatic plant tissues with sophisticated optical instruments and other "mysterious apparatuses" and procedures will reveal the cause. In fact, detailed laboratory examinations often merely show further symptom development at the tissue, cellular, and physiological levels without an indication of the causal agent. For example, microscopic observation of cells in necrotic tissue of a leaf spot does not aid in finding the cause of the leaf spot.

Field Disease Signatures

Field diagnosing is based on the fact that plant diseases caused by different agents follow certain patterns: the field disease signatures. Specific causes of plant diseases can be grouped into many general categories for diagnostic purposes. Those within a category will follow similar patterns, e.g., field distribution, time of appearance, rapidity of symptom development, plant species involved, and plant symptoms. For example, leaf spots caused by pathogens are usually host specific, and weeds or other plant species in the area are not normally affected. Leaf spots caused by drifting droplets of a contact herbicide (e.g., paraquat) are not host specific, and a variety of plant species in the area would show similar leaf spotting. Other differences would be apparent: distribution of affected plants in the field, position of affected leaves on the plant, and the rapidity of occurrence. After determining the field characteristics of the leaf spot disease, it would be easy to diagnose the cause of the problem as either an airborne pathogen or a chemical. The next step is to determine what pathogen or chemical is responsible; but in either case, time is not wasted in searching for a microorganism or chemical when the symptoms are caused by the other.

Other agents of plant disease fall into diagnostic groups including air pollution, vectorborne viruses, seedborne and soilborne diseases, nematodes, lightning, fire, hail, drought, drowning ("wet feet"), soil herbicide carryover, herbicide drift, spray injuries, and nutritional problems. The field disease signatures of these groups are given in Table 4.1.

Field Distribution of Diseases and Possible Causes

The distribution of affected plants in a field (Plate 1) can provide much information on a probable cause of the problem. For example, diseases caused by soilborne pathogens are usually host specific, have a strong interaction with soil characteristics (i.e., low, wet; high, dry; and texture, structure), and are rarely uniformly present over an entire field. Characteristically, these soilborne diseases involve only the crop and are often distributed in large, circular to irregular areas that may reflect soil type, cropping history, herbicide usage, or topographic features (Fig. 4.1 and Tables 4.1 and 4.2).

Some plant species germinate or grow poorly in association with or following other plant species (Plate 1E). This may be caused by soilborne pathogens (e.g., root-knot nematodes) that affect both species or by toxic substances released from one plant (allelopathy) or produced by the decomposition of crop residues. (See Appendix 4.2 for a listing of some problem plant associations or sequences and Appendix 4.3 for plant species susceptible to juglone injury.)

Air pollution, on the other hand, is generally fairly uniform over a large area and usually affects numerous plant species. Thus, symptomatic plants would involve the crop and other sensitive species in the area, including some weeds and trees. However, there may be differences in the intensity of symptoms over the field because plant growth and degree of sensitivity or susceptibility are not usually uniform. Figure 4.1 and Table 4.2 summarize various field distribution patterns and possible causes.

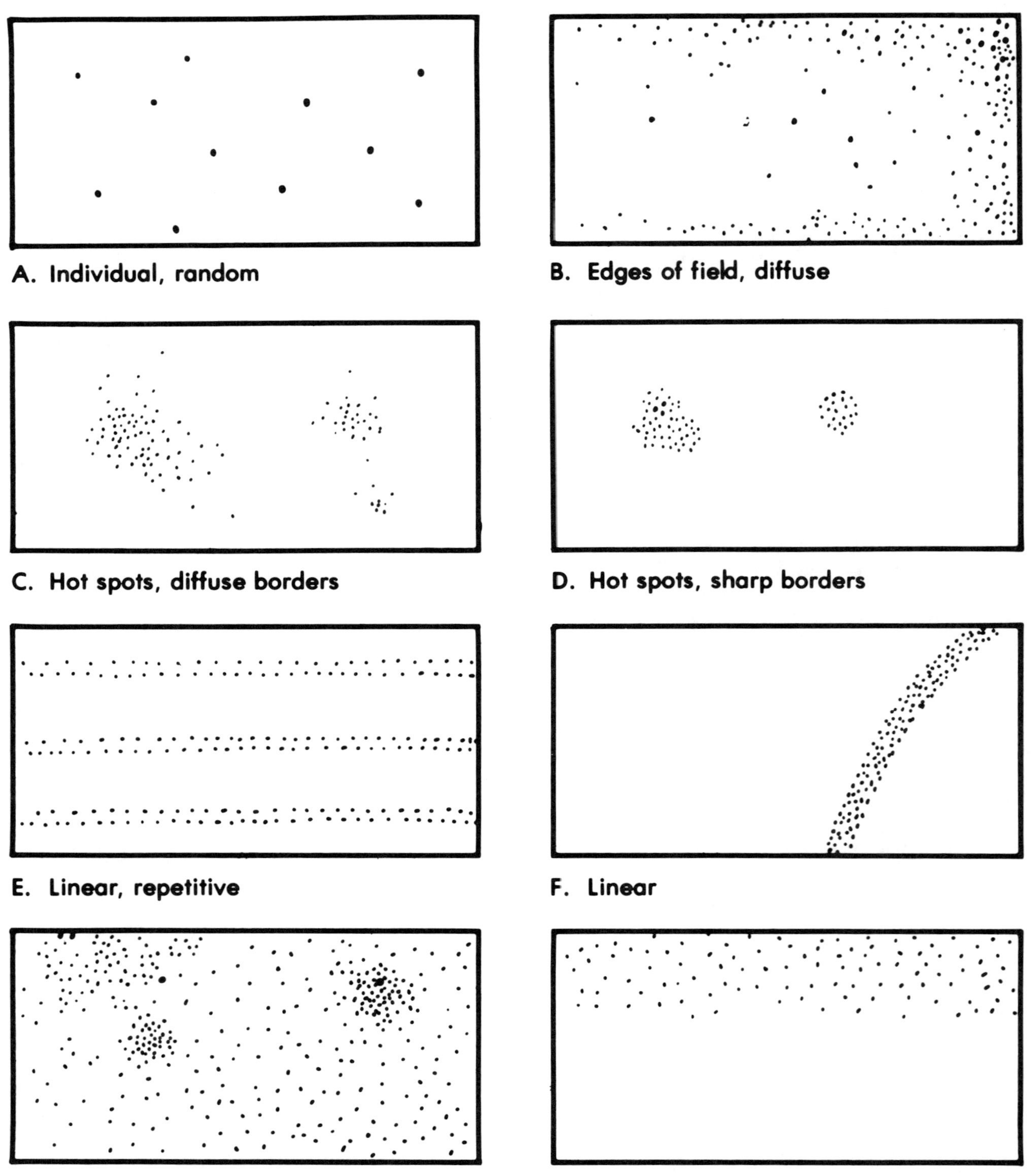

Fig. 4.1. Initial field distribution patterns of disease plants (in annual row crops). See Table 4.2 for possible causes of these patterns.

Table 4.1. Groupings of plant disease agents and typical disease signatures (see also Figure 4.1)

Plant disease agent	Typical field signature				
	Number of hosts	Time of appearance	Symptomatic tissues	Field distribution	Comments
Pathogens					
Airborne	1	Gradual, late	Older shoots	Uniform, topographic	High weather interaction
Seedborne	1	Gradual, early	Older shoots	Random plants, topographic	...
Soilborne	1	Gradual, early	Roots, stems	Topographic, soil type	...
Insect-vectored	1	Gradual, late	Shoots	Field edges, random plants	...
Abiotic agents					
Air pollution	Several to many	Sudden, fast	One age of foliage; interveinal	Topographic, uniform	Look for indicator plants
Toxic droplet drift	Many	Sudden	Exposed shoots	Uniform	Greater towards source
Toxic vapors	Many	Sudden	Shoots	Uniform, islands	...
Soluble salts	Many	Slow	Lower leaves, leaf margins, interveinal	Uniform, topographic, soil type	Worse during dry periods and in sandy soils
Nutritional deficiency or excess	Many	Slow	Variable	Uniform, soil type	Worse during wet periods and in sandy soils
Spray injury from improper pesticide	Few to many	Sudden	Exposed shoots	Uniform	...
Soil herbicide	Many	Fast	Variable	Uniform, topographic, soil type	Worse in sandy soil
Cold injury	Many	Fast	Tender, exposed	Low areas	Shaded foliage on ground may escape
Heat injury	Many	Fast	Tender, exposed	Uniform	Worse during sunny, dry periods
Drought	Many	Slow	Older leaves, roots	Uniform, topographic	Worse in sandy soil
Malfunctioning equipment	1	Sudden	Variable	Regular pattern	Often in streaks and ends of rows

Table 4.2. Initial field distribution of diseased plants and possible causes (see also Figure 4.1)

Distribution	Possible cause	Distribution	Possible cause
Random, individual plants	Seedborne disease Long-distance spore showers Insect-vectored pathogen	Linear, often repetitive	Malfunction of equipment (depth of seeder, adjustment or plugging of a fertilizer or pesticide applicator) Infectious disease spread by contact (pattern consistent with cultural operations) Carryover of soil herbicide Soilborne disease from a previous crop
Random "hot spots"	Herbicide injury Seedborne disease Insect-vectored pathogen Nematodes (concentrated) Soilborne disease	Linear	Old roadbed or chemical spill Movement of soilborne pathogen from high to lower ground
Edges of field	Insect-vectored disease Windborne pathogens from adjacent fields Pesticide drift from adjacent fields Improper lime, fertilizer, or pesticide application to soil Shaded areas	Uniform	Spore shower Air pollution injury Volatile herbicide injury Herbicide injury Excessive fertility or a nutrient deficiency Seedborne disease (seed badly infested or infected)
Low areas	Soilborne disease Herbicide concentration Cold injury Excess moisture	Limited to one species	Infectious disease
High areas	Moisture stress High soil soluble salts level Lightning	Several species involved	Air pollution injury Herbicide injury
Soil type	Nutrient problem often associated with soil pH after liming (high pH in sandy soils, low pH in soils high in clay or organic matter) Herbicide injury (more severe in sandy soils low in organic matter) High soil soluble salts level (more severe in sandy soils or soils low in organic matter) Soil nematodes (often more severe in sandy soils)	Many species involved	Herbicide injury Lightning (restricted area up to 20–30 ft across) Soil toxicant injury Cold injury
		Large, linear section of field	Field sections have different cropping history (often a carryover herbicide problem)
		Ends of rows or symptoms more pronounced in turn-around areas	Excess or deficiency of some chemical applied by the grower Soil compaction

Seasonal Occurrence of Plant Disease and Possible Causes

In most geographic areas, there is a general, predictable, yearly progression of disease occurrence and types of specimens submitted to plant disease clinics. This phenomenon results from plant growth stages, weather patterns, crop husbandry practices, and plant shipments. Recently, the reasons for yearly seasonal occurrence and intensity of disease have been vigorously investigated in order to improve disease control strategies, e.g., disease forecasting and integrated disease and pest management.

An awareness of the normal seasonal occurrence of plant diseases and factors (especially weather) conducive to disease development is valuable to the plant disease diagnostician. For example, late blight of potato is likely to occur after a month or so of cool, wet weather. If the weather has been hot, above 32°C (90°F), and dry, late blight can usually be eliminated as a possible cause. Information on the requirements for disease occurrence and development is given in many books and bulletins that describe diseases by crop.

A useful table was prepared by the Cooperative Extension Service of Illinois for predicting the occurrence of diseases of various crops on the basis of spring and summer rainfall (Table 4.3). Although the table is specific for the Midwest, it has application to most areas with similar weather patterns and crops. Similar charts for other areas can be prepared by recording the date of disease occurrence on a crop in the area served. Because seasons may be early or late, it is useful to include a growth stage of the host or a common plant as a time reference. Flowering of the crop or of certain plants (e.g., dogwood, apple, or peony) is especially convenient. A particular disease may occur over a 2-week period just after flowering, which usually occurs on or close to a certain date. However, when spring is late, flowering may be delayed 10 days or more, and the disease may then be anticipated about 2 weeks later. Table 4.4 illustrates a disease progression chart, initially developed by R. K. Jones for ornamentals in North Carolina. This region is characterized by well-defined seasons (spring, summer, autumn, and winter), abundant rainfall with intervening dry periods, and varied soil types, topography, and elevations. Only those diseases that usually occur at a particular time of year or at a particular growth stage of the plant are listed.

There are problems in using disease occurrence charts. Local grower practices, such as irrigation and fertilization, may promote or influence disease occurrence at unexpected times and locations. This is especially true for greenhouse-grown crops and others with high management inputs. Also, disease symptoms may develop initially as anticipated, but a symptom such as a gall or canker may persist for months.

Plant Parts Affected and Possible Causes

The location of symptomatic tissue on a plant is often a poor indicator of the cause of an abnormality. Frequently, the source of the injury or entry of the causal agent is at some distant point on the plant. The causal agent may produce no symptoms at the place of entry. For example, a canker on a tree trunk may cause the leaves on one or several branches above the canker to discolor, wither, and die. High concentrations of soluble salts in the soil may cause undetectable symptoms on roots but cause marginal and interveinal lesions on older leaves with whitish deposits commonly occurring on the soil surface. The same is true for many nutritional deficiencies and injuries from soil

Table 4.3. Potential damage from some common infectious plant diseases in Illinois, depending on the rainfall during the spring and summer

Crop	Disease	Spring and summer conditions		
		Wet	Normal	Dry
All plants	Crown and root rots, leaf spots and blights, seed rot, damping-off	Severe	Light to moderate	Doubtful
Small grains	Foliar rusts, scab, glume blotch	Severe	Light to severe	Doubtful
	Yellow dwarf (virus)	Doubtful	Light to severe	Light to severe
Corn (maize)	Leaf blights, stalk rots, ear rots	Moderate	Light	Doubtful
Soybeans	Foliar diseases, Phytophthora root and stem rot, pod and stem blight	Severe	Light	Doubtful
Alfalfa	Foliar diseases, Phytophthora root rot	Moderate to severe	Moderate	Light
Beans, garden	Foliar diseases (bacterial blights, rust), white mold, damping-off, root and stem rot, wilt	Severe	Moderate	Light
Potato, tomato, and vine crops	Blights, leaf spots, anthracnoses, fruit/tuber rots, white mold	Severe	Light	Doubtful
Apple and pear	Scab, fire blight, cedar rusts, fruit spots and rots	Severe	Moderate to severe	Light to moderate
Grape	Black rot, downy mildew, fruit rots	Severe	Light to severe	Light to moderate
Strawberry	Foliar diseases, fruit rots, root rots	Severe	Light to moderate	Light
Raspberry	Anthracnose, cane blights, fruit rots	Severe	Light to moderate	Light
Trees and shrubs	Leaf spots, blights, anthracnoses, rusts	Severe	Light to moderate	Doubtful
Evergreens	Needle blights, Sphaeropsis tip blight	Severe	Moderate	Light
	Browning, needle drop	Doubtful	Light	Moderate to severe
Lawns and other turf areas	Leaf spots, melting-out, slime molds, dollar spot	Severe	Light to moderate	Doubtful
	Rusts, summer patch, powdery mildew	Doubtful	Light to moderate	Moderate to severe

Table 4.4. Annual progression of some plant diseases in North Carolina[a]

Plant	Marker / Disease	Jan	Feb	Mar	Apr	May	Jun	Jul	Aug	Sep	Oct	Nov	Dec	Probable area[b]
Alfalfa	Dogwood flowering				✿✿✿✿									
	Anthracnose						★	★	★	★				
	Sclerotinia blight	★	★	★	★						★	★	★	
Apple	Flowering				✿✿✿									
	Bitter rot							★	★	★	★			1, A
	Black rot (leaves)				★	★	★							1, A
	Black rot (fruit)								★	★	★			1, A
	Cedar rusts					★	★	★						1, A
	Fire blight				★	★	★							1, A
	Powdery mildew				★	★	★							1, A
	Scab			★	★	★	★	★	★	★	★	★		1, A
Asparagus	Harvest				✿✿✿									
	Cercospora leaf spot						★	★	★	★	★	★		1, CP
Azalea	Flowering				✿✿✿✿✿✿	✿								
	Leaf gall				★	★								1, A
	Ovulinia petal blight				★	★								1, A
Bean	Flowering					✿✿✿✿✿	✿		✿✿✿					
	Seed rot					★	★	★						1, A
	Damping-off						★	★	★					1, A
	Root rot						★	★	★	★				1, A
	Anthracnose						★	★	★					1, A
	Botrytis blight					★	★			★				1, A
	Rust						★	★	★	★				1, A
	Powdery mildew							★	★	★				1, A
Cabbage	Heading					✿✿✿				✿✿✿				
	Black rot			★	★	★	★			★	★	★		2, A
	Blackleg			★	★	★	★			★	★	★		2, A
	Sclerotinia watery soft rot				★	★	★				★			3, M, P
	Bacterial head rot						★	★	★					3, A

(continued on next page)

Table 4.4. (*continued*)

Plant	Marker / Disease	Jan	Feb	Mar	Apr	May	Jun	Jul	Aug	Sep	Oct	Nov	Dec	Probable area[b]
Cabbage (*continued*)	Clubroot				★	★	★	★		★				2, NW
	Downy mildew				★	★	★	★	★	★	★	★		1, A
	Alternaria head rot						★	★	★	★	★	★		1, A
Carrot	Harvest						✿				✿	✿		
	Southern blight						★	★		★				2, P, CP
	Leaf spots								★		★	★		1, A
Corn (maize)	Seeding				✿	✿								
	Anthracnose					★		★	★					1, A
	Cercospora leaf spot								★	★				1, A
	Southern leaf blight						★	★	★					1, A
Cucurbits	Seeding				✿	✿		✿						
	Anthracnose							★	★	★				1, A
	Bacterial wilt						★	★	★	★	★			1, A
	Gummy stem blight						★	★	★	★	★			1, A
	Downy mildew						★	★	★	★	★			1, A
	Powdery mildew						★	★	★	★	★			1, A
	Damping-off (Pythium)					★	★	★	★					1, A
	Damping-off (Rhizoctonia)							★	★	★				1, A
Dogwood	Flowering				✿									
	Anthracnose				★	★	★							1, A
	Septoria leaf spot						★	★	★					1, A
	Leaf scorch (abiotic)							★	★					1, A
Flowering annuals	Dogwood flowering				✿									
	Botrytis gray mold						★	★	★	★	★			2, A
	Rhizoctonia stem rot						★	★						2, A
Peanut	Seeding					✿								
	Sclerotinia blight								★	★	★			2, CP (N)
	Southern blight									★	★			2, CP
	Cercospora leaf spot								★	★	★			1, CP
	Cylindrocladium root rot						★	★	★	★	★			2, CP

Host	Stage / Disease	Code
Pine	Flowering	
	Eastern gall rust	1, P
	Fusiform rust	1, P, CP
	Hypoderma needle cast	1, A
	Needle rust	3, A
Rose	Flowering	
	Black spot	1, A
	Botrytis blight	1, A
	Powdery mildew	1, A
Soybean	Seeding	
	Septoria brown spot	1, A
	Powdery mildew	1, A
	Downy mildew	1, A
	Diaporthe pod and stem blight	1, A
	Brown spot	1, A
	Phytophthora root rot	2, P
	Rhizoctonia damping-off, root rot	2, P, CP
	Southern stem blight	2, P, CP
Strawberry	Flowering	
	Fruiting	
	Gray mold	
	Red stele	
	Powdery mildew	
	Anthracnose	
	Leaf spots and blights	2, P, CP
	Leather rot	2, P, CP
Tobacco	Transplanting	
	Blue mold in plant bed	4, A
	Blue mold in field	4, A
	Black shank	2, A
	Bacterial wilt	2, P, CP
	Mosaic (tobacco mosaic virus)	1, P, CP

(continued on next page)

Table 4.4. (*continued*)

Plant	Marker / Disease	Jan	Feb	Mar	Apr	May	Jun	Jul	Aug	Sep	Oct	Nov	Dec	Probable area[b]
Tomato (field)	Transplanting				✿	✿								
	Septoria leaf spot					★	★	★	★					2, A
	Early blight						★	★	★	★				1, A
	Late blight						★	★	★	★	★			4, M, P
	Bacterial leaf spot						★	★	★	★	★			2, A
	Bacterial wilt						★	★	★					2, P, CP
	Mosaic (tobacco mosaic virus)						★	★	★					2, A
Turf (fescue and rye)	Dogwood flowering				✿									
	Brown patch (hot weather)					★	★	★	★					2, A
	Brown patch (cold weather)	★	★									★	★	2, A
	Dollar spot			★	★	★	★	★	★	★				2, A
	Fairy ring		★	★	★	★	★	★	★	★	★	★	★	2, A
	Fusarium patch	★	★									★	★	2, A
	Helminthosporium leaf spot			★	★	★	★	★	★	★	★			2, A
	Helminthosporium crown rot				★	★	★	★	★					2, A
	Nematodes						★	★	★					2, A
	Powdery mildew				★	★	★	★	★					2, A
	Pythium blight					★	★	★	★					2, A
	Pythium root rot				★	★	★	★	★	★	★			2, A
	Red thread				★	★	★	★	★					2, A
	Rust			★	★	★	★	★	★	★	★			2, A
	Slime mold					★	★	★	★	★				2, A
	Southern blight						★	★	★					2, A
Wheat	Dogwood flowering				✿									
	Septoria glume blotch				★	★	★							2, A
	Powdery mildew			★	★	★	★							1, A
	Rust			★	★	★	★							2, A
	Gaeumannomyces take-all		★	★	★	★	★	★					★	4, A

[a] ✿ = Planting, transplanting, flowering, fruiting, heading, or seeding; ★ = disease probability (the size of the star represents the degree of probability of disease).

[b] 1 = Most plantings, usually; 2 = fairly common; 3 = infrequent; and 4 = rare. Areas in North Carolina: A = all areas; CP = Coastal Plain; M = mountains; N = north; NW = northwest; P = Piedmont.

herbicides and soil toxins. By recognizing these limitations and carefully examining the symptoms and other parts of the plant, one can use the position of symptomatic tissue on the plant to help categorize the cause of the problem. Tables 4.5–4.8 relate symptom position with possible causes.

Plant Symptoms and Possible Causes

Details of symptoms on individual plants are extremely useful in diagnosing causes of plant diseases. Each causal agent, biotic or abiotic, will produce specific symptoms. These symptoms, of course, can vary within limits depending on exposure intensity, growth stage of the plant, and environmental conditions during and after exposure. In fact,

many keys for identifying plant diseases on the basis of symptomology have been prepared for specific plants or groups of plants, such as turf and ornamentals. Those that relate to a single host tend to be most useful, and the diagnostician is encouraged to collect and use them. Most general plant diagnostic keys based on symptomology are inadequate because they include only a narrow group of causal agents (e.g., insects, air pollutants, pathogens, herbicides, or nutritional problems) and omit causal agents in other groups that may produce similar symptoms. However, once the diagnostician determines the general nature of the causal agent, these keys become useful. See Tables 4.6–4.8 and Figures 4.2 and 4.3 for specific symptoms and possible causes.

Key for Diagnosing Disorders of Annual Row Crops

Detailed observations of field distribution, plant parts affected, time of occurrence, and plant symptoms can be useful in determining the general cause of disease in annual row crops. Table 4.7 is a key to field disease signatures with general causes. Within limits, the key can be used for perennial and tree crops.

Air Pollution Injury

Air pollution is an increasing problem worldwide, causing plant disease especially in industrial and urban regions but also in some rural areas. There are many natural and human sources of air pollutants, which can be categorized in two general ways:

1. Point source pollutants are those emanating from one to a few closely located sources (e.g., a factory or volcano) and affecting plants downwind, usually within a few miles. These include sulfur dioxide, nitrogen oxides, chlorine, ammonia, hydrogen fluoride and chloride, mercury, ethylene, volatile herbicides such as 2,4-D, and particulate matter.
2. Non-point source or general atmospheric pollutants are those arising from many small sources (e.g., internal combustion engines) over an extensive area and at times contaminating the atmosphere over significant portions of continents. These include ozone and peroxyacetyl-nitrates (PAN) (Table 4.9).

Air pollution injury varies depending on the type of pollutant and its duration and concentration (often less than 1 ppm) and on the plant species (in some species, even on plant cultivar or variety), growth stage, and previous growing conditions. *Chronic injury* results from long-term exposure to low levels of pollutants, typically ozone and other oxidant types, the most common non-point source pollutants. These plants may show reduced growth, unusual color, or premature senescence. *Acute injury* results from high levels of pollutants. Plant symptoms are fairly distinct, usu-

Table 4.5. Plant parts affected and possible causes

Part affected	Possible causes
Roots	Soilborne disease Asphyxiation (waterlogged soil) Toxic substance in soil, such as excess soluble salts, aluminum, ammonia, nitrites, herbicides Insects (usually larval forms such as grubs and wireworms)
Trunk	Mechanical injury Cold or heat injury (usually side toward afternoon sun; bark often split or cambium cankered) Insects (usually immature forms such as borers) Infectious disease (usually soilborne or endemic with perennials) Injury to trunk or disease in stems or roots Herbicide or chemical injury (some herbicides can penetrate the bark)
Stems	Infectious disease Insects (usually immature forms such as borers) Cold or heat injury (see Trunk)
Terminal shoots	Infectious disease Nutritional problems Insects Eriophyid mites Herbicide injury (growth-regulator type) Cold injury
Young leaves	Infectious disease Cold injury Insects, mites Spore shower Heat injury Moisture stress (older leaves more symptomatic) Herbicide injury, from both soil and topical exposure Air pollution
Old leaves	Infectious disease, systemic (e.g., wilts) Infectious disease, localized (e.g., leaf spot or blight) Herbicide injury Soluble salts Moisture stress
Entire above-ground portion	Cold injury Herbicide (especially contact type) or chemical spray injury Root, stem, or crown rot Lightning Infectious disease, may be systemic (e.g., wilts)

Table 4.6. Plant disease symptoms, characteristics, and possible causes (see glossary for definitions)

Symptom	Possible causes	Symptom	Possible causes
Aborted flowers	Poor pollination (fertilization) Ethylene air pollution	Dieback	Infectious disease (e.g., fungal or bacterial infection) Cold or heat stress Insects Mechanical injury (e.g., girdling wires or twine) Gas main leak Change in soil grade Lightning Poor planting (e.g., too shallow or deep, girdling roots)
Adventitious root development	Interference with translocation at or below ground Prolonged water stress Root or collar (crown) rot Ethylene air pollution Nematode injury to roots Phenoxy-type herbicide		
Blast	Freezing injury Very high temperatures Drought Severe infectious disease	Dieback (sudden) of terminal shoots with a sharp transition to healthy tissue; dead leaves are retained	Cold, frost, or chemical injury Infectious disease (e.g., stem canker, fire blight)
"Blisters" on succulent tissues	Eriophyid mites Toxic chemical injury Nutritional deficiency Infectious disease (e.g., *Taphrina*) Edema	Dieback of terminal shoots in spring with gradual transition to healthy tissue; cracking of bark and wood, often one-sided on larger plants	Winter injury
Brooming (witches'-brooms)	Eriophyid mites Virus or mycoplasma infection Fungal infection Aphids Mistletoes Salt spray damage	Dying (gradual) of a branch in tree or shrub	Canker Shoot blight
		Edema (oedema)	Excess intercellular water accumulation resulting from rapid uptake by roots and low transpiration Occasionally insects or chemical injury
Canker	Infectious disease (Fig. 4.3c) Localized exposure to temperature extremes or toxins Often follows mechanical damage	Enation	Viral infection
		Epinasty	Hormone accumulation at base of petiole Verticillium wilt Bacterial wilt (*Pseudomonas solanacearum*) Ethylene (air pollutant) Herbicide (e.g., 2,4-D)
Chlorosis (see also Yellows)	Nutritional deficiency or excess Herbicide injury (chlorophyll inhibitor) Air pollution (e.g., ethylene) Infectious disease with release of systemic toxins (e.g., *Fusarium* or *Verticillium*)		
		Erineum (gall from plant hair)	Eriophyid mites
		Etiolated growth (may be induced en route to clinic)	Light deficiency Excessively warm conditions
		Flagging	Loss of cell turgor pressure Vascular wilt disease; infection confined to a radial section of the vascular system Stylet-feeding insect injury
Damping-off	Soilborne fungi (e.g., *Pythium*) Insects (e.g., lesser cornstalk borer, seed corn maggot, cutworms) Very high soil soluble salts levels High soil temperature injury mimics damping-off		
Death (gradual) of all or major portion of tree or shrub	See Wilt	Premature fruit drop	Petiole- or pedicel-rotting fungus Formation of abscission layer caused by ethylene air pollutant, poor pollination, infectious disease with production of auxins, or pesticide application (e.g., carbaryl) Insect damage
Decline (gradual) of woody shoots and retention of dead leaves; margin between affected and healthy tissue is irregular and sunken; small pimple-like fruiting bodies over surface of dead bark; limited growth of branches indicated by closeness of the terminal bud scars	Infectious disease (usually a fungal canker)		
		Galls	Insects and mites (Fig. 4.3d) Infectious disease (e.g., crown gall, clubroot of crucifers)
		Gummosis	Mechanical injury Insect injury Infectious disease
		Leaf bronzing, silvering	Air pollution (ozone) Mites Thrips
Decline (sudden) of tree or shrub or large parts of it	Toxic substance in soil Severe weather change (e.g., drought or freezing) Mechanical injury to roots (e.g., from heavy machinery) or trunk Change in the soil grade Gas main leak	Leaf burn or scorching of margins and between veins (dead tissue may drop out and adjacent tissue may be chlorotic)	Accumulation of toxic substances through transpiration (Figs. 4.2e and 4.3a) Drought Infectious disease (e.g., fungal or bacterial infection)

(*continued on next page*)

Table 4.6. (*continued*)

Symptom	Possible causes	Symptom	Possible causes
Leaf coating (sticky) followed by a sooty mold	Insects (especially aphids, scales, mealybugs, and whiteflies) Eriophyid mites	Leaf yellowing in bands along larger veins in contrast to dark green bands associated with nutrient deficiency problems	Viral infection (Fig. 4.2h)
Leaf fall	Formation of abscission layer at base of petiole (a natural occurrence) Ethylene (air pollutant) Infectious disease such as certain bacterial leaf spots (e.g., *Xanthomonas campestris* pv. *vesicatoria*) and anthracnose Plant under stress (e.g., drought)	Mold on foliage, green stems, fruits, and occasionally flowers	
		White mold with underlying tissue green	Powdery mildew (Fig. 4.2d)
Leaf mottling	Air pollution Mites Thrips Viral infection Herbicide injury	Black mold with underlying tissue sticky	Sooty mold
		Black, dry mold, usually in large spots; tropical	Black mildew
Leaf russeting	Air pollution Mites Thrips	Gray white, yellow, or purple mold on turf that easily brushes off	Slime mold
		Olive drab mold	Cladosporium leaf mold (or early apple scab)
Leaf scald	Excess heat Air pollution	Tan or gray mold	Botrytis or gray mold
		Green or blue mold	Penicillium rot
Leaf scorch	Usually caused by rapid and excessive loss of water or exposure to toxins (Figs. 4.2e and 4.3a) Excess heat and wind, especially under moisture stress (usually interveinal and marginal) Drought Application of toxic substance (e.g., excess fertilizer or pesticide, improper pesticide formulation, salt) Exposure to toxic gas (e.g., ammonia; SO_2; fluorine compounds; flue fumes, especially from fuels with high sulfur levels) Moisture stress (youngest and oldest leaves are more resistant; usually interveinal or marginal) Damage to roots, stem, trunk, or branches Infectious disease (e.g., root rot) Reduced root system Spray injury Leafhopper injury	Cottony mold	Pythium, Rhizopus, or Sclerotinia rot
		Mosaic	Viral infection (Fig. 4.2f and i)
		Mottle	Viral infection Nutritional problems Air pollution
		Needle cast, blight, death, or dieback (scattered); more frequent on older needles; fruiting bodies often present	Fungal needle disease (Fig. 4.3b)
		Needle drop	
		Old needles	Normal and seasonal; new needles produced Stress, especially drought
		Mostly young needles	Fungal disease (needle cast) Air pollution Nutritional deficiency Toxic substance in soil Herbicide injury
		Needle tip burn	Air pollution (Fig. 4.3a) Soil herbicide Soil soluble salts Drought Freezing
Leaf spots		Ooze	Bacterial or fungal infection
Variable size, round or oval, usually elongate on stems; different color or texture zones may develop in necrotic tissue giving a bull's-eye effect; often chlorotic halo around the spot; mold or fruiting structures in older necrotic tissues	Fungal and bacterial diseases (Fig. 4.2a and b)	Phyllody	Mycoplasma infection
		Pustule	Fungal or bacterial infection
		Ring spot	Viral infection
		Rosetting, rosetted shoots	Nutritional deficiency or excess Viral or mycoplasma infection Insects or eriophyid mites
Angular and limited by small leaf veins; tissue at first "oily" or water soaked, translucent and papery after drying; often lower leaf margins with a very narrow band of necrotic tissue; margins sharp	Bacterial leaf spot (Fig. 4.2b)	Rot	Fungal or bacterial infection (may be secondary to injury)
		Russetting of fruit	Improper spray Mite injury Cold injury Mechanical injury Powdery mildew
Evenly distributed; uniform in size and color; margins sharp but adjacent tissue may be chlorotic	Toxic chemical droplet drift (Fig. 4.2c)	Transformation (metamorphosis)	Infectious disease Some herbicides Mechanical injury

(*continued on next page*)

Table 4.6. (*continued*)

Symptom	Possible causes	Symptom	Possible causes
Vein banding	Viral infection (Fig. 4.2g) Herbicides	Wilt (*continued*)	Roots pruned (by nematodes, herbicides, insects) Cold soil (usually reversible) Vascular disease Hot, dry, and windy conditions (usually temporary) Waterlogged soil Insect injury Drying en route to clinic
Vein clearing	Viral infection (Fig. 4.2h) Herbicide in soil		
Warts on leaves, especially along veins on undersurface	Edema (oedema) Insect galls (Fig. 4.3d)		
Wilt	Dry soil High soluble salts in soil water Stem or trunk damage (e.g., canker, mechanical, etc.) Root rot	Yellows (see also Chlorosis)	Viral or mycoplasma infection Fungal infection Nutritional deficiency or excess

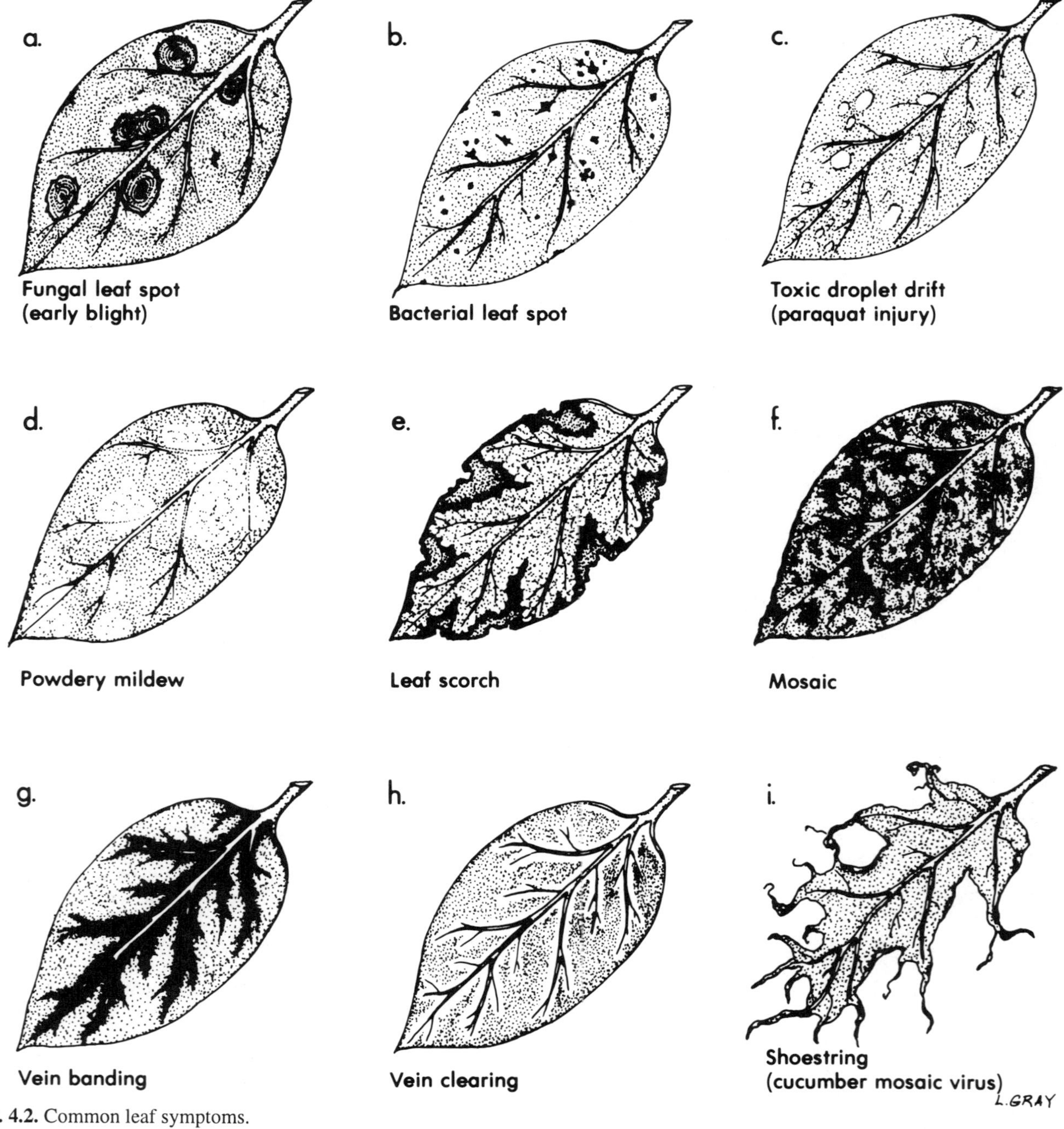

Fig. 4.2. Common leaf symptoms.

ally developing within a week of the exposure. Often the injury is limited to a certain age group of leaves on the plant.

Diagnosis of air pollution injury is based on plant symptoms (Plate 4 and Table 4.9), the spectrum of plant species involved (Table 4.10), knowledge of pollutants in the atmosphere, and in a few cases, a chemical assay of symptomatic plants (e.g., for hydrogen fluoride injury) in specialized laboratories. Concentrations of non-point source pollutants (especially ozone and PAN) are monitored in many areas by state and federal agencies, and records may be available in newspapers or by contacting the area meteorologist. When pollutant levels exceed air quality standards or known injurious levels, sensitive plants should show air pollution injury a week or so later (Tables 4.9 and 4.10).

Diagnosing air pollution injury on individual plants received in clinics is often not possible because symptoms (Plate 4) mimic those caused by other agents such as mites,

chemical spray injury, high levels of soluble salts in the soil, residual soil herbicides, and improper soil fertility, including water deficiency. In the clinic, however, it is useful to eliminate pathogens (fungi, bacteria, viruses, and nematodes), high soluble salts levels, and improper nutrition (from soil and tissue analyses reports) as possible causal agents. Diagnosing air pollution injury to plants is done mostly in the field. The following procedure is suggested:

1. Examine the area and determine

   a. the presence of one or more potential point sources of pollution (consider power plants, factories, smelters, ignited trash heaps, and chemical spills);

   b. terrain characteristics and wind direction at the time of exposure (in long, narrow valleys, plant injury from point source pollutants will be concentrated downwind on the floor and sides of the valley);

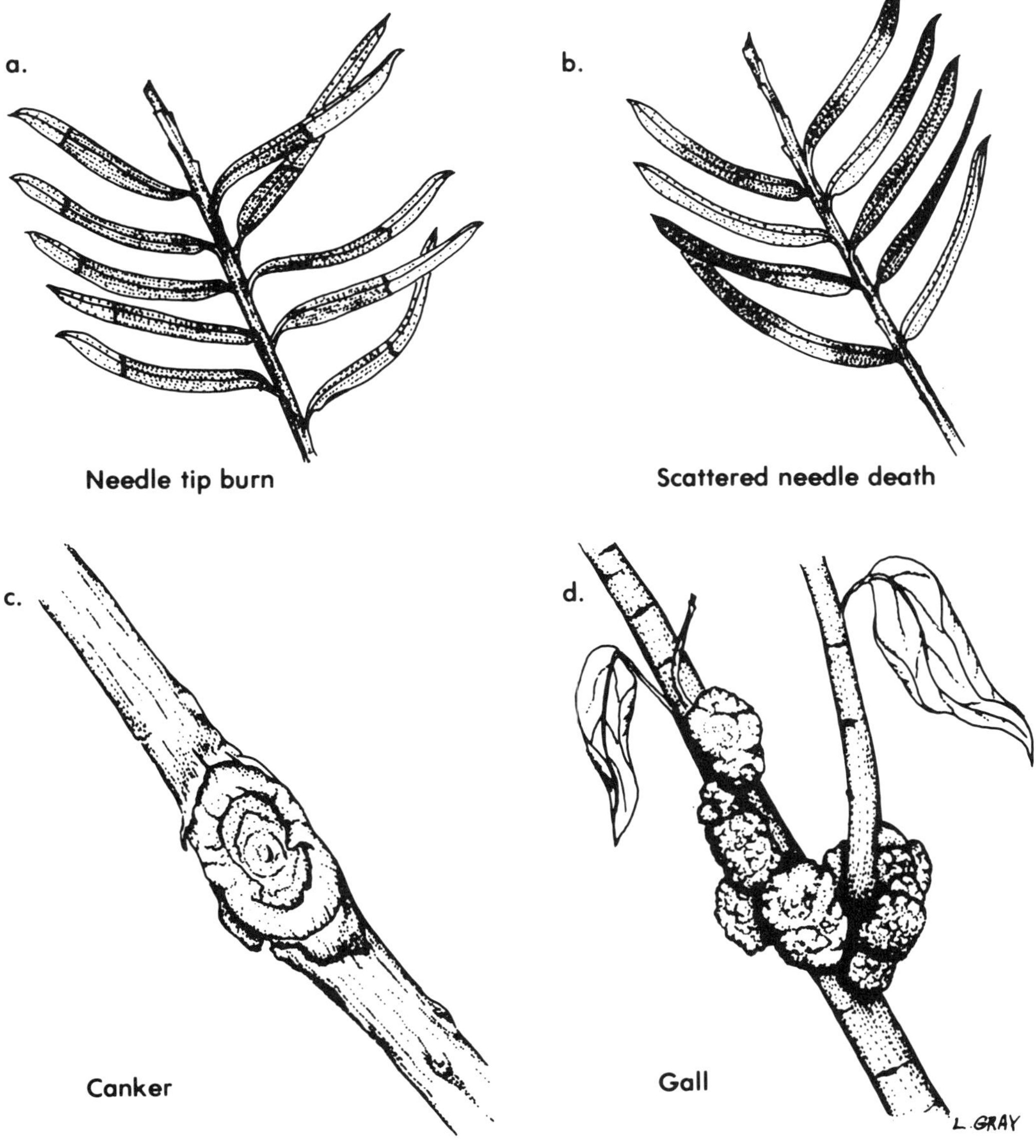

Fig. 4.3. Needle and stem symptoms.

Table 4.7. Diagnosing common disorders of annual row crops (mostly vegetables) in the field

Symptom or condition	Diagnosis
I. Plants in seedling stage or prior to emergence (consider II)	
A. Emergence poor (consider I.B)	
1. Growing conditions normal	
a. Seed did not germinate, rotted, or seedling failed to emerge	Cold soil
	Seed old, injured, or nonviable
	Seed planted deeper than twice its diameter
	Seed planted very shallow; soil dry
	Soil compacted or crusted
	Fertilizer or herbicide placed near seed or in excessive amount
	Seed eaten by birds or other animals (e.g., insects)
2. Growing conditions wet	
a. Field flooded, or standing water longer than 1 day	Asphyxiation, seed decay
b. Weather cool and overcast	
(1) Seeds appear normal	Cold soil (problem temporary)
(2) Seeds rotted	Preemergence damping-off
(3) Emergence scattered, organic soil	Seed corn maggot
3. Seed germinates but soon rots or stems of seedlings rot; more prevalent in wet weather (cold or hot) than in dry weather	
a. Cottony growth on stems in morning	Pythium, Sclerotinia, or Sclerotium damping-off
b. Dry rot, often brownish	Rhizoctonia or Fusarium damping-off
c. Small, dark dots appear in lesions	Various fungi
d. None of the above	Fertilizer injury, high soil salts
B. Emergence normal but growth poor	
1. Condition uniform over entire field or sections of it (consider section I.B.2)	
a. Seedlings twisted or distorted, or leaves mottled with light green, or leaves nearly white, or base of hypocotyl swollen	Residual herbicide in soil
	Improper depth placement of herbicide, fertilizer, or seed
b. Flecking, interveinal necrosis, or marginal burn on leaves	Air pollution, wind burn, or chemical burn
c. Marginal and tip leaf burn; roots may be brownish with smaller ones rotted	High soil salts (excessive fertilizer)
d. Plants stunted and wilted; white maggots on roots, crown, bulbs, etc.	Insect damage, e.g., cabbage or onion maggot
e. None of the above	Improper soil pH; fertilizer leached or poorly placed
	Soil herbicide or residue from previous crop
2. Condition usually scattered starting in areas 3–30 m (10–100 ft) in diameter with few plants affected in intervening areas; "hot spots" or "foci"	
a. Condition associated with topographic or soil features (low areas, deep sands, ridges, etc.)	
(1) Water stands in area; roots rotted	Asphyxiation of roots; residual herbicide
(2) Well-drained site	Nutrients leached from root zone; insufficient moisture
(3) Site of old house, dump, or roadbed	Soil toxicant (e.g., lime, zinc, etc.); soil compaction
(4) Sandy soil	Soil herbicide; high soil pH after liming
b. Condition not associated with topographic or soil features or only slightly so	
(1) Seedlings clipped near ground line	Cucumber beetles, cutworms
(2) Roots with galls or knots	Root-knot nematode, crown gall
(3) Crucifer plant with galls or clubs	Clubroot
(4) Roots rotted	Root-rotting fungi, very high soil soluble salts
(5) Seedlings constricted at ground level, have fallen over or died	
(a) Cottony growth on stems in morning	Sclerotinia or Pythium damping-off
(b) Dry rot, often brownish	Fusarium or Rhizoctonia damping-off, blight
(c) Small, dark dots appear in lesions	Various fungi causing damping-off
(6) None of the above	Residual herbicide, fertilizers leached from soil, fertilizer burn, or soil salt injury
II. Plants well established	
A. Condition in areas 3–30 m (10–100 ft) in diameter; "hot spots" with few plants affected in intervening areas; may follow topographic or cultural patterns associated with wet weather (consider sections II.B and C)	
1. Stem rots (consider II.A.2, 3, 4, and 5)	
a. Pith compressed into ladderlike disks	Lightning
b. White moldy growth on diseased tissue	
(1) Cool weather, moldy growth only on stems or aerial portions, black sclerotia in fungus mat	Sclerotinia blight
(2) Warm weather	
(a) Fungal growth on stem and adjacent soil; small, round, light brown sclerotia in fungus mat	Southern stem blight (*Sclerotium rolfsii*)
(b) Fungal growth not visible; rot soft, wet, often with bad odor; rot often dark	Bacterial soft rot
2. Fruit rots (consider II.A.3, 4, and 5)	
a. Fruit touching ground or subject to splashing water from soil	
(1) Cottony growth visible, especially on cucumber	Pythium fruit rot, "leak"
(2) Dry, crusty, reddish brown lesions on underside of cucumber	Rhizoctonia fruit rot, "belly rot"
(3) Firm rot with concentric growth rings on tomato	Phytophthora rot, "buckeye rot"

(continued on next page)

Table 4.7. (*continued*)

Symptom or condition	Diagnosis
II. A. 2. (*continued*)	
b. Fruit rot not associated with position on plant	
(1) Soft, smelly, and watery rot	Bacterial soft rot, often associated with insect or mechanical injury
(2) Sunken, decayed, circular spots on fruit, often with tiny black dots	Many different fungi
(3) Fungal growth on fruit during cool weather	Sclerotinia blight or Trichothecium rot
(4) Short, fuzzy, tan gray fungus on rot	Botrytis blight
(5) Soft rot on blossom end, covered with black mold on squash, pumpkin, okra, pepper, southern pea, and pea	Choanephora wet rot
(6) Dark, dry rot on blossom end of tomato, pepper, squash, pumpkin, and watermelon often associated with drought	Blossom-end rot (calcium deficiency in tissue)
3. Leaf spots or leaf blights; green stems may also be affected; usually more severe on older leaves (consider II.A.4 and 5)	
a. Short, fuzzy, tan gray fungus on dead tissue, usually associated with cool, damp nights	Botrytis blight
b. Circular to angular leaf spots with concentric rings and yellow border, usually on older leaves	Alternaria leaf spot and others
c. Circular leaf spots with light-colored centers	Cercospora leaf spot and others
d. Downy, bluish, fungal growth on undersides of leaves visible before dew dries; spots initially with yellowing but no necrosis; usually during prolonged wet weather	Downy mildew (e.g., on crucifers and cucurbits) Late blight of tomato or potato
e. Small, yellowish orange spots with little leaf necrosis initially; yellow area will rub off as rust-colored powder (note: do not confuse with spider mite injury)	Rust (beans)
f. Crucifer leaves with V-shaped yellow areas extending inward along the veins from margin (veins are black)	Black rot
g. Lower tomato leaves with V-shaped yellow areas extending inward from tips	Verticillium wilt
4. Plants wilt; off color	
a. Roots in good condition, at least initially; vascular system in stem is brown	
(1) Weather warm	
(a) Sudden wilt (often after a wet period)	Bacterial wilt
(b) Slow wilt; lower leaves turn yellow	Fusarium wilt
(2) Weather cool; little wilting, leaves turn yellow	Verticillium wilt
b. Root system in poor condition; vascular system not usually discolored more than 3–4 in. above ground	
(1) Field flooded or saturated for more than 24 h	Asphyxiation of roots (tomato plants are very susceptible)
(2) Roots pruned or rotted	
(a) Soil well drained	Nematode or salt (fertilizer) injury
(b) Soil poorly drained	Fungus root rot or salt (fertilizer) injury
(3) Galls on roots	Root-knot nematode
(4) Galls or clubs on roots of crucifer plants	Clubroot
5. Plants off-color; new growth abnormal; leaves with mosaic, twisted, or spindly; growth slightly stunted; distribution of affected plants greater on edge of a field with normal plants interspersed	Insectborne virus
B. Condition random in field or around borders; root system in good condition; insects or remains often evident (consider II.C)	
1. Much plant-to-plant variation	
a. Wilting or breaking of plants; stems tunneled; frass; webbing, pupal cases or larvae present	Borers (e.g., lesser cornstalk, lima bean vine, European corn, squash vine)
b. Plants cut off at base; sudden damage usually after warm, moist evenings	Cutworms
c. Leaf spots, blights, or mosaic	Insectborne diseases, windborne pathogens from adjacent fields
2. Little plant-to-plant variation	
a. Plants stunted, wilted, chlorotic; white maggots in roots or crown	Cabbage and other maggots
b. Vegetation affected, insect evidence	
(1) Leaves mined with twisting tunnels	Leafminer
(2) Leaves chewed or ragged at margins; frass, feces, or insects and webbing present	Larvae of corn earworm, cabbage looper, armyworm, etc.
(3) Leaves chewed at margins; beetles present	Beetles (adult cucumber, Colorado potato, or bean leaf, etc.)
(4) Leaves skeletonized between veins	Japanese beetle, Mexican bean beetle or larvae
(5) Leaves mottled or chlorotic; may be some wilting as the result of sucking insects (usually in colonies)	Sucking insects (aphids, leafhoppers, harlequin bugs, stink bugs, etc.)
c. Fruit (etc.) affected by insects	
(1) Fruit with holes or tunneling; feces or insects present	Insect larvae
(2) Fruit misshapen or with needlelike punctures	Stink bugs
(3) Underground roots or tubers with small to large holes	Beetle larvae (white grub or wireworm, etc.)
d. Leaf spots, blights, or mosaic	Wind- and insectborne diseases

(*continued on next page*)

Table 4.7. (*continued*)

Symptom or condition	Diagnosis
II. B. (*continued*)	
3. Abnormal fruit not associated with insects	
a. Fruit set normal	
(1) Yellow squash turns green or mottled green	Virus (watermelon mosaic and others)
(2) Tomato fruit puffy with open locules and few seeds	Poor pollination or fertilization
(3) Tomato, eggplant, and pepper with fasciated or cat-faced fruit	Cool, humid weather at flowering or phenoxy-type herbicide injury
(4) Off-shaped fruit; seeds partially missing	Poor pollination; viral infection
(5) Blossom end turns black (more prevalent in dry weather)	Blossom-end rot
(6) Fruit pointed at blossom end	Low nitrogen; dry weather
(7) Fruit constricted at stem end	Low K or poor pollination
b. Fruit set poor	
(1) Flowering sparse	Excessive nitrogen or shading
(2) Flowering normal; bees and other pollinating insects in low numbers	Poor pollination
(3) Flowering normal; cold conditions	Poor pollination or fertilization
C. Condition uniform over field; very little plant-to-plant variation	
1. Spots on leaves, lesions on stems, or blights on foliage; usually associated with wet weather and older leaves	Seedborne disease caused by bacteria or fungi
2. White, powdery, circular, moldy growth starting on lower leaves; may be serious in dry weather	Powdery mildew
3. Gradual yellowing, bronzing of older leaves; very fine webbing and minute mites present on undersides of leaves	Spider mites
4. Flecking, marginal burning, or interveinal tissue necrosis (note: may be more pronounced on the side of the field closer to the source of pollutant; often appears suddenly)	Air pollution, herbicide drift, windburn, chemical burn, cold injury, frost, etc.
5. Abnormal growth, such as leaf twisting, terminal necrosis, and mosaic patterns on leaves (note: may be more pronounced on the side of the field closer to the source of the pollutant and exhibit a sudden, uniform appearance)	Air pollution, herbicide drift
6. Plants growing slowly, foliage off color or with necrotic areas; soil pH normal (5.5–6.8)	
a. Symptoms more pronounced on older (lower) leaves, at least initially	
(1) Foliage with purple cast, especially on underside; stems thin	Phosphorus (P) deficiency
(2) Older leaves light green with major veins purplish	Nitrogen (N) deficiency
(3) Foliage with bronze cast; necrotic spotting or marginal burning; roots brown; stems narrow	Potassium (K) deficiency
(4) Yellowing between veins; brittle leaves, leaf margins turn upward; usually on acid soils below pH 5.5	Magnesium (Mg) deficiency
b. Young leaves chlorotic	
(1) Veins tend to retain green color, usually on alkaline soil (or on over-limed soil)	Iron (Fe) deficiency
(2) Light green or yellowish color in interveinal area; veins remain green to the smallest branches, usually on soil above pH 6.8	Manganese (Mn) deficiency
c. Terminal bud abnormal or absent	
(1) Pale margins on young leaves; leaves remain turgid and curl up	Calcium (Ca) deficiency
(2) Retarded growth and interveinal chlorosis of young leaves; leaves and stems brittle; loss of color at base of bud leaf	Boron (B) deficiency
(3) Terminal buds missing	Cabbage webworms or cutworms on crucifers, crusty soil on snap beans
d. Color normal, permanent mottling of upper leaves, usually on soil with high organic matter	Copper (Cu) deficiency
e. Interior of storage roots of cole crops, especially just under the crown, is brownish; stems hollow	Boron (B) deficiency
f. Fruit turns black on distal (flower) end and rots; no moldy growth seen (especially watermelon and tomatoes)	Calcium (Ca) deficiency

    c. the distribution of affected plants;

    d. the plant species affected (grasses, herbaceous plants, deciduous trees, and conifers may all show injury to various degrees);

    e. the type of plant symptoms and age group of affected foliage (Table 4.9 and Fig. 4.2);

    f. the presence of pathogens or other possible causal agents, e.g., drought, winter kill, insects, or herbicide spray drift;

    g. the cropping history, especially for application of herbicides (some herbicides are more persistent in the soil than others, and plants vary in their sensitivity to herbicides [see next section]);

    h. movement of soil herbicides from a nearby area or field by wind or water.

  2. Construct a sensitivity chart on the basis of field observations of symptomatic and nonsymptomatic plants. Table 4.10 gives the expected sensitivity reaction of many plants to various pollutants. Write in a + for sensitive plants, +− for an intermediate reaction, and − for tolerant plants. Leave blank spaces for unlisted reactions. A hypothetical example is given in Figure 4.4 for a

Table 4.8. Abiotic diseases or disorders of interiorscape plants grown in pots or other containers in homes or on greenhouse benches[a]

Problem	Probable causes
Rapid leaf drop	Extreme changes in temperature; hot or cold drafts; extreme changes in light intensity; low humidity (especially during the winter); exposure to gas or furnace fumes; root loss from transplanting; overwatering and/or poor drainage; lack of water
Gradual leaf drop	Insufficient light; lack of fertilizer; overwatering and/or poor drainage; lack of water
Sudden wilting of the entire plant	Exposure to cold; excess fertilizer; overwatering and/or poor drainage; lack of water; high salts in water; plant is pot bound; root or crown rot; nematodes (especially root-knot nematode)
Loss of leaf color	Insufficient or too much light; lack of fertilizer; overwatering and/or poor drainage; lack of water; buildup of spider mites (check for webbing)
Yellowing or death of lower leaves	Nutrient deficiency, usually nitrogen; plant is pot bound
Yellowing of all leaves and poor growth	Overwatering and/or poor drainage; insufficient or excess fertilizer; improper light; improper temperature; soil pH not right; root or crown rot; nematodes (especially root-knot nematode)
Spotted leaves (yellow, yellowish brown, brown, or white spots)	Excess sunlight; cold water on leaves; air pollution; plant-cleaning product; aerosol product; hot grease in kitchen area; spider mites (check for webbing); fungal or bacterial infection
Brown leaf tips and margins	Low humidity (hot, dry air); soil pH not right; air pollution (especially fluorides); salt accumulation; insufficient fertilizer; excess fertilizer; overwatering and/or poor drainage; drafts; lack of water; insect or mite feeding; objects rubbing against leaves
Lack of vigor; stunted	Insufficient or excess fertilizer; overwatering and/or poor drainage; lack of water; viral infection; root or crown rot
Plant distortion (thickening, curling, or drooping of leaves); deformation and dropping of leaves and flowers	Air pollution (gas fumes); lack of water; pesticide injury; heavy infestation of spider or cyclamen mites
Pale and spindly new growth with unnaturally small leaves; loss of flowers	Insufficient light; not enough fertilizer; lack of water; viral infection
New growth wilted or blackened	Cold or hot drafts; excess sunlight; too warm or too cold (e.g., chilling injury during winter transport); excess fertilizer; lack of water
Pale white or yellow leaf blotches that appear speckled; buds may fail to open normally	Spider mites or thrips (check for webbing or frass); cold water on leaves (especially African violets)
Cottony masses on stems and leaf axils; plants may become stunted and distorted and eventually die	Mealybugs
White or brown bumps on leaves or stems	Scale insects; galls; edema
Sticky spots on undersides of leaves or on young shoots; poor growth with distorted leaves	Aphids clustered on leaves, buds, and tender shoots; scale crawlers
Whitish, crusty or crystalline growth on soil and/or pot surface	Salt accumulation (from hard or softened water); excess fertilizer
White or pinkish creamy growth on soil surface	Saprophytic soil fungus
Torn leaves	Damage by animals; mechanical injury

[a] In a home, apartment, commercial building, or well-managed greenhouse, most plant problems are caused by unfavorable growing conditions and not by infectious agents or pathogens.

long, north-south oriented valley with a city, a generating plant, an aluminum processing plant, and a smelter to the north. Air pollution injury is suspected on crops and other plants to the south. After the area was examined, it was concluded that the cause of the injury was sulfur dioxide (SO_2) from the generating plant (Plate 4K–O). The case would be strengthened by examining plants on the hillsides, recent meteorological records (for inversion layers and wind direction), and plants for pathogens. In this case, only four pollutants were considered because the others seemed highly improbable as causes of the injury, considering the symptoms and widespread distribution of affected plants. Notice that complete information was available for only three of the plant species.

Herbicide Injury

Herbicide injury to plants has become a major problem in recent years as an ever-increasing array of new herbicides are used by commercial agriculture, industry, and homeowners. At times, 20% or more of the specimens received by clinics are diagnosed as having probable herbicide injury. Compounding the problem for many diagnosticians is a lack of experience with herbicides, their uses and injury symptoms, and a lack of familiarity with good reference materials. In many instances, growers find it easier and more profitable to blame herbicides for plant injury than to investigate other causes, e.g., soil salts (Plate 5Q) and foliar pesticide (Plate 5K). Diagnosticians are cautioned to handle herbicide injury problems with special care.

Herbicide injury is suspected under the following conditions:

1. Plant symptoms appear uniformly and simultaneously over the entire field, in large portions of it, in turn-around areas at edges of fields, in overlaps at turn rows, or in bands reflecting width (or overlap) of the spray boom; in the case of drift, plant symptoms decrease with distance from the source (Plate 5A–P). Variations are often associated with soil type, topography, and cropping or husbandry practices. All susceptible crop plants and

Table 4.9. Sources, symptoms, and modes of action of air pollutants

Pollutant Type	Sources	National air-quality standards and bioindicators	Typical symptoms and mode of action
Ozone (O_3) Non-point source, oxidant	Internal combustion engines; NO_x and CHO interaction with sunlight; stratosphere; lightning	0.12 ppm, 1 h; okra, petunia[a] (cultivar White Cascade), pinto bean, forest trees, tobacco (cultivar Bell-W-3)	Enters through stomata causing light to dark stippling or mottling on upper surfaces of one age group of leaves; premature senescence
PAN[b] Non-point source, oxidant	Internal combustion engines; interactions of gasoline vapor with ozone and NO_x	0.12 ppm, 1 h; okra, petunia[a] (cultivar White Cascade), pinto bean, forest trees, tobacco (cultivar Bell-W-3)	See Ozone. White to bronze spots on undersides of leaves, i.e., "silver leaf"; common near metropolitan areas with smog and inversion layers
Nitrogen oxides (NO_x) Point source, oxidant	Internal combustion engines and furnaces; bacteria; nitric acid spills; rockets	0.05 ppm, annual mean	Plant growth suppression; leaf bronzing, bleaching, and chlorosis; reduced plant growth from chronic exposure
Sulfur dioxide (SO_2) Point source	Furnaces and smelters that use fuels with high sulfur content; volcanoes	0.14 ppm, 24 h; alfalfa, aspen, bracken fern, beans, lichens,[c] ragweed, white birch	Chronic exposure causes general chlorosis; acute exposure causes interveinal bleaching
Carbon monoxide (CO) Point source	Internal combustion engines	9 ppm, 8 h; maximum of one episode per year	Plants are not sensitive to this widely distributed pollutant; at concentrations of 500 ppm, epinasty may develop
Particulate matter (various dusts) Point source	Furnaces, manufacturing smelters, nonpaved roads	260 mg/m³, 24 h; maximum of one episode per year	Dusty or crusty layer on foliage, depending on type of particle; foliage may become chlorotic or develop necrotic spots, especially after being wetted by dew or rain
Fluorides (HF, F) Point source	Ore or oil processing; phosphate plants; glass, aluminum, and ceramic manufacturing	None established (0.1 ppb may be toxic); gladiolus (cultivar Elizabeth the Queen), iris, ponderosa pine	Leaf margins of dicots and leaf tips of monocots turn tan to dark brown; premature leaf fall
Oxidant (O_3, PAN, and NO_x) Non-point source	Products from combustion interacting with hydrocarbons and moisture in sunlight	See individual pollutant; bluegrass, petunia (cultivar White Cascade)	See individual pollutant
Chlorine (Cl_2), hydrogen chloride (HCl) Point source	Manufacture or incineration of plastics, glass, paints, and stains; refineries; chemical spills	None established (Russia: 0.01 ppm, 29 µg/m³, 24 h for Cl_2 and 0.009 ppm, 13 µg/m³, 24 h for HCl)	Bleached and necrotic areas between veins and along margins of leaves; premature leaf fall
Ethylene ($CH_2=CH_2$) Point source	Incomplete fuel combustion (e.g., unvented gas heaters in greenhouse); internal combustion engines; ripening fruit; plant response to injury and exposure to ozone	None established (0.05 ppm may be toxic); orchids, African marigold	Leaf curl; reduced leaf expansion; premature bud break; premature senescence; reduced flower and fruit set
Phenoxy herbicide (e.g., 2,4-D) Point source	Use of highly volatile ester type of 2,4-D in gardens and farms	None established; tomato, grape, redbud, and tobacco; grasses and conifers are resistant	Leaves become twisted and elongated, "shoestring," "fan shaped"; epinasty
Ammonia (NH_4) Point source	Accidental spills of liquid ammonia fertilizer; manufacturing	None established (Canada: 5 ppm for 30 min); sunflower, lamb's-quarters	Color change in leaves; dull, dark green spots, typically on leaf margins, that turn dark brown later
Hydrogen sulfide (H_2S) Point source	Paper mills	None established; the odor is very offensive at 7.2 ppm	Plants are not affected at H_2S levels found in a very polluted atmosphere; at very high levels (>86 ppm), youngest leaves and growing points may be killed

[a] Some cultivars. Other cultivars are likely to be sensitive.
[b] Peroxyacetylnitrates.
[c] Lichen species that are especially susceptible include *Buellia stillingiana*, *Candelaria concolor*, *Centraria ciliaris*, *Lecidea nylanderi*, *Parmelia rudecta*, *P. trabeculata*, *P. orbicularis*, *Ramalina papillata*, and *Xanthoria polycarpa*.

weeds in the affected area usually show little plant-to-plant variability in symptomology. Depending on the herbicide, the appearance of plant injury is usually uniform; it may be sudden, in the case of drift, or gradual if roots are growing into soil horizons contaminated with a herbicide. Many herbicides are slow in their action, and plants, especially perennials, may show a slow decline leading to dieback or death.

2. Other plant species (weeds) in the crop and in adjacent areas (weeds and brush) show similar symptoms.

3. Suspected herbicides (on the basis of symptoms) have been used on the farm or in the community.

4. Plants show some of the following symptoms: stem and foliar twisting; elongated leaves; tendency toward parallel venation in foliage of dicot species (Plate 5A); leaf veinal, interveinal, or marginal chlorosis or necrosis (Plate 5E and L); upward or downward cupping of expanding leaves; swelling at the base of the hypocotyl; root pruning; premature leaf or flower fall; abnormal arrangement of leaves and flowers on the stems; and

Table 4.10. Sensitivity[a] of commonly grown plant species to various air pollutants[b,c]

Plant	O_3	PAN	SO_2	F	NO_x	2,4-D	Cl_2	NH_4	HCl	Hg	H_2S	CH_2
Acacia (*Acacia* sp.)												T
Adonis (*Adonis* sp.)									T			
Ageratum (*Ageratum* sp.)												I
Alder (*Alnus* sp.)	I		S									
European or black alder (*A. glutinosa*)									I			
Alfalfa (*Medicago sativa*)	S	I	S	T	S		S					
Aloe (*Aloe* sp.)												T
Andromeda, Japanese (*Pieris japonica*)	T											
Anthurium (*Anthurium* sp.)		T										
Apple (*Malus sylvestris*)		T	S	I	S	S						
Apricot (*Prunus armeniaca*)	T											
Chinese (*P. armeniaca* var. Chinense)	I			S								
Moorpark (*P. armeniaca* var. Moorpark)				I								
Royal (*P. armeniaca* var. Royal)				S								
Tilton (*P. armeniaca* var. Tilton)				I								
Arborvitae, American (*Thuja occidentalis*)	T	T	T	TI			T		T			I
Ash (*Fraxinus* sp.)	S					T	S					
Red or green (*F. pennsylvanica*)	S	T	S	I								
Velvet (*F. velutina*)			T									
White (*F. americana*)	S	T	S									
Asparagus (*Asparagus officinalis*)				T	T							
Aspen												
Big-tooth or large-tooth (*Populus grandidentata*)			S									
Quaking or trembling (*P. tremuloides*)	S		S	I								
Aster (*Aster* sp.)		S	S	I		I						
Avocado (*Persea americana*)	T											
Azalea (*Rhododendron* spp.)		T		I	S		T			I		
Campfire (*R. kaempferi* 'Campfire')	S											
Chinese (*R. molle*)	T											
Hino (*R. hinodegiri*)	S											
Korean (*R. poukhanense*)	S											
Snow (*R. mucronatum*)	S											
Bachelor's-button or cornflower (*Centaurea cyanus*)			S						S			
Barley (*Hordeum vulgare*)	S	I	S		S							
Young				S								
Mature				I								
Basil, common or sweet (*Ocimum basilicum*)		S										
Basswood, American (see Linden) (*Tilia americana*)		T								I		
Bean (*Phaseolus* sp.)	S	S	S							S		
Bush (*P. vulgaris* var. *humilis*)				I	T							
Field (*P. vulgaris*)				I								
Lima (*P. lunatus*)		T										
Pinto (*P. vulgaris*)					S		I		I			
Scotia (*P. vulgaris*)							I					
Snap, string, garden, or wax (*P. vulgaris*)				T								
Beech												
American (*Fagus grandifolia*)			S									
European (*F. sylvatica*)	T											
Purple (*F. sylvatica atropurpurea*)				T								
Beet, sugar and table or garden (*Beta vulgaris*)	T	I	S									
Begonia (*Begonia* sp.)	I	T					T			I		
Tuberous hybrid (*Begonia × tuberhybrida*)					S							
Birch (*Betula* sp.)	S		SI			S						
Cutleaf (*B. pendula* var. *gracilis*)				T								
European white (*B. pendula*)	T	T		T	S							
White, paper, or canoe (*B. papyrifera*)			S	T								
Blackberry (*Rubus* sp.)								S				
Blueberry (*Vaccinium* sp.)			S									
Bougainvillea, Brazil (*Bougainvillea spectabilis*)				S								
Box elder (*Acer negundo*)	SI		TI	S		S	S					
Boxwood, Japanese (*Buxus microphylla* var. *japonica*)	T											
Brazilian pepper-tree (*Schinus terebinthifolius*)			S									
Bridal-wreath (*Spiraea prunifolia*)	S			T								
Broccoli (*Brassica oleracea* var. *botrytis*)	I	T	S									
Bromelia (*Bromelia* sp.)		T										
Browallia (*Browallia* sp.)	S											
Buckwheat, common or notch-seeded (*Fagopyrum esculentum*)	S		S				I	T				
Burdock (*Arctium* sp.)				T								
Butterfly-weed or Indian paintbrush (*Asclepias tuberosa*)										S		

(*continued on next page*)

Table 4.10. (*continued*)

Plant	O_3	PAN	SO_2	F	NO_x	2,4-D	Cl_2	NH_4	HCl	Hg	H_2S	CH_2
Cabbage or savoy (*Brassica oleracea* var. *capitata*)	I	T	T	T	T	T	...	...	...	...	...	...
Cacti, Cactaceae	...	T	...	...	...	...	T	...	...	...	...	...
Cajeput or weeping tea-tree (*Melaleuca leucadendra*)	...	...	...	...	I	...	...	...	...	...	...	...
Calendula (*Calendula* sp.)	...	T	...	...	...	...	...	...	...	...	...	T
Camellia (*Camellia* sp.)	...	T	...	T	...	...	...	...	...	I	...	...
Cape jasmine or common gardenia (*Gardenia jasminoides*)	...	...	...	...	I	...	...	...	...	...	...	...
Carissa, karanda or perunkila (*Carissa carandas*)	...	...	...	...	T	...	...	...	...	...	...	...
Carnation (*Dianthus caryophyllus*)	I	T	...	...	...	...	...	...	...	...	...	I
Carrot (*Daucus carota* subsp. *sativus*)	I	I	S	T	S	...	...	...	...	...	...	S
Catalpa, southern or common (*Catalpa bignonioides*)	SI	...	SI	...	...	...	...	...	...	...	...	...
Cauliflower (*Brassica oleracea* var. *botrytis*)	...	T	...	T	...	...	...	...	...	...	...	...
Cedar (*Cedrus* sp.)	...	...	T	...	I	...	...	...	...	...	...	...
Incense cedar (*Calocedrus decurrens*)	I	...	...	...	...	...	...	...	...	...	...	...
Celery (*Apium graveolens*)	...	S	T	T	I	...	...	...	...	...	...	S
Cheese-weed or little mallow (*Malva parviflora*)	...	I	...	...	I	...	I	S	...	...	T	...
Chenopodium or goosefoot (*Chenopodium* spp.)	...	...	...	...	...	...	T	...	...	...	...	...
Cherry (*Prunus* sp.)	...	...	...	...	...	I	...	...	...	T	...	...
Bing (*P. avium* var. Bing)	S	...	...	...	...	...	...	...	...	...	...	...
Black, wild (*P. serotina*)	...	...	...	...	...	...	I	...	I	...	...	...
Chokecherry, common (*P. virginiana*)	...	...	...	I	...	I	...	...	...	...	...	...
Flowering, oriental or Japanese (*P. serrulata*)	...	...	...	TI	...	...	...	...	...	...	...	...
Ground or European (*P. fruticosa*)	...	S	...	...	...	...	...	...	...	...	...	...
Lambert (*P. avium* var. *Lambert*)	I	...	...	...	...	...	...	...	...	...	...	...
Sweet (*P. avium*)	...	...	T	I	...	...	...	...	...	...	...	...
Chickweed, common or stitch-wort (*Stellaria media*)	...	S	S	S	I	...	S	I	...	...	T	...
China aster (*Callistephus chinensis*)	...	...	...	...	...	...	...	...	...	...	...	T
Chrysanthemum (*Chrysanthemum* sp.)	I	T	...	T	...	...	T	...	I	...	...	...
Cinquefoil (*Potentilla* sp.)	...	...	...	...	...	...	...	...	...	S	...	...
Citrus (*Citrus* sp.)	S	...	T	...	...	...	...	...	...	...	...	...
Clover (*Trifolium* spp.)	...	S	...	...	...	...	...	...	...	...	...	...
Crimson or Italian (*T. incarnatum*)	...	...	...	I	S	...	...	...	...	...	...	...
Red (*T. pratense*)	S	...	S	...	S	...	...	...	...	...	...	...
Sweet-clover (*Melilotus* sp.)	...	...	...	I	...	...	...	...	...	...	...	...
Yellow sweet-clover (*M. officinalis*)	...	...	S	...	...	...	...	...	...	...	...	...
Coleus (*Coleus* sp.)	S	T	...	...	...	...	...	T	...	...	...	...
Columbine (*Aquilegia* sp.)	...	...	...	...	...	...	...	...	...	S	I	...
Corn or maize (*Zea mays*)	...	T	T	...	I	I	I	...	...	...	...	...
Field corn	I	...	...	SI	...	...	T	...	...	...	...	...
Sweet corn	S	...	...	SI	...	...	...	...	...	...	...	...
Cosmos (*Cosmos* sp.)	...	...	S	...	...	...	...	...	...	I	...	...
Cotoneaster (*Cotoneaster* sp.)	...	...	...	...	...	...	...	...	...	I	...	...
Rock (*C. horizontalis*)	S	...	...	...	...	...	...	...	...	...	...	...
Spreading (*C. divaricata*)	S	...	...	...	...	...	...	...	...	...	...	...
Cotton (*Gossypium* sp.)	T	T	S	T	...	...	...	...	...	...	...	I
Cow pea, common or black-eyed pea (*Vigna unguiculata*)	I	...	...	...	...	...	T	...	...	...	...	I
Crabapple, Siberian (*Malus baccata*)	SI	...	S	...	...	...	S	...	...	...	...	...
Croton, variegated (common florist) (*Codiaeum variegatum*)	...	...	...	...	T	...	...	...	...	T	...	...
Cucumber (*Cucumis sativus*)	I	T	S	T	...	...	...	I	...	...	...	S
Currant (*Ribes* sp.)	...	...	...	T	...	...	...	...	...	...	...	...
Cyclamen (*Cyclamen* sp.)	...	T	...	...	...	...	...	...	...	...	...	...
Dahlia (*Dahlia* sp.)	...	S	...	I	I	...	I	...	...	...	...	T
Daisy, ox-eye (*Chrysanthemum leucanthemum*)	...	...	...	...	T	...	...	...	...	S	...	...
Dandelion, common (*Taraxacum officinale*)	...	...	...	...	I	.	I	I	...	...	T	...
Descurainia or tansy mustard (*Descurainia pinnata*)	T	...	...	...	...	...	...	...	...	...	T	...
Dock (*Rumex* sp.) (see also Sorrel)	...	...	S	T	...	...	...	...	...	...	...	...
Curly dock or yellow dock (*R. crispus*)	...	I	...	...	...	...	...	...	...	...	...	...
Dogwood (*Cornus* sp.)	T	T	...	T	...	S	S	...	...	...	...	...
Gray or panicled (*C. racemosa*)	T	...	...	...	...	...	...	...	...	...	...	...
Flowering (*C. florida*)	T	...	...	...	...	...	...	...	...	...	...	...
Douglas fir (*Pseudotsuga menziesii*)	T	S	I	S	...	...	...	...	...	...	...	...
Eggplant (*Solanum melongena* var. *esculentum*)	...	...	...	T	...	T	...	...	...	...	...	...
Elderberry (*Sambucus* sp.)	...	...	...	T	...	S	...	...	...	...	...	...
Elm (*Ulmus* sp.)	...	...	T	T	T	...	...	...	...	...	...	...
American or white (*U. americana*)	...	...	S	T	...	...	...	...	...	...	...	...
Siberian or Chinese (*U. pumila*)	...	...	...	T	...	...	...	...	...	...	...	...
Endive or escarole (*Cichorium endivia*)	I	S	...	...	...	...	...	...	...	...	...	S
Euonymus (*Euonymus* sp.)	T	...	...	...	...	...	...	...	...	...	...	...
Winged spindle-tree (*E. alatus*)	T	...	...	...	...	...	...	...	...	...	...	...
Dwarf euonymus (*E. nanus*)	T	...	...	...	...	...	...	...	...	...	...	...

(*continued on next page*)

Table 4.10. (*continued*)

Plant	O_3	PAN	SO_2	F	NO_x	2,4-D	Cl_2	NH_4	HCl	Hg	H_2S	CH_2
Fern (various genera)												
Boston fern (*Nephrolepis exaltata* var. *bostoniensis*)											S	
Bracken or pasture brake (*Pteridium aquilinum*)			S									
Holly fern (*Cyrtomium falcata*)											S	
Fir												
Balsam (*Abies balsamea*)	T	T			I		T		T			
Big-cone Douglas fir (*Pseudotsuga macrocarpa*)	I											
Grand or lowland white (*A. grandis*)				I								
White (*A. concolor*)	IT	T	S									
Firethorn, scarlet (*Pyracantha coccinea*) (see also Pyracantha)				T								
Laland (*P. coccinea* cv. Lalandei)	T											
Forget-me-not (*Myosotis* sp.)											T	
Forsythia (*Forsythia* sp.)				S						I		
Cultivar Lynwood Gold (*F. intermedia spectabilis*)	I											
Four-o'clock, common (*Mirabilis jalapa*)			S									
Fuchsia (*Fuchsia* sp.)	T	S			I					I		
Gardenia, common or cape jasmine (*Gardenia jasminoides*)					I					I		
Geranium (*Geranium* sp.)	T			I			T			I		
Ginkgo or maidenhair-tree (*Ginkgo biloba*)			T	T								
Gladiolus (*Gladiolus* sp.)	T		S	S	T	I						
Gloxinia, violet-slipper (*Sinningia speciosa*)	T											
Goldenrod (*Solidago* sp.)				I					T			
Gomphrena, globe amaranth (*Gomphrena globosa*)						S						
Goosefoot, white or lamb's-quarters (*Chenopodium album*)			I	T		I	I			S		
Grape (*Vitis* sp.)	S				S							
American grape, Concord (*V. labrusca*)	S			I								
European wine (*V. vinifera*)				S								
Thompson seedless (*Vitis* sp.)	I											
Grapefruit (*Citrus paradisi*)				I								
Grasses (various spp.)												
Annual bluegrass (*Poa annua*)		S	S		I		I	S			T	
Bent grass, creeping (*Agrostis stolonifera*)	S											
Brome grass, smooth (*Bromus inermis*)	S											
Crabgrass, smooth (*Digitaria ischaemum*)	S			S								
Johnson grass (*Sorghum halepense*)				I								
Kentucky bluegrass (*Poa pratensis*)					T		T	I			T	
Orchard grass (*Dactylis glomerata*)	S											
Gum, black or sour (*Nyssa sylvatica*)	T						I					
Hazelnut or filbert (*Corylus* sp.)			S									
Heath (*Erica* sp.)				T								
Hemlock (*Tsuga* sp.)						I						
Eastern (*T. canadensis*)	TS	T	S	T			T					
Hibiscus (*Hibiscus* sp.)					S							
Hickory (*Carya* sp.)			S			S						
Holly							T					I
American (male and female) (*Ilex opaca*)	T									I		
Chinese or horned (*I. cornuta*)										T		
Hetz's Japanese (*I. crenata*)	T											
Honeysuckle, blue-leaf (*Lonicera korolokowi* var. *zabeli*)	I											
Hornbeam, European (*Carpinus betulus*)					T							
Horse-chestnut, common or European (*Aesculus hippocastanum*)						S						
Hydrangea (*Hydrangea* sp.)										S		
Hypericum (*Hypericum* sp.)	I											
Impatiens (*Impatiens* sp.)	T											
Iris (*Iris* sp.)				S					T			
Ivy, English (*Hedera helix*)	T	T								T		
Ixora, flame-of-the-woods (*Ixora coccinea*)					I							
Jerusalem cherry (*Solanum pseudocapsicum*)	T			SI								
Jimsonweed or common thornapple (*Datura stramonium*)		S										
Juniper (*Juniperus* sp.)	T		T	T	T		S					
Shore (*J. conferta*)						T						
Sierra (*J. occidentalis*)	T											
Kohlrabi (*Brassica oleracea* var. *gongylodes*)			S		T							
Lamb's-quarters (see also Goosefoot)	T	I		SI	T	S	T	S			S	
Larch (*Larix* sp.)	S		S									
European (*L. decidua*)	SI	T	S		S							
Japanese (*L. kaempferi*)	SI	T			S							
Western (*L. occidentalis*)				S								
Larkspur (*Delphinium* sp.)									T			I

(*continued on next page*)

Table 4.10. (*continued*)

Plant	O_3	PAN	SO_2	F	NO_x	2,4-D	Cl_2	NH_4	HCl	Hg	H_2S	CH_2
Laurel, mountain (*Kalmia latifolia*)	T			T								
Leek (*Allium porrum*)					S							
Lemon (*Citrus limon*)	T			I								
Lettuce												
Garden (*Lactuca sativa*)	T	S	S		S							S
Prickly (*L. serriola*)			S									
Ligustrum or glossy privet (*Ligustrum lucidum*)				I								
Lilac (*Syringa* sp.)			T	I								
Chinese (*S. chinensis*)	S											
Common (*S. vulgaris*)	S	T										
Lily (*Lilium* and other genera)		T										I
Calla-lily, common (*Zantedeschia aethiopica*)										I		
Easter or Bermuda Easter (*L. longiflorum* var. *eximium*)										I		
Lily-of-the-valley (*Convallaria majalis*)					T				T			
Plantain lily, fragrant (*Hosta plantaginea*)					T				T			
Linden (*Tilia* sp.)			TI	I	S	S						
American basswood (*T. americana*)	T			T			T					
Little-leaf or European (*T. cordata*)	T	S	I	T			I					
Lobelia (*Lobelia* sp.)												T
Locust												
Black (*Robinia pseudoacacia*)	ST			T	T							
Honey (*Gleditsia triacanthos*)	S	T										
London planetree (*Platanus acerifolia*)				T		S						
Lupine (*Lupinus* sp.)									T			
Maple (*Acer* sp.)	T		T	S								
Hedge or field (*A. campestre*)				I								
Japanese (*A. palmatum*)					I		T			I		
Norway (*A. platanoides*)	T	T		I	I	S	S		I			
Red or swamp (*A. rubrum*)	T		S									
Silver or white (*A. saccharinum*)	S	T	T	I			S					
Sugar or rock (*A. saccharum*)	T	T	T				S		I			
Marigold (*Tagetes* sp.)	T											
African or Aztec (*T. erecta*)												S
Mignonette, common (*Reseda odorata*)							T					
Milo (*Sorghum bicolor*) (see also Sorghum)			SI									
Mimosa or silktree (*Albizia julibrissin*)	T										S	
Mimulus (*Mimulus* sp.)		S										
Mint (*Mentha* sp.)	T	S										
Mock-orange (*Philadelphus* sp.)				T								
Sweet (*P. coronarius*)	I											
Mountain-ash												
American (*Sorbus americana*)				T								
European (*S. aucuparia*)	S		S	IT		I						
Mulberry (*Morus* sp.)			S	I		I	S					
Red or American (*M. rubra*)				I								
Muskmelon or netted melon (*Cucumis melo* var. *reticulatus*)	S		T									
Mustard, cultivated and wild (*Brassica* sp.)		S	S		S		S	S			T	
Myrtle (*Myrtus* sp.)							T					
Narcissus (*Narcissus* sp.)		T		I								
Nasturtium, garden (*Tropaeolum majus*)							I					
Nettle, burning or dog (*Urtica urens*)		S										
Nightshade, black (*Solanum nigrum*)			S	T								
Oak (*Quercus* sp.)	T		T	TI	T		T			I		
Black or yellow-barked (*Q. velutina*)	I					S						
Bur or mossy-cup (*Q. macrocarpa*)	T											
English or truffle (*Q. robur*)	T	T										
Gambel (*Q. gambelii*)	S											
Northern red (*Q. rubra*)	T	T										
Pin or Spanish (*Q. palustris*)	SI	T	S			I	S					
Red (*Q. rubra*)	T					I	T			T		
Scarlet (*Q. coccinea*)	SI											
Shingle or laurel (*Q. imbricaria*)	T											
White (*Q. alba*)	S	T										
Oat, common (*Avena sativa*)	S	S	S			ST						
Young and mature				I								
Wild oat (*A. fatua*)		S										
Oleander or rosebay (*Nerium oleander*)					S							
Olive, Russian or oleaster (*Elaeagnus angustifolia*)				T								
Onion (*Allium cepa*)	S	T	T		T			I				

(*continued on next page*)

Table 4.10. (*continued*)

Plant	O₃	PAN	SO₂	F	NOₓ	2,4-D	Cl₂	NH₄	HCl	Hg	H₂S	CH₂
Orange, sweet (*Citrus sinensis*)				I	I							
Oregon grape (*Berberis aquifolium*)					S							
Orchid, Orchidaceae		T										S
Oxalis (*Oxalis* sp.)							T			S		
Pachysandra, Japanese spurge (*Pachysandra terminalis*)	T											
Pagoda-tree, Japanese (*Sophora japonica*)	T											
Palmer's pigweed (*Amaranthus palmeri*)			S									
Parsley (*Petroselinum crispum*)	I				S							
Parsnip (*Pastinaca sativa*)	I											
Pea, common or garden (*Pisum sativum*)	I		S	T	S							
Peach (*Prunus persica*)	T		I			I	I			I		
Fruit				S								
Foliage				I								
Peanut, common or groundnut (*Arachis hypogaea*)	I											
Pear, common (*Pyrus communis*)			S	T	ST	T						S
Cultivar Bartlett	T											
Pecan (*Carya illinoensis*)			S									
Peony (*Paeonia* sp.)				I		T						
Pepper, black (*Piper nigrum*)	I	S					T					I
Pepper, bell or sweet (*Capsicum frutescens* cv. grossum)				T								
Periwinkle, common (*Vinca minor*)	T	T										
Persimmon, common or American (*Diospyros virginiana*)										I		
Petunia (*Petunia* sp.)	S	S		T	I		I					
Pheasant's-eye pink (*Dianthus* sp.)									T			
Phlox												
Tufted (*Phlox caespitosa*)										S		
Summer or perennial (*P. paniculata*)									T			
Piggyback-plant or pickaback plant (*Tolmiea menziesii*)	T											
Pigweed, red-root or rough (*Amaranthus retroflexus*)		S		I	T		T	S			T	
Pine (*Pinus* sp.)				S	T		T					
Austrian or black (*P. nigra*)	S	T	S						T			
Coulter or big-cone (*P. coulteri*)	S	S										
Digger (*P. sabiniana*)	T											
Eastern white (*P. strobus*)	S		S	S			S					
White			S						I	I		
Jack (*P. banksiana*)	S		S					I				
Jeffrey (*P. ponderosa* var. *jeffreyi*)	S	S										
Knob-cone (*P. attenuata*)	I			S								
Loblolly (*P. taeda*)			S					I				
Lodgepole (*P. contorta* var. *latifolia*)	I			ST								
Monterey (*P. radiata*)	S	S										
Mugo or Swiss mountain (*P. mugo*)				T								
Pitch (*P. rigida*)	SI	T										
Ponderosa (*P. ponderosa*)	S	S	S	S								
Red or Norway (*P. resinosa*)	T	T	S									
Scotch or Scot's (*P. sylvestris*)	SI	T	S	S								
Short-leaf (*P. echinata*)							I					
Single-leaf pinyon or stone (*P. monophylla* var. *monophylla*)	T											
Slash (*P. elliottii*)				S			I					
Sugar (*P. lambertiana*)	I											
Torrey or soledad (*P. torreyana*)	I											
Virginia (*P. virginiana*)	S	T	S									
Western white (*P. monticola*)	S		S	S								
Pittosporum, Japanese (*Pittosporum tobira*)					I							
Plantain cooking banana (*Musa paradisiaca*)				T								
Plum												
Common or garden (*Prunus domestica*)				S								
Myrobalan or cherry (*P. cerasifera*)			I	ST								
Poinsettia (*Euphorbia pulcherrima*)	I											
Polygonum (*Polygonum* sp.)							T					
Poplar (*Populus* sp.)	S		I									
Balsam (*P. balsamifera*)				T								
Carolina (*P. canadensis*)				I								
Hybrid (*Populus* sp.)	S	T										
Lombardy or black (*P. nigra* cv. Italica)			S	I								
Poppy, Oriental (*Papaver orientale*)										S		
Potato (*Solanum tuberosum*)	S		T	T	I	I						
Primrose (*Primula* sp.)		S										

(continued on next page)

Table 4.10. (*continued*)

Plant	O_3	PAN	SO_2	F	NO_x	2,4-D	Cl_2	NH_4	HCl	Hg	H_2S	CH_2
Privet (*Ligustrum* sp.)	...	...	T	T	...	I	...	...	...	SI	...	...
Amur cultivar North (*L. amurense*)	T	...	...	...	...	...	...	...	...	...	...	...
Common (*L. vulgare*)	I	...	...	...	...	...	...	...	...	...	...	...
Lodense (*L. vulgare* var. *pyramidale*)	S	...	...	...	...	...	...	...	...	...	...	...
Purslane, common (*Portulaca oleracea*)	...	...	...	T	...	...	...	...	...	...	...	...
Pyracantha (*Pyracantha* sp.) (see also Firethorn)	...	...	...	...	S	...	...	...	...	...	...	...
Radish (*Raphanus sativus*)	S	T	S	...	...	...	S	...	...	...	...	...
Ragweed, common (*Ambrosia artemisiifolia*)	...	...	S	...	...	I	...	...	...	...	...	...
Ranunculus (*Ranunculus* sp.)	...	S	...	...	...	...	...	...	...	...	...	...
Raspberry, red or European red (*Rubus idaeus*)	...	...	...	T	...	...	...	...	...	...	...	...
Redbud, eastern (*Cercis canadensis*)	SI	...	...	...	...	...	...	S	...	...	...	...
Redwood or coast redwood (*Sequoia sempervirens*)	T	T	...	...	...	...	...	...	...	...	...	...
Rhododendron (*Rhododendron* sp.)	...	...	...	I	...	I	...	...	...	...	...	...
Carolina (*R. carolinianum*)	T	...	...	...	...	...	...	...	...	...	...	...
Catawba or mountain rosebay (*R. catawbiense*)	...	...	...	...	I	...	...	...	...	...	...	...
R. catawbiense album	I	...	...	...	...	...	...	...	...	...	...	...
R. nova zembla	I	...	...	...	...	...	...	...	...	...	...	...
R. roseum elegans	I	...	...	...	...	...	...	...	...	...	...	...
Rhubarb or wineplant (*Rheum rhabarbarum*)	...	T	S	...	S	T	...	...	...	...	...	...
Rice (*Oryza sativa*)	T	...	...	...	...	...	...	...	...	...	...	...
Rose, tea (*Rosa odorata*)	...	...	...	I	S	I	S	...	...	...	...	I
Rye (*Secale cereale*)	S	...	...	...	I	...	...	...	...	...	...	...
Mature and young	...	...	...	I	...	...	...	...	...	...	...	...
Safflower (*Carthamus tinctorius*)	S	...	...	...	...	...	...	...	...	...	...	...
Salvia (*Salvia* sp.)	...	...	...	...	...	...	...	...	...	I	...	...
Sarcococca (*Sarcococca* sp.)	...	...	...	...	...	...	...	...	...	T	...	...
Saxifrage (*Saxifraga* sp.)	...	...	...	...	...	...	...	...	...	I	...	...
Sequoia, giant (*Sequoiadendron giganteum*)	T	T	...	...	...	...	...	...	...	...	...	...
Serviceberry, Saskatoon or western (*Amelanchier alnifolia*)	...	...	S	I	...	...	...	...	...	...	...	...
Silverberry (*Elaeagnus commutata*)	I	...	...	...	...	...	...	...	...	...	...	...
Smartweed or knotweed (*Polygonum* sp.)	S	...	...	SI	...	...	...	...	...	...	...	...
Snapdragon, common or garden (*Antirrhinum majus*)	T	S	...	...	S	...	...	...	...	...	...	...
Snowberry, common (*Symphoricarpos* sp.)	S	...	...	...	...	...	...	...	...	...	...	...
Sorghum (*Sorghum bicolor*) (see also Milo)	I	T	...	...	...	T	...	...	...	...	...	...
Sorrel (*Rumex* sp.) (see also Dock)	...	...	...	...	...	S	...	...	...	...	...	...
Sow-thistle (*Sonchus* sp.)	...	...	S	...	...	...	...	...	...	...	...	...
Soybean (*Glycine max*)	S	I	...	IT	...	...	...	...	...	...	...	...
Spinach (*Spinacia oleracea*)	S	I	S	T	...	...	...	...	...	...	...	...
Spirea (*Spiraea* sp.)	T	...	...	...	...	...	...	...	...	...	...	...
Spruce (*Picea* sp.)	...	...	...	...	I	I	...	...	...	...	...	...
Black Hills (*P. glauca* var. *densata*)	T	T	...	...	...	...	...	...	...	...	...	...
Blue (*P. pungens*)	...	T	...	SI	...	...	...	...	...	...	...	...
Colorado blue (*P. pungens* cv. Glauca)	T	...	...	SI	...	...	...	...	...	...	...	...
Engelmann (*P. engelmannii*)	...	...	S	T	...	...	...	...	...	...	...	...
Norway (*P. abies*)	T	T	...	...	...	...	...	S	T	...	...	...
White or cat (*P. glauca*)	T	T	...	I	...	...	...	...	...	...	...	...
Squash												
Winter or summer (*Cucurbita pepo*)	...	T	S	...	...	...	...	I	...	...	...	...
Bush or bush pumpkin (*C. pepo* var. *melopepo*)	...	...	...	T	...	...	...	...	...	...	...	...

(*continued on next page*)

inappropriate formation of roots on stems (Plate 5C). Swelling of the hypocotyl associated with root pruning is a common symptom induced by some soil-applied herbicides.

Other agents such as air pollutants, high rates of other pesticides, nutritional or moisture problems, and infection by pathogens can mimic these symptoms with similar field patterns (Plates 1P; 4A–T; 5K and Q; and 6C, E, G, H, and Q–S). They must also be considered. For example, injury symptoms of 2,4-D on corn ("buggy whip") often is confused with genetic unfurling or infection by *Pseudomonas andropogonis* (bacterial stripe and leaf spot) or *Sclerophthora macrospora* (crazy-top).

All plants are susceptible to herbicide injury if sufficient toxicant enters actively growing plant tissues. The selectivity of herbicides is based on precisely specified methods of application. If any of the application specifications are violated, the probability of plant injury is high (Plate 5O). Common causes of herbicide injury include

1. movement of herbicide from nearby fields by means of droplet or vapor drift or wind-blown deposition of treated soil;
2. use of contaminated equipment, fertilizers, supplies, and irrigation water;
3. planting in a field that has incompatible herbicide residues in the soil from a previous crop (carry-over);

Table 4.10. (*continued*)

Plant	O_3	PAN	SO_2	F	NO_x	2,4-D	Cl_2	NH_4	HCl	Hg	H_2S	CH_2
St.-John's-wort, common (*Hypericum perforatum*)	...	...	...	S	...	...	...	...	...	...	...	...
Stevia (*Piqueria trinervia*)	I	...	...	...	...	...	...	...	...	...	...	...
Strawberry (*Fragaria* sp.)	T	T	...	I	...	...	...	...	...	I	...	...
Sugarcane (*Saccharum officinarum*)	...	...	...	T	...	...	...	...	...	...	...	...
Sumac (*Rhus* sp.)	S	...	...	...	...	S	...	...	...	...	...	...
Smooth or scarlet (*R. glabra*)	...	...	...	I	...	...	...	...	...	...	...	...
Staghorn or Virginian (*R. typhina*)	...	...	T	I	...	...	...	...	...	...	...	...
Sunflower (*Helianthus* sp.)	...	...	...	I	S	...	S	S	...	S	T	...
Sweetgum, American or red gum (*Liquidambar styraciflua*)	I	T	...	T	...	I	S	...	...	...	...	...
Sweet pea (*Lathyrus odoratus*)	...	...	S	...	S	...	...	...	...	...	...	...
Sweet potato or yam (*Ipomoea batatas*)	T	...	S	SI	...	...	...	...	...	...	...	...
Sweet-William (*Dianthus barbatus*)	...	...	S	I	...	...	...	...	T	...	...	...
Swiss chard (*Beta vulgaris* subsp. *cicla*)	...	S	S	...	...	...	...	...	...	...	...	...
Sycamore or planetree (*Platanus* sp.)	...	...	S	T	...	...	...	...	...	...	...	...
American or buttonwood (*P. occidentalis*)	S	...	...	...	...	...	...	...	...	...	...	...
Tangerine or Mandarin orange (*Citrus reticulata*)	...	...	...	I	...	...	...	...	...	...	...	...
Timothy, common (*Phleum pratense*)	I	...	...	...	...	...	...	...	...	...	...	...
Tobacco (*Nicotiana tabacum*)	S	I	...	T	S	S	S	...	...	I	...	...
Tomato (*Lycopersicon esculentum*)	S	S	S	I	I	S	I	T	S	I	...	I
Tree-of-heaven or ailanthus (*Ailanthus altissima*)	S	...	...	T	...	S	S	...	...	...	...	...
Tulip (*Tulipa gesneriana*)	...	...	S	S	...	...	S	...	...	...	...	...
Tulip-tree or tulip-poplar (*Liriodendron tulipifera*)	S	ST	S	...	...	...	...	...	S	...	...	...
Turnip (*Brassica rapa*)	I	...	S	...	...	...	...	...	...	...	...	...
Vetch, common or tare (*Vicia sativa*)	...	...	...	...	S	...	...	...	...	...	...	...
Viburnum (*Viburnum* sp.)	T	...	...	...	...	...	...	...	...	I	...	...
Korean spice (*V. carlesii*)	T	...	...	...	...	...	...	...	...	...	...	...
Linden (*V. dilatatum*)	I	...	...	...	...	...	...	...	...	...	...	...
Tea or Chinese (*V. setigerum*)	I	...	...	...	...	...	...	...	...	...	...	...
Vinca (*Vinca* sp.) (see also Periwinkle)	...	...	...	...	...	...	...	...	...	I	...	...
Violet (*Viola* sp.)	...	...	S	I	...	...	...	...	...	...	...	T
Virginia creeper or woodbine (*Parthenocissus quinquefolia*)	T	...	...	T	...	...	S	...	...	...	...	...
Walnut (*Juglans* sp.)	S	...	...	...	...	...	...	...	...	...	...	...
Black (*J. nigra*)	T	...	...	I	...	...	...	...	...	...	...	...
English or Persian (*J. regia*)	SI	...	...	I	...	...	...	...	...	...	...	...
Wandering-Jew or inchplant (*Zebrina pendula*)	...	...	...	...	...	...	S	...	...	...	...	...
Wheat, common or bread (*Triticum aestivum*)	S	I	S	I	I	...	...	...	...	...	...	...
Willow (*Salix* sp.)	...	...	S	T	...	...	...	...	...	S	...	...
Weeping (*S. babylonica*)	SI	...	S	T	...	...	...	...	...	...	...	...
Wisteria (*Wisteria* sp.)	...	...	...	...	...	S	...	...	...	...	...	...
Yellow-wood or virgilia (*Cladrastis lutea*)	...	...	...	...	...	S	...	...	...	...	...	...
Yew, English (*Taxus baccata*)	T	...	...	I	T	I	T	...	...	...	...	...
Dense (*T. cuspidata* var. *densiformis*)	T	...	...	...	...	...	...	...	...	...	...	...
Hatfield's pyramidal (*T. media* var. *hatfieldi*)	T	...	...	...	...	...	...	...	...	...	...	...
Japanese (*T. cuspidata*)	...	...	...	I	...	...	...	...	...	...	...	...
Zinnia (*Zinnia* sp.)	T	...	S	...	...	S	S	...	...	...	...	I

[a] S = sensitive, I = intermediate, and T = tolerant.

[b] Pollutants: O_3 = ozone, PAN = peroxyacetylnitrate, SO_2 = sulfur dioxide, F = fluorides, NO_x = oxides of nitrogen, 2,4-D = 2,4-D phenoxy herbicide, Cl_2 = chlorine, NH_4 = ammonia, HCl = hydrogen chloride, Hg = mercury vapor, H_2S = hydrogen sulfide, and CH_2 = ethylene.

[c] Adapted from Lacasse, N. L., and Treshow, M., 1976, Diagnosing Vegetation Injury Caused by Air Pollution, Applied Science Association, Air Pollution Training Institute, Research Triangle Park, NC.

4. herbicide application, for example, inappropriate selection of herbicide, rate for soil type, timing (age of plant), or placement. The following, suggested by herbicide specialists to avoid herbicide injury, may serve as a checklist for diagnosing application problems:

    a. Carefully select the herbicide rate given on the label.

    b. Calibrate the equipment accurately.

    c. Apply the herbicide accurately and uniformly.

    d. Use nozzle tips of the same type and size.

    e. Adjust boom height and spray patterns for uniform application.

    f. Permit thorough mixing of the herbicide in the tank prior to making an application. The sprayer must provide adequate and continuous agitation during mixing and application.

    g. Avoid skips and overlaps during application.

    h. Maintain uniform speed during application.

    i. Use proper equipment for uniform dispersal and incorporation of the herbicide.

    j. Avoid crop contact with a postemergence herbicide.

    k. Shut off the sprayer when turning at field ends.

    l. Avoid spills and overflows when filling the applicator.

Herbicide injury to ornamental and vegetable crops is especially frequent because of the widespread use of herbicides by growers and homeowners, the close proximity of many plant species, and the rapid growth of plants with surfaces susceptible to injury. For example, a volatile growth regulator herbicide may be applied to a lawn for control of broadleaf weeds and by vapor drift injure azalea, bean, redbud, tomato, and other plants growing in the homeowner's or a neighbor's yard.

Unusual exposure may result in atypical herbicide injury symptoms. For example, the contact herbicide paraquat causes localized necrotic foliar spots in drift situations (Plate 5N) or foliar desiccation when properly sprayed on the foliage. However, if the nutrient solution of a hydroponic growing system becomes even slightly contaminated with paraquat, the herbicide will enter the transpiration stream through the roots and cause systemic browning along the leaf veins followed by rapid death of the plant. This symptom is rarely seen in the field because paraquat is tightly held by mineral soils.

Symptoms of herbicide injury on plants may also deviate significantly from those expected as a result of environmental factors, the growth stage and age or condition of the plant, the presence of other chemicals, the exposure court (parts of the plant exposed), the herbicide concentration, the plant species, and in some cases the plant cultivar. Knowledge of herbicide properties and proper application is useful in diagnosing herbicide problems. Consider some of the properties of the herbicide: solubility (movement in water), volatility (vapor drift), longevity in soil, plant specificity under normal conditions, translocation in the plant, and physiological action on the plant. Some of these properties are given in Tables 4.11 and 4.12. For a more comprehensive discussion of specific herbicides, consult the container

FIELD OBSERVATION

AREA OBSERVED: Along valley floor.   DATE 9/22/89

CROP: wheat and oats   . SYMPTOMS FIRST NOTICED: 9/12/89

PREVAILING WINDS WEEK EARLIER: Variable, mostly from NW.

FOLIAGE AFFECTED: Mostly in mid-part of plant.

FIELD NOTES		PLANT SENSITIVITY TO POLLUTANTS											
PLANT NAME	SYMPTO-MATIC	O3	PAN	SO2	Fl	NOx	24D	Cl	NH3	HCl	HG	HS	ETH
Wheat	+	+	+-	+	-								
Cucumber	+	+-	-	+	-								
Oats	+	+	+	+									
Onion	-	+	-	-									
White pine	+			+									
Corn	-	-+			+								
Chrysanthemum	-	+	-		-								
Alfalfa	+	+	-+	+	-								
Apple	+		-	+	+-								

Fig. 4.4. Sensitivity chart based on field notes taken after examination of an area for a potential air pollution problem.

label or a recent text on herbicides (see references at the end of this chapter).

As with other pesticides, each herbicide has three names: the common (preferred) name (e.g., simazine); the often-used trade name(s) (e.g., Princep); and an unwieldy chemical name (e.g., 2-chloro-4,6-bis[ethylamino]-*s*-triazine). Herbicides are usually grouped into families with similar chemistry. Those within a family usually have similar properties and induce similar plant symptoms. A useful classification for diagnosing plant injury is mode of action with similar symptoms. These include 1) inhibitors of photosynthesis or chlorophyll synthesis (Plate 5E–I); 2) inhibitors of plant pigments (carotenoids); 3) inhibitors of cell division and meristems (growth suppressors) (Plate 5J–L, O, and P); 4) growth regulators; and 5) contact or desiccant poisons (Plate 5N and Tables 4.11 and 4.12).

The procedure for diagnosing herbicide injury is as follows:

1. Rule out other possible causal agents. Air pollution, high soil soluble salts, nutritional deficiencies, and virus-induced diseases can usually be determined by carefully observing the field signature (see other sections in this chapter and Plates 1M; 4; 5M and Q; and 6C–E, G, H, R, and S).
2. Check that the distribution of affected plants is uniform in the affected area with little plant-to-plant variation in symptomology.
3. Ascertain from the grower that the timing of symptom

development was uniform. Depending on the herbicide and application method, it may be sudden or gradual. The intensity of symptoms may vary in the field because of soil characteristics and the source of the herbicide.

4. Examine other plant species in the affected area for symptoms (or absence of symptoms) characteristic of each plant type resulting from the suspected herbicide (Tables 4.11 and 4.12).
5. Determine which herbicides were used in the planting or in the community.
6. Ask the grower to review application procedures if herbicides were used.

    a. Check the label (or consult with a herbicide specialist) to see whether the application method, timing, environmental conditions, and rates were appropriate.

    b. Take a look at the equipment, and check for leaky hoses and connections, proper and uniform height of booms and drop nozzles, and uniformity of spray pattern and rate for each nozzle.

    c. Question the grower on the source of irrigation water and herbicides used previously and those applied to the previous crop.

    d. Question the grower regarding the use of other pesticides, which may interact with herbicides and sometimes cause unusual injury symptoms.

7. Depending on the herbicide or herbicides used, the soil type, amount of rainfall, and temperature may be important.
8. Assays can be run in critical cases.

Table 4.11. Plant symptoms produced by major herbicide families

| | Herbicide mode of action | | | | |
Symptoms	Photosynthesis inhibitor	Pigment inhibitor	Metabolic inhibitor	Growth regulator	Contact[a]
Foliar discoloration					
Veinal chlorosis	+	−	−	−	−
Interveinal chlorosis	+	−	−	−	−
Marginal chlorosis	+	−	+	−	−
Albinism	−	+	−	−	−
Leaves dark green	−	−	+	−	−
Purpling of leaves	−	+	+	−	−
Abnormal shoot growth					
Plant stunting	−	−	+	+	−
Leaf curl	−	−	+	+	−
Leaf strapping	−	−	−	+	−
Leaf crinkling	+	−	+	+	−
"Fiddlenecking"	−	−	−	+	−
Onion leaf	−	−	−	+	−
Buggy whip	−	−	−	+	−
Epinasty	−	−	−	+	−
Abnormal root growth					
Root pruning	−	−	+	−	−
Root swelling, tip enlargement	−	−	+	−	−
Hypocotyl basal swelling	−	−	+	−	−
Brace roots on stems	−	−	−	+	−
Necrosis					
Round leaf spots	−	−	+	−	+
Angular leaf spots	−	−	−	−	−
Sudden desiccation or necrosis of entire plant	−	−	−	−	+
Vascular browning	−	−	−	−	+

[a] Contact herbicides are so designated on the basis of their extremely rapid effects on plants. The specific modes of action for these herbicides are listed in Table 4.12.

Table 4.12. Herbicide injury symptoms, groups, common and trade names, soil persistence, usual method of application, plant translocation, and comments[a]

Herbicide mode of action Primary symptoms	Herbicide group Common and trade names	Persistence in soil[b]	Application, translocation,[c] selectivity, and comments
Photosynthesis or chlorophyll inhibitors			
Crinkling and rough leaf margins; plant stunting; twisted and bent leaves in grasses	Anilides propanil, Stam	1–3 days	Foliar application; apoplastic translocation; injury common to rice when organophosphate insecticides are used in conjunction with propanil
Mottled chlorosis and necrosis on contacted leaves; stunting and scorch of older leaves	Benzothiodiazoles bentazon, Basagran	<6 wk	Foliar application; minimal translocation; most broad-leaf species and some grasses susceptible
Leaf chlorosis beginning on leaf margins and then along veins followed by necrosis across veins	s-Phenylureas chloroxuron, Tenoran diuron, Karmex linuron, Lorox methazole, Probe tebuthiuron, Spike	5–6 mo 1–2 yr 2–5 mo 1–2 mo 12–15 mo[t]	Soil application; in oats and barley, chlorosis is white; in cereals, chlorosis starts in midleaf extending to the tip; broad-leaf crops especially sensitive
Interveinal and marginal chlorosis, especially on older leaves	s-Triazines ametryn, Evik atrazine, Aatrex cyanazine, Bladex dipropetryn, Sancap prometon, Pramitol prometryn, Caparol propazine, Milogard simazine, Princep terbutryn, Igran	1–3 mo 12 mo[t] 2–4 wk[t] 1–3 mo[t] 8–12 mo 1–3 mo 6–10 mo 30 days[t] 4–6 mo	Soil and foliar application; symplastic translocation; selectivity varies among herbicides in this group
Leaf marginal and later interveinal chlorosis and necrosis	Triazinones hexazinone, Velpar metribuzin, Lexone and Sencor	1–6 mo[t] 1–2 mo[t]	Soil and foliar application; injury dependent on soil type and rainfall
Similar to those of s-phenylureas; veinal chlorosis	Uracils bromacil, Hyvar terbacil, Sinbar	5–6 mo[t] 5–6 mo[t]	Soil application; rapid apoplastic translocation; nonselective; industrial uses, especially along rights-of-way
Pigment inhibitors (carotenoids)			
Severe chlorosis and bleaching; pink coloration on younger leaves; albino appearance	. . . amitrole, Weedazol clomazone, Command fluridone, Sonar norflurazon, Zorial	 2–4 wk 1–3 mo 8–12 mo 6–15 mo	Soil and foliar application; rapid apoplastic and symplastic translocation; all green tissue susceptible
Cell division and metabolism inhibitors			
Similar to those of amides; stunting; rough and crinkled leaf margins; twisted and bent grass leaves	Acetanilides alachlor, Lasso diethatyl, Antor metolachlor, Dual propachlor, Ramrod	6–10 wk 6–10 wk 2–7 wk 2–3 mo	Soil application; apoplastic translocation; selectivity varies considerably among herbicides in this group
Stunting; leaf cupping; chlorosis; epinasty; shoot and root stunting; reduction of wax on leaves	Aliphatic acids dalapon, Dowpon	<1 mo	Soil and foliar application; readily leached; apoplastic and symplastic translocation; grasses susceptible
Shoot, bud, and root tip stunting; root pruning with bulbous tips; leaves dark green with marginal and interveinal chlorosis; grass coleoptiles stunted and swollen; emergence delayed	Amides diphenamid, Enide pronamide, Kerb	3–6 mo 2–9 mo	Soil and foliar application; limited translocation; grasses and cereals especially susceptible
Plant stunting and foliar discoloration; some leaf distortion, spots, and necrosis	Amino acids glyphosate, Roundup and Kleenup	None	Foliar application; readily translocates symplastically and apoplastically; all green plant tissue susceptible; symptoms often not visible for 7–10 days
Growth cessation; wilting; mottled chlorosis; grass foliage may take on a rust brown or purple color	Aryloxphenoxy protonates diclofop, Hoelon fenoxaprop, Whip and Acclaim fluazifop, Fusilade haloxyfop, Verdict and Gallant quizalofop, Assure	 1 mo NL 2 mo 1–3 mo 1–2 mo	Foliar application; symplastic and apoplastic translocation; hormonal growth regulation; inhibits lipid synthesis; grasses very susceptible

(continued on next page)

Table 4.12. (*continued*)

Herbicide mode of action Primary symptoms	Herbicide group Common and trade names	Persistence in soil[b]	Application, translocation,[c] selectivity, and comments
Cell division and metabolism inhibitors (*continued*)			
Internal tissue browning; leaf chlorosis and necrosis	Cyclohexanediones clethodim, Select sethoxydim (ISO), Poast tralkoxydim, Achieve	2 days[t] 4–11 days[t] NL	Foliar application; symplastic translocation; grasses susceptible; broad-leaf plants resistant
Similar to those of amides; stunting and swell- ing of roots, particularly laterals; hypocotyl swelling; grass leaves fail to unroll	Dinitroanilines benefin, Balan ethalfluralin, Sonalan fluchloralin, Basalin isopropalin, Paarlan oryzalin, Surflan pendimethalin, Prowl trifluralin, Treflan	4–8 mo 4–6 mo 4–6 mo 6–12 mo 4–12 mo 3–6 mo 6–12 mo	Soil application; inhibits root meristems; little trans- location; broad-leaf species susceptible; injury common in cucurbits and solanaceous crops
Dwarfing; yellowing of new growth followed by necrosis and death; symptoms not notice- able for several days after application	Imidazolinones imazameth, Cadre imazamethabenz, Assert imazapyr, Arsenal imazaquin, Scepter imazethapyr, Pursuit	NL NL 3–24 mo >11 mo NL	Foliar and soil application; apoplastic and symplastic translocation; selectivity varies greatly among chemicals in this group; imazapyr used along rights-of-way; legumes tolerant of imazaquin and imazethapyr; most grasses and broad-leaf weeds susceptible
Slow growth; stunting; leaf chlorosis	. . . maleic hydrazide, MH-30	<1 mo	Foliar application; symplastic and apoplastic trans- location; inhibits cell division but not cell enlarge- ment
Stunting; marginal chlorosis on leaves; seed germination inhibited; root stunting and thickening; exposed bark may develop le- sions at base of trunk	Nitriles bromoxynil, Brominal and Buctril dichlobenil, Casoron	1–2 mo 2–12 mo	Soil and foliar application; soluble; limited trans- location; some noncrop uses; young plants sus- ceptible; bromoxynil inhibits photosynthesis
Severe stunting and root pruning; purpling in corn	Organophosphorus bensulide, Prefar and Betasan	4–6 mo[t]	Soil application; little apoplastic translocation; grasses and some broad-leaf species susceptible
Similar to those of amides; stunting; chlorosis, especially at leaf tips	Phenylcarbamates barban, Carbyne chlorpropham, Furloe phenmedipham, Betanal propham, Chem-Hoe	<3 wk 2 mo[t] 25 days[t] 5–15 days[t]	Soil and foliar application; apoplastic translocation; absorbed by seed, shoots, and to a lesser extent by roots; selectivity varies among herbicides in this group
General plant stunting; root inhibition; stem cracking; negative geotropism	Phthalamates chlorthal-dimethyl, Dacthal naptalam, Alanap	2–3 mo[t] 3–8 wk	Growth inhibitor; soil application; inhibits germi- nating seed and roots; many grasses and some broad-leaf species susceptible
Dwarfing; yellowing of new growth followed by necrosis and death; symptoms not notice- able for several days after application	Sulfonylureas chlorimuron, Classic chlorsulfuron, Glean metsulfuron, Ally nicosulfuron, Accent primisulfuron, Beacon prosulfuron, Peak rimsulfuron, Titus sulfometuron, Oust thifensulfuron, Harmony and Pinnacle	NL 4–6 wk[t] 1–6 wk[t] NL NL NL NL NL NL	Soil and foliar application; apoplastic and symplastic translocation; selectivity varies greatly among chemicals in this group; some right-of-way use
Leaf cupping and necrosis; hypocotyl swell- ing; young leaves may stick to older leaves; in cereals, kinking near base of stem; shiny, thick leaves in soybean	Thiocarbamates butylate, Sutan+ diallate, Avadex EPTC, Eptam pebulate, Tillam vernolate, Vernam	2–3 wk[t] 6–8 wk 1 wk[t] 2 wk[t] 10 days[t]	Soil application; affects hormone distribution and cell elongation; uptake through seeds, roots, and shoots; broad-leaf species and grasses susceptible
Growth regulators			
Leaf epinasty; cupping; "fiddlenecking"; abnormal brace roots	Benzoic acids clopyralid, Stinger and Reclaim chloramben, Amiben and Vegiben dicamba, Banvel	2–10 wk 6–8 wk 3–12 wk	Foliar application; symplastic and apoplastic trans- location; hormonal growth regulation; seedling grass and broad-leaf weeds susceptible

(*continued on next page*)

Table 4.12. (*continued*)

Herbicide mode of action Primary symptoms	Herbicide group Common and trade names	Persistence in soil[b]	Application, translocation,[c] selectivity, and comments
Growth regulators (*continued*)			
Abnormal growth and morphology; growth stops; dicots develop narrow leaves, mottled chlorosis, parallel veins, epinasty, stem splitting, bending, twisting, callus tissue, and wilting; abnormal brace roots	Phenoxy 2,4-D, various 2,4-DB, Butyrac Dichlorprop, mixtures with 2,4-D MCPA, Rhonox MCPB, Thistrol	 1–4 wk NL NL 1–6 mo NL	Foliar application; symplastic translocation; rapidly growing broad-leaf plants highly susceptible; injury to grasses and corn not unusual; may penetrate bark of trees and shrubs
Tissue proliferation; leaf epinasty and "fiddle-necking"	Pyridines picloram, Tordon triclopyr, Garlon	 6–18 mo 46 days[t]	Foliar and soil application; symplastic and apoplastic translocation; absorbed by foliage, stems, and roots; slowly degraded in plants; grasses resistant; annual and perennial broad-leaf weeds and woody species susceptible
Contact herbicides			
Necrotic spots on leaves; rapid desiccation of entire green shoot; vascular browning in potato tubers	Bipyridiliums diquat, Diquat paraquat, Paraquat	 1 day 1 day	Foliar application; apoplastic translocation; root uptake causes veinal browning and chlorosis in leaves; mode of action is inhibition of photosynthesis; all green plant tissue susceptible
Foliar flecking; necrotic spots; desiccation	Diphenyl ethers acifluorfen, Blazer bifenox, Modown fomesafen, Reflex lactofen, Cobra oxyfluorfen, Goal	 2–9 wk[t] NL NL NL 4–6 wk[t]	Soil and foliar application; little translocation; mode of action is inhibition of photosynthesis; broad-leaf species and some grasses susceptible
Necrotic spots on leaves; desiccation of entire green shoot; vascular browning in potato tubers	Organic arsenicals DSMA, various MSMA, Weed-Hoe	 NL NL	Foliar application; mode of action is metabolic inhibition; grasses, pigweed, and cocklebur susceptible; cotton resistant
Necrotic spots on leaves; desiccation of entire green shoot; vascular browning in potato tubers	Petroleum oils kerosene, Varsol, Stoddard solvent	NL	Foliar application; little translocation; selectivity varies
Necrotic spots on leaves; desiccation of entire green shoot; vascular browning in potato tubers; wilting within several hours of application	Phenolics dinoseb, Premerge	2–4 wk	Foliar application; apoplastic translocation; mode of action is metabolic inhibition; all grass plant tissue susceptible

[a] We acknowledge the critical reviews and assistance of Janet L. Shurtleff, North Carolina Department of Agriculture, and A. R. Bonanno and T. J. Monaco, North Carolina State University, in the preparation of this section.

[b] Persistence of injurious levels of herbicide, as listed in Weed Science Society of America Herbicide Handbook, 6th ed., 1989. NL = the herbicide was registered after the publication of the sixth edition or persistence was not given, and t = information represents the half-life of the herbicide.

[c] Translocation or movement within the plant. Apoplastic translocation is upward movement through nonliving xylem and the transpiration stream. Symplastic translocation is downward movement through living protoplasts of cells and the phloem stream.

a. Obtain appropriate plant or soil samples and send them to a chemical assay laboratory for analysis (not a plant disease clinic). This analysis is slow and expensive. The herbicide to be assayed and a list of other pesticides and chemicals applied must be specified.

b. For soil herbicide residue situations, bioassays with sensitive plants can be run. The general procedure is to take representative soil samples from both problem and nonproblem areas with similar soil types. A sample of nonproblem soil is "spiked" with the suspected herbicide, and a soil dilution series is prepared with "clean" soil as the diluent. Sensitive plants with predictable responses are grown in each and observed for symptom development (Fig. 4.5). There are many pitfalls, and bioassays should be done by an experienced weed specialist or in close cooperation with one.

9. When possible, document the problem with color photographs.

Crop-Association Problems

Some plant species germinate or grow poorly in association with or following other plant species. This may result from many factors such as soilborne pathogens (e.g., root-knot nematodes, root-rotting fungi, and Verticillium wilt) that affect both species, from the release of toxic substances from one plant species that inhibits another species (allelopathy), or from toxins produced by the decomposition of crop residues. In other cases, one species might have "reduced growth" because of competition for limited nutrients or water. In situations involving decomposition of a large amount of biomass in the soil, inhibition may be

Fig. 4.5. Bioassay to determine the amount of suspected herbicide residue in soil. Untreated or "clean" soil was added to the pot on the far left, and increasing amounts of a benzoic acid herbicide were mixed with the soil in pots to the right. A pot of soil suspected of containing a residue of the same herbicide can be planted and compared with this series to estimate the amount of residue present.

caused by altered C:N ratios, and the problem will most likely be transitory. In many of these situations, several plant species in the field may show symptoms. See Appendix 4.2 for a listing of some problem plant species associations. No doubt other detrimental plant species associations exist and need to be properly documented. The diagnostician should use Appendix 4.2 with discretion and, with each specimen received, attempt to find the cause of the problem. If none can be found, the diagnostician may state that the cause of the problem was not diagnosed but that the plant association has been reported as a problem and perhaps give some possible causes that may be of use to the grower. If the problem occurs frequently, the diagnostician should seek the assistance of a research plant pathologist or other appropriate scientist.

BLACK WALNUT INJURY

Black walnut (*Juglans nigra*) and, to a lesser extent, Persian or English walnut (*J. regia*), butternut or white walnut (*J. cinerea*), Japanese walnut or onigurumi (*J. sieboldiana*), shagbark or shellbark hickory (*Carya ovata*), pecan (*C. illinoensis*), *Pterocarya stenoptera*, and other related plant species produce juglone in roots, leaves, bark, and husks. Juglone, a slightly water-soluble naphthaquinone, is fungicidal, insecticidal, and phytotoxic. Injury occurs when roots of sensitive plants grow into the root zone of walnuts. Sensitive plants with shallow root systems may escape injury because the roots do not make contact with the deeper-rooted walnuts. Symptoms include stunting, wilting, lack of flowering, root rot, and death. Internally, symptoms can include tyloses in the xylem vessels and browning in the vascular region. Symptoms may resemble Fusarium and Verticillium wilts on some plants. Typically, the problem appears gradually or suddenly after midseason on small farms and gardens where susceptible crops are growing near walnut trees, usually within a distance equal to the height of the trees. Plant injury does not necessarily occur every season because environmental conditions are not always conducive to release of juglone from the tree roots. Injury is more probable in poorly drained soils than in well-drained soils. Injury can occur on the site for 3 years after removal of the walnut trees.

See Appendix 4.3 for plant species susceptible to juglone injury.

Postharvest Disease Problems

Commodities such as fruits, ornamentals, and vegetables are often subject to significant losses caused by postharvest diseases and normal deterioration. For example, losses of highly perishable fruits, vegetables, and flowers in grocery stores average 15% or more. Occasionally, losses during shipment and storage approach 100%.

Diagnostic Procedure

1. If possible, obtain descriptions of common postharvest diseases in publications. See the references at the end of this chapter.
2. Obtain from the grower a complete record of field conditions and disease-control practices prior to harvesting, shipping, and storage. Temperatures, humidities, ventilation, and duration of each are especially important.
3. Examine diseased tissue for microorganisms, especially fungi and bacteria. However, their presence does *not* necessarily establish them as the causal agents (see the following paragraphs).
4. Look up the recommended storage conditions and normal longevity of the commodity (many of these are listed in the following paragraphs and tables).

Pathogens, especially fungi (e.g., species of *Aspergillus, Botrytis, Mucor, Penicillium, Rhizopus,* and *Sclerotinia* [*Monilinia*]) and bacteria (*Erwinia* and *Pseudomonas* spp.) are frequent causes of spots and rots. Infection usually occurs through the inevitable injuries associated with harvesting and handling in the field, shipping, and storage and marketing facilities. In some cases, infection by fungi and bacteria starts during the growing season in the field (e.g., bacterial spot of pepper, nailhead spot of tomato, pox of sweet potato, and scab of potato). In any case, the diagnosis is based on observing and identifying the pathogen.

Unfortunately, injuries caused by abiotic and physiogenic factors may be confused with pathogen-induced diseases, because the damaged tissue is often rapidly colonized by soft-rotting fungi and bacteria.

Abiotic diseases are often the results of improper storage and/or handling conditions. Symptoms induced by abiotic factors are quite variable, depending on the type of exposure and the commodity and its maturity. The damaging effects of intermittent exposure to adverse storage environments on fresh produce are usually cumulative. Most commodities have well-defined storage conditions and maximum storage lives; these are clearly stated in numerous federal, state, and commodity association publications (see references at the end of this chapter). Unfortunately, injuries

Table 4.13. Desirable storage temperatures and storage lives of ornamentals

Commodity	Storage temperature (°F)	Storage life	Commodity	Storage temperature (°F)	Storage life
Cut flowers			Bulbs, corms, and roots (*continued*)		
Acacia (*Acacia* spp.)	40	3–5 days	Iris		
Alstroemeria, Peruvian lily			Spanish (*Iris xiphium*)	68–70	4–12 mo
(*Alstroemeria* spp.)	40	2–3 days	English (*I. xiphioides*)	63	Unknown
Anemone (*Anemone* spp.)	40–45	1–2 days	*I. reticulata*	63	Unknown
Anthurium, tailflower (*Anthurium* spp.)	56	2–4 wk	Ixia (*Ixia* spp.)	68–77	Unknown
Bird-of-paradise (*Strelitzia reginae*)	45	1–3 wk	Lily, gloriosa (*Gloriosa rothschildiana*)	50–63	4–12 mo
Camellia (*Camellia* spp.)	45	3–6 days	Lily-of-the-valley (*Convallaria majalis*)	25–28	1 yr
Clarkia (*Clarkia* spp.)	50	7 days	Muscari, grape- and tassel-hyacinth		
Eucharis (*Eucharis grandiflora*)	45–50	7–10 days	(*Muscari* spp.)	63	2–4 mo
Ginger (*Zingiber officinale*)	55	4–7 days	Narcissus (*Narcissus* spp.)	55–63	2–4 mo
Gladioli (*Gladiolus* spp.)	35–42	6–8 days	*Ornithogalum thyrsoides*	77	Unknown
Gloriosa-lily (*Gloriosa rothschildiana*)	45	4–7 days	*Oxalis adenophylla*	63–68	Unknown
Heliconia (*Heliconia psittacorum*)	55	10 days	Peony, common garden (*Paeonia lacti-*		
Orchids			*flora*)	33–35	5 mo
Cattleya spp.	45–50	2 wk	Primrose (*Primula* spp.)	45–50	Unknown
Cymbidium spp.	31–40	2 wk	*Puschkinia scilloides* (Striped squill)	60–68	Unknown
Vanda spp.	55	5 days	Ranunculus, buttercups (*Ranunculus* spp.)	50–55	Unknown
Poinsettia (*Euphorbia pulcherrima*)	50–60	4–7 days	*Scilla nonscripta*	63–68	Unknown
Sweet-William (*Dianthus barbatus*)	45	3–4 days	Siberian or tartar-lily (*Ixiolirion tataricum*)	68	Unknown
Florist greens			Snowdrop (*Galanthus nivalis*)	55–60	Unknown
Anthurium (*Anthurium* spp.)	40–50	Unknown	Star-of-Bethlehem (*Ornithogalum*		
Chamaedorea (*Chamaedorea* spp.)	45	2–3 wk	*umbellatum*)	68	Unknown
Cordyline (*Cordyline terminalis*)	45–50	2–3 wk	Squills (*Scilla* spp.)	63–73	Unknown
Dieffenbachia (*Dieffenbachia* spp.)	55	Unknown	Taro (*Colocasia esculenta*)	45	Unknown
Fern			Toad or checkered-lily		
Maidenhair (*Adiantum* spp.)	32–40	Unknown	(*Fritillaria meleagris*)	48–55	Unknown
Staghorn (*Platycerium* spp.)	55	Unknown	Trillium (*Trillium* spp.)	33–35	Unknown
Mistletoes (*Phoradendron* spp.)	32	3–4 wk	Triplet-lily (*Brodiaea laxa*)	68–77	Unknown
Palms (many genera)	45	Unknown	Tuberose (*Polianthes tuberosa*)	40–45	4 mo
Podocarpus (*Podocarpus macrophyllus*)	45	Unknown	Tulips (*Tulipa* spp.)	63	2–6 mo
Bulbs, corms, and roots (few tolerate			For forcing	40–50	2–4 mo
refrigeration temperatures)			For outdoors	31–32	5–6 mo
Achimenes (*Achimenes grandiflora*)	45–50	Unknown	Wand-flower (*Sparaxis tricolor*)	77	Unknown
Acidanthera (*Acidanthera* spp.)	45–50	Unknown	Herbaceous perennials		
Alstroemeria (*Alstroemeria* spp.)	40–50	Unknown	Hollyhock (*Alcea rosea*)	33–40	Unknown
Amaryllis (*Amaryllis belladonna*)	38–45	5 mo	Iris (*Iris* spp.)	33–40	Unknown
Anemone, poppy (*Anemone coronaria*)	45–55	3–4 mo	Lady's-slipper (root mat)		
Begonia (*Begonia* spp.)	35–45	3–5 mo	(*Cypripedium* spp.)	33–40	Unknown
Bletilla orchid (*Bletilla* spp.)	35–40	Unknown	Lupine (*Lupinus albus*)	33–40	Unknown
Caladium (*Caladium* spp.)	70	Unknown	Rose mallow (fleshy roots) (*Hibiscus* spp.)	33–40	Unknown
Calla or water-arum (*Calla palustris*)	36–40	Unknown	Transvaal daisy (*Gerbera jamesonii*)	33–40	Unknown
Camassia (*Camassia* spp.)	63–68	Unknown	Nursery stock		
Canna (*Canna* spp.)	40–50	Unknown	Asparagus rhizomes (*Asparagus* spp.)	30–32	3–4 mo
Colchicum (*Colchicum autumnale*)	Unknown	Unknown	Blueberries (wood, unrooted)		
Crocus (*Crocus* spp.)	48–63	2–3 mo	(*Vaccinium* spp.)	30	5 mo
Crown imperial (*Fritillaria imperialis*)	73–77	Unknown	Christmas trees (genera unspecified)	22–32	6–7 wk
Dahlia (*Dahlia* spp.)	40–48	5 mo	Conifer seedlings (genera unspecified)	32–35	4–6 mo
Erythronium (*Erythronium* spp.)	48–63	Unknown	Flower seedlings (genera unspecified)	32–35	2–6 wk
Freesia (*Freesia* spp.)	86	3–4 mo	Rose (*Rosa* spp.)		
Galanthus, snowdrop (*Galanthus nivalis*)	55–60	Unknown	Budwood	28–31	1–2 yr
Garlic, giant (*Allium giganteum*)	73–77	Unknown	Bushes	32	4–5 mo
Gladioli (*Gladiolus* spp.)	45–50	5–8 mo	Strawberry plants (*Fragaria* spp.)	30–32	8–10 mo
Gloxinia (*Sinningia speciosa*)	41–50	5–7 mo	Tomato plants (*Lycopersicon esculentum*)	50–55	10 days
Hemerocallis, daylily (*Hemerocallis* spp.)	50	1 month	Trees and shrubs (genera unspecified)	32–36	4–5 mo
Hyacinth (*Hyacinthus orientalis*)	63–68	2–5 mo	Bedding plants	40–55	2–4 wk
Hymenocallis, spider-lily			Woody ornamentals	32–36	4–5 mo
(*Hymenocallis* spp.)	60–70	Unknown	Herbaceous perennials	31–35	3–7 mo

Table 4.14. Desirable storage temperatures of fruits and vegetables, storage lives, and injury symptoms

Commodity	Storage temperature (°F)	Storage life	Symptoms
Amaranth (*Amaranthus* spp.)	34–36	10–14 days	. . .
Anise (*Pimpinella anisum*)	34–36	2–3 wk	. . .
Apples (some) (*Malus* spp.)	36–38	4–5 wk	Internal browning; brown core; soggy breakdown; soft scald
Apricot (*Prunus armeniaca*)	31–32	3–5 wk	Skin pitting and scalding; flesh grayish brown; fails to soften when ripe
Artichoke			
Globe (*Cynara cardunculus*)	32	2–3 wk	Decay
Jerusalem (*Helianthus tuberosus*)	31–32	4–5 mo	Shriveling; rot
Asparagus (*Asparagus officinalis*)	32–36	2–3 wk	Dull green and limp tips
Asparagus bean or yard-long bean			
(*Vigna unguiculata* subsp. *sesquipedalis*)	40–45	7–10 days	. . .
Atemoya (*Annona cherimola* × *squamosa*)	55–60	4–6 wk	. . .
Avocado (*Persea americana*)	40–55	2–8 wk	Grayish brown discoloration of flesh; pitting; bitter taste
Babaco (*Carica pentagona*)	55–60	1–3 wk	. . .
Balsam-pear, la-kwa (*Momordica charantia*)	53–55	2–3 wk	. . .
Bananas (green or ripe) (*Musa* spp.)	53–56	. . .	Dull color when ripe; dark brown streaks under epidermis of peel
Barbados cherry (*Malpighia glabra*)	32	7–8 wk	. . .
Beans			
Dry (*Phaseolus* spp.)	40–50	6–10 mo	. . .
Snap, string, wax (*P. vulgaris*)	40–45	7–10 days	Pitting; bronzing; russeting
Bean sprouts (several genera)	32	2–3 wk	Darkening of radicle, cotyledons, and hypocotyl; sliminess; rot; mold
Beets (*Beta vulgaris*)	32		Wilting; shriveling
Garden		10–14 days	
Topped		1–3 mo	
Blackberry (*Rubus allegheniensis*)	31–32	2–3 days	Not adapted to long storage
Black sapote (*Diospyros ebenaster*)	55–60	2–3 wk	. . .
Blood orange (*Citrus* sp.)	40–44	3–8 wk	. . .
Blueberry (*Vaccinium* spp.)	31–32	3–5 wk	Off flavor; decay
Bok choy, pak-choi (*Brassica chinensis*			
var. *chinensis*)	34–36	3 wk	. . .
Boniato or air-potato (*Dioscorea bulbifera*)	55–60	4–5 mo	. . .
Breadfruit (*Artocarpus altilis*)	55–60	2–6 mo	. . .
Broccoli (*Brassica oleracea* var. *botrytis*)	32	10–14 days	Off odors and flavors; leaf discoloration; buds yellow and fall; softening; Botrytis mold; rots
Brussels sprouts (*Brassica oleracea* var. *gemmifera*)	32	3–5 wk	Black specking; loss of bright green color; wilt, rot, and discoloration of cut stem
Cabbage (*Brassica oleracea* var. *capitata*)	32		Wilt; black flecks; off flavor; rots
Early		3–6 wk	
Late		3–4 mo	
Cactus leaves, cactus pear, or Indian fig			
(*Opuntia ficus-indica*)	36–40	3 wk	. . .
Cainito or star-apple (*Chrysophyllum cainito*)	50–55	3 wk	. . .
Calabaza, winter or summer squash			
(*Cucurbita pepo*)	50–55	2–3 mo	. . .
Calamondin (*Citrofortunella mitis*)	48–50	2 wk	. . .
Canistel (*Pouteria campechiana*)	55–60	3 wk	. . .
Carambola or star-fruit (*Averrhoa carambola*)	48–50	3–4 wk	. . .
Carrot (*Daucus carota* subsp. *sativus*)	32		Decay; softening; sprouting; bitterness; surface browning
Bunch		2 wk	
Mature		7–9 mo	
Immature		4–6 wk	
Cashew or caju (*Anacardium occidentale*)	32–36	5 wk	. . .
Cassava (*Manihot esculenta*)	32–41	1–2 mo	Vascular streaking; grayish pulp
Cauliflower (*Brassica oleracea* var. *botrytis*)	32	3–4 wk	Leaf wilt, yellowing, and drop; decay and soft rot; browning and rotting of head
Celeriac (*Apium graveolens* var. *rapaceum*)	32	6 wk	Wilting; pithiness; blanching; soft rot
Celery (*Apium graveolens* var. *dulce*)	32	6 wk	Wilting; pithiness; blanching; soft rot
Chayote (*Sechium edule*)	45	4–6 wk	. . .
Cherimoya, custard-apple			
(*Annona cherimola*)	55	2–4 wk	. . .
Cherry			
Sour (*Prunus cerasus*)	32	3–7 days	Unsuitable for long storage
Sweet (*P. avium*)	30–32	14 days	Off flavor; loss of brightness; surface pitting
Chicory, common or witloof			
(*Cichorium intybus*)	32	2–4 wks	Marginal leaf browning; wilting
Chinese cabbage (*Brassica pekinensis*)	32	2–3 mo	. . .
Coconut palm (*Cocos nucifera*)	32–35	1–2 mo	Loss of weight
Corn or maize (*Zea mays* var. *saccharata*)	32	4–8 days	Loss of kernel tenderness and sweetness; denting (wilting) of kernels

(continued on next page)

Table 4.14. (*continued*)

Commodity	Storage temperature (°F)	Storage life	Symptoms
Cranberry, cultivated or large			
(*Vaccinium macrocarpon*)	36	8–16 wk	Rubbery texture, red flesh
Cucumber (*Cucumis sativus*)	45–50	10–14 days	Pitting; water-soaked spots; decay
Currant, northern red or white (*Ribes rubrum*)	31–32	1–4 wk	. . .
Custard-apple (*Annona reticulata*)	41–45	4–6 wk	. . .
Daikon or Chinese radish (*Raphanus sativus*			
cv. longipinnatus)	32–34	4 mo	. . .
Date palm (*Phoenix dactylifera*)	32	6–12 mo	Molds; darkening; loss of flavor
Dewberry, American (*Rubus flagellaris*)	31–32	2–3 days	Not adapted to storage
Durian (*Durio zibethinus*)	39–42	6–8 wk	. . .
Eggplant (*Solanum melongena*)	45–50	7 days	Surface scald; Alternaria rot; bronzing; blackening of seed
Elderberry, American (*Sambucus canadensis*)	31–32	1–2 wk	. . .
Endive and escarole (*Cichorium endivia*)	32	2–3 wk	Wilting
Feijoa or pineapple guava (*Feijoa sellowiana*)	41–50	2–3 wk	. . .
Fig, common (*Ficus carica*)	31–32	7–10 days	Rots
Garlic (*Allium sativum*)	32–35	6–7 mo	Loss of weight; sprouting; root growth; mold growth
Ginger (*Zingiber officinale*)	50	6 mo	Weight loss; moldy growth; skin peeling; shriveling; softening; purple discoloration
Gooseberry, European (*Ribes grossularis*)	31–32	3–5 wk	. . .
Granadilla (*Passiflora edulis* f. *flavicarpa*)	50	3–4 wk	. . .
Grape			
European wine (*Vitis vinifera*)	30–31	1–6 mo	Molds; rots; loss of brightness; flaccid texture
American (*Vitis* spp.)	31–32	2–8 wk	Red cultivars become gray purple; green types become gray green to brown
Grapefruit (*Citrus paradisi*)	50–60	6–8 wk	Scald; pitting; watery breakdown; brown stain
Greens (collards, kale, rape, Swiss chard,			
beet, turnip)	32	10–21 days	Wilting, rot
Guava, common or apple (*Psidium guajava*)	40	2–3 wk	Pulp injury; decay
Haricot bean (*Phaseolus vulgaris*)	40–45	7–10 days	. . .
Horseradish (*Armoracia rusticana*)	30–32	10–12 mo	. . .
Jaboticaba (*Myrciaria cauliflora*)	55–60	2–3 days	Not adapted to long storage
Jackfruit (*Artocarpus heterophyllus*)	55	2–6 wk	. . .
Jaffa or sweet orange (*Citrus sinensis*)	46–50	8–12 wk	. . .
Jerusalem artichoke (*Helianthus tuberosus*)	31–32	4–5 mo	. . .
Jicama or yambean (*Pachyrhizus erosus*)	55–65	4–8 wk	Surface decay; discoloration
King orange, tangerine, satsuma orange or			
Mandarin orange (*Citrus reticulata*)	40	2–4 wk	. . .
Kiwano (*Cucumis melo* var. *deliciosa*)	50–60	6 mo	. . .
Kiwi or Chinese gooseberry			
(*Actinidia chinensis*)	31–32	3–5 wk	. . .
Kohlrabi (*Brassica oleracea* var. *gongylodes*)	32	2–3 mo	Wilt; soft rot
With leaves attached		2 wk	
Kumquat, oval (*Fortunella margarita*)	40	2–4 wk	. . .
Langsat (*Lansium domesticum*)	52–58	2 wk	. . .
Leek (*Allium porrum*)	32	1–3 mo	At higher temperatures: yellowing, elongation, curving, and rot
Lemon (*Citrus limon*)	55–60	4–16 wk	Pitting; membrane staining; red blotch
Lettuce (*Lactuca sativa*)	32	2–3 wk	Russet spotting; pink rib; butt discoloration; brown stain; rot
Lima bean (*Phaseolus lunatus*)			
Pods	34–40	5–7 days	Pitting; russeting; rusty brown areas
Shelled beans	40–45	7–10 days	Pitting; russeting; rusty brown areas
Lime (*Citrus aurantifolia*)	48–50	6–8 wk	Pitting, turning tan with time
LoBok (*Raphanus sativus* var. *niger*)	32–35	2–4 mo	. . .
Loganberrry or boysenberry			
(*Rubus loganobaccus*)	31–32	2–3 days	Not adapted to long storage
Longan or lungan (*Euphoria longana*)	35	3–5 wk	. . .
Loquat or Japanese medlar			
(*Eriobotrya japonica*)	32	3 wk	Internal browning; shriveling; rot; weight loss
Lychee or litchi (*Litchi chinensis*)	35	3–5 wk	Darkening; decay; desiccation
Malanga (*Xanthosoma* spp.)	45	3 mo	. . .
Mamey (*Calocarpum sapota*)	55–60	2–6 wk	. . .
Mango (*Mangifera indica*)	50–55	2–3 wk	Grayish, scaldlike discoloration of skin; uneven ripening; poor flavor
Mangosteen (*Garcinia mangostana*)	55	2–4 wk	. . .
Melon			
Cantaloupe (*Cucumis melo*			
var. *cantalupensis*)	36–41	2 wk	Pitting; surface decay
Casaba (*C. melo* var. *inodorus*)	45–50	3 wk	Pitting; poor ripening
Crenshaw (*C. melo* var. *inodorus*)	45–50	2 wk	Pitting; poor ripening

(*continued on next page*)

Table 4.14. (*continued*)

Commodity	Storage temperature (°F)	Storage life	Symptoms
Melon (*continued*)			
Honeydew (*C. melo* var. *inodorus*)	50	3 wk	Pitting; poor ripening; tarnish discoloration
Persian (*C. melo* var. *reticulatus*)	45–50	2 wk	Pitting; poor ripening
Mushrooms (many genera and types)	32	3–4 days	Darkening; wilt; rot; stalk elongation; opening of cap
Nectarine (*Prunus persica* var. *nucipersica*)	31–32	2–4 wk	. . .
Okra or gumbo (*Abelmoschus esculentus*)	45	7–10 days	Discoloration; water-soaked areas; pitting; decay
Olive, common (fresh) (*Olea europaea*)	41–50	4–6 wk	Internal browning around seed; shriveling
Onion (*Allium cepa*)			
Dry	32–35	1–8 mo	Decay; sprouting; root growth; water-soaked scales; yellowing and leaf decay
Green	32	2–4 wk	
Sets	32–35	6–8 mo	. . .
Orange, sweet (*Citrus sinensis*)	38	3–8 wk	Pitting; brown stain (Arizona and California)
Papaya or pawpaw (*Carica papaya*)	45	1–3 wk	Pitting; failure to ripen; off flavor; decay
Parsley (*Petroselinum crispum*)	32	4–6 wk	Wilting; loss of green color
Parsnip (*Pastinaca sativa*)	32	2–6 mo	Surface browning or yellowing; soft rots
Passion-fruit (*Passiflora edulis*)	45–50	3–5 wk	. . .
Pea			
Common or garden (*Pisum sativum*)	32	1–2 wk	Loss of flavor; shriveling
Snow or sugar (*P. sativum* var. *macro-carpon*)	32–34	1–2 wk	. . .
Southern or cowpea (*Vigna unguiculata*)	40–41	6–8 wk	Loss of flavor and green color; pod yellowing
Peach (*Prunus persica*)	31–32	2–4 wk	Failure to ripen; off flavor; flesh dry, mealy, or wet, dull
Pear, common (*Pyrus communis*)	29–31	2–7 mo	Off flavor; dry texture; core breakdown; rot
Pepino or melon pear (*Solanum muricatum*)	40	1 mo	. . .
Pepper			
Sweet or bell (*Capsicum frutescens* cv. grossum)			
Green	45–50	2–3 wk	Sheet pitting; Alternaria rot on pods and calyxes; poor color; darkening of seed
Ripe	41		
Chili (*Capsicum frutescens* var. *longum*), dry	32–50	6 mo	. . .
Persimmon, Japanese (*Diospyros kaki*)	30	3–4 mo	. . .
Pineapple (*Ananas comosus*)	45–50	2–4 wk	Dull green when ripened
Plantain (*Musa* spp.)	56–58	1–5 wk	. . .
Plum, common and prune (*Prunus domestica*)	31–32	2–5 wk	Off flavors; flesh browning; mealiness; shriveling
Pomegranate (*Punica granatum*)	41	2–3 mo	Pitting; external and internal browning
Potato, Irish (late crop) (*Solanum tuberosum*)	38–40	4–6 mo	Mahogany browning (cultivars Chippewa and Sebago); sweetening
Pummelo or shaddock (*Citrus maxima*)	45–48	12 wk	. . .
Pumpkin and winter squash (*Cucurbita pepo*)	50–55	2–4 mo	Decay, especially Alternaria rot
Quince (*Cydonia oblonga*)	31–32	2–3 mo	Similar to symptoms on apple and pear
Raddichio or radichetta (*Cichorium intybus* var. *foliosum*)	32–34	2–3 wk	. . .
Radish (*Raphanus sativus*)	32	2–3 wk	Black rot; softening; regrowth of tops and roots
Rambutan (mamon chino) (*Nephelium lappaceum*)	54	1–3 wk	. . .
Raspberry, red or European red (*Rubus idaeus*)	31–32	2–3 days	Not adapted to long storage
Rhubarb (*Rheum rhabarbarum*)	32	2–4 wk	Moldy growth; petiole splitting; moisture loss
Rutabaga or swede (*Brassica napus* var. *napobrassica*)	32	4–6 mo	Moisture loss; shriveling; internal breakdown; water soaking; light browning; sprouting
Salsify			
Vegetable-oyster (*Tragopogon porrifolius*)	32	2–4 mo	Moisture loss; shriveling
Black or Spanish (*Scorzonera hispanica*)	32–34	6 mo	. . .
Santol (*Sandoricum koetjape*)	45–48	3 wk	. . .
Sapodilla or sapotilla (*Manilkara zapota*)	60–68	2–3 wk	. . .
Shallot (*Allium ascalonicum*)	32–35	6–8 mo	. . .
Soursop (*Annona reticulata*)	55	1–2 wk	. . .
Spinach (*Spinacia oleracea*)	32	10–14 days	Wilting; yellowing
Squash, bush or bush pumpkin (*Cucurbita pepo* var. *melopepo*)	45–55	7–10 days	Pitting; poor color; decay
Strawberry, beach or Chilean (*Fragaria chiloensis*)	32	5–7 days	Off flavor; dull; shriveling; rot
Sugar-apple or sweetsop (*Annona squamosa*)	45	4 wk	. . .
Sweet potato (*Ipomoea batatas*)	55	4–8 mo	Decay; pitting; internal darkening; hard core after cooking
Tamarind (*Tamarindus indica*)	45	3–4 wk	. . .
Tangerine, Mandarin orange, and related citrus (*Citrus reticulata*)	40	2–4 wk	. . .
Taro or dasheen (*Colocasia esculenta*)	45–50	4–5 mo	. . .

(*continued on next page*)

Table 4.14. (*continued*)

Commodity	Storage temperature (°F)	Storage life	Symptoms
Tomatillo (*Physalis peruviana*)	55–60	3 wk	. . .
Tomato (*Lycopersicon esculentum*)			
Ripe	60–65	4–7 days	Water soaking; softening; decay
Mature green	60–65	1–4 wk	Poor color when ripe; Alternaria rot
Tomato-tree or tree-tomato			
(*Cyphomandra betacea*)	37–40	10 wk	Surface pitting; discoloration; rots
Turnip (*Brassica rapa*)	32	4–5 mo	. . .
Ugli fruit (*Citrus* hybrid)	40	2–3 wk	. . .
Water-chestnut, Chinese			
(*Eleocharis tuberosa*)	32–36	10 mo	Molds
Watercress (*Nasturtium officinale*)	32	2 wk	Wilting; yellowing of leaves; sliminess
Watermelon (*Citrullus lanatus*)	40–50	2–3 wk	Pitting; objectionable flavor; pale red color
White sapote or Mexican apple			
(*Casimiroa edulis*)	67–70	2–3 wk	. . .
Winged bean (*Psophocarpus tetragonolobus*)	50	4 wk	. . .
Yam, Chinese or cinnamon vine			
(*Dioscorea batatas*)	61	7–8 mo	Discolored pulp; softening; waterlogging; decay
Yuca, cassava, or manioc (*Manihot esculenta*)	32–41	1–2 wk	. . .

caused by abiotic factors such as bruising, chilling, freezing, overheating, asphyxiation, and exposure to ethylene, ammonia, and other gases usually are not immediately apparent. For example, bananas bruised during the late green stage of maturity may develop softening and darkening of the seed cavity after the fruit turns yellow.

Despite the many problems associated with relating symptoms to abiotic causes, some generalizations can be made.

1. Low atmospheric humidity causes weight loss, shriveling, and softness in many fruits, flowers or other ornamental plants, and vegetables. For most fruits and vegetables, the relative humidity should be 85–90% and for most ornamentals, 90–95%. Higher humidities may be desirable, but the risk of water condensation and infections by pathogens usually increases.

   Most cut flowers and florist greens are best stored at a relative humidity of 90–95% with a temperature near the freezing point (32–40°F [0–4.4°C]). However, there are important exceptions. See Table 4.13 for a list of desirable storage temperatures and expected storage life of ornamentals.

2. Asphyxiation symptoms are ill defined because they develop gradually and the tissues are rapidly colonized by soft-rotting pathogens, especially anaerobic bacteria. Large, fleshy commodities, such as potato and sweet potato, are especially susceptible when stored at relatively high temperatures in airtight chambers. Potato tubers commonly show internal browning. In sweet potato roots, the typical white latex flow is not evident on freshly cut surfaces, the surface is dull, and there is a characteristic "ferment" odor. With most root and tuber commodities, the lenticels often become prominent. Waterlogged soil prior to harvest is a frequent cause of asphyxiation.

3. Chilling often induces skin pitting, surface blemishes, internal discoloration, and failure of the commodity to ripen evenly when it is returned to normal ripening temperatures (Table 4.14). Affected commodities are often more susceptible to infection by rot organisms. Chilling effects are usually cumulative with a time-temperature interaction. Thus, low temperatures in the field and during transit add to the total effect of chilling that may occur during storage. Chilling injury is often seen in plant clinics because homeowners, handlers, and growers are ignorant of the fact that many commodities are susceptible to this type of injury at surprisingly high temperatures (in some cases 55°F [12.8°C]). For many commodities, chilling injury symptoms are apparent only after ripening.

4. Freezing injury to postharvest commodities occurs at somewhat below the freezing point of water; the exact temperature depends on the commodity, the state of maturity, and the exposure time. Freeze-injured tissue appears water soaked, the result of ice crystals forming in the tissue and rupturing the cells. Frozen or partially frozen commodities are especially susceptible to injury by movement and should not be handled until after they are slowly thawed. Fast thawing increases the risk of freeze injury. When undisturbed, most fruits and vegetables can be cooled one to several degrees below their normal freezing points without injury. This phenomenon is called supercooling or undercooling. However, if the commodity is bumped or jolted, injury occurs. Commodities can be grouped by tolerance to one or more light freezings (Table 4.15).

5. The storage life of fresh fruits, vegetables, and ornamentals under ideal conditions is limited to days or weeks, depending on the commodity. Longer storage periods result in rapid deterioration of the commodity including softening, pitting, decay, and the presence of microorganisms. These may be confused with infectious diseases and injuries.

6. Ethylene gas may cause premature senescence of foliage and ripening of fruit. For example, cucumber and celery turn yellow, while lettuce turns brown. Leafy vegetables and susceptible plants, e.g., potted ornamentals, may be exposed to ethylene gas while in mixed storage with commodities, especially fruit, that produce ethylene.

Mycotoxins and Mycotoxicoses

Mycotoxins are diverse, extracellular substances (secondary metabolites) produced by more than 200 fungi that may be toxic or carcinogenic to animals and humans by producing a disorder called mycotoxicosis. Under some conditions, feed and food invaded by certain fungi may contain mycotoxins. Although the effects of mycotoxins have been known since the Middle Ages (e.g., "holy fire" or ergotism), the breadth of the problem became apparent in 1960 with the death of 100,000 turkey poults in England after they ingested feed containing peanut meal with aflatoxins, a type of mycotoxin. Methods for assaying plant tissues for mycotoxins is beyond the scope of this text. Private and state-supported laboratories are available for information and to assay food or feed for aflatoxins and other mycotoxins (Appendix 4.1).

The diagnostician should remember that the presence of the fungus in infected tissue does not necessarily mean that the food or feed contains mycotoxins. The production of mycotoxins is usually dependent on the strain of the fungus and exact moisture, humidity, and temperature conditions in the field and during transportation and storage. Important fungi and their respective toxins are briefly discussed below; for additional details see the references at the end of this chapter.

1. *Acremonium coenophialum* in tall fescue and *A. typhinum* (*Epichloë typhina*) in perennial rye grass: lolitrems "summer syndrome" (related to the tremorgenic mycotoxins, e.g., aflatrems, penitrems, and janthitrems, that cause trembling in affected animals).
2. *Aspergillus flavus, A. glaucus, A. parasiticus,* and other species: aflatoxins common in cereals, legumes, peanut, cottonseed, fishmeal, Brazil nuts, copra, and other seeds and nuts grown in humid, warm-to-hot regions; ochratoxin in grains; citreoviridin, citrinin, luteoskyrin, and cyclochlorotine in stored rice, barley, corn (maize), and dried fish; tremorgenic toxins in stored and refrigerated foods, grains, and cereals; penicillic acid (a carcinogen) in moldy cereal grains, feeds, tobacco, and other products; and patulin (a carcinogen) in bread, apples, and other fruits.
3. *Cephalosporium* sp.: trichothecene in feed.
4. *Claviceps purpurea* and other species: various ergot alkaloids of many grasses and cereals, especially rye.
5. *Fusarium graminearum, F. moniliforme, F. oxysporum, F. roseum,* and other species: zearalenone (F-2 myco-

Table 4.15. Freezing sensitivity of fruits and vegetables

Commodity	Sus.	Mod.	Tol.
Apple (*Malus sylvestris*)	. . .	+	. . .
Apricot (*Prunus armeniaca*)	+	. . .	. . .
Asparagus (*Asparagus officinalis*)	+	. . .	. . .
Avocado (*Persea americana*)	+	. . .	. . .
Banana (*Musa acuminata*)	+	. . .	. . .
Beans (*Phaseolus vulgaris*)	+	. . .	. . .
Beet, common (no tops) (*Beta vulgaris*)	. . .	. . .	+
Berries (except cranberries)	+	. . .	. . .
Broccoli (sprouting) (*Brassica oleracea* var. *botrytis*)	. . .	+	. . .
Brussels sprouts (*Brassica oleracea* var. *gemmifera*)	. . .	. . .	+
Cabbage (*Brassica oleracea* var. *capitata*)			
New	. . .	+	. . .
Old	. . .	. . .	+
Carrot (no tops) (*Daucus carota* subsp. *sativus*)	. . .	+	. . .
Cauliflower (*Brassica oleracea* var. *botrytis*)	. . .	+	. . .
Celery (*Apium graveolens*)	. . .	+	. . .
Cranberry, American (*Vaccinium macrocarpon*)	. . .	+	. . .
Cucumber (*Cucumis sativus*)	+	. . .	. . .
Date palm (*Phoenix dactylifera*)	. . .	. . .	+
Eggplant (*Solanum melongena*)	+	. . .	. . .
Grape, European wine (*Vitis vinifera*)	. . .	+	. . .
Grapefruit (*Citrus paradisi*)	. . .	+	. . .
Kale (*Brassica oleracea* var. *acephala*)	. . .	. . .	+
Kohlrabi (*Brassica oleracea* var. *gongylodes*)	. . .	. . .	+
Lemon (*Citrus limon*)	+	. . .	. . .
Lettuce (*Lactuca sativa*)	+	. . .	. . .
Lime (*Citrus aurantifolia*)	+	. . .	. . .
Okra (*Abelmoschus esculentus*)	+	. . .	. . .
Onion, dry (*Allium cepa*)	. . .	+	. . .
Orange, sweet (*Citrus sinensis*)	. . .	+	. . .
Parsley (*Petroselinum crispum*)	. . .	+	. . .
Parsnip (*Pastinaca sativa*)	. . .	. . .	+
Pea (*Pisum sativum*)	. . .	+	. . .
Peach (*Prunus persica*)	+	. . .	. . .
Pear (*Pyrus communis*)	. . .	+	. . .
Pepper, bell or sweet (*Capsicum frutescens* cv. grossum)	+	. . .	. . .
Plum (*Prunus domestica*)	+	. . .	. . .
Potato (*Solanum tuberosum*)	+	. . .	. . .
Radish (no tops) (*Raphanus sativus*)	. . .	+	. . .
Rutabaga (*Brassica napus*)	. . .	. . .	+
Salsify (*Tragopogon porrifolius*)	. . .	. . .	+
Spinach (*Spinacia oleracea*)	. . .	+	. . .
Squash			
Winter or autumn (*Cucurbita maxima*)	. . .	+	. . .
Summer or winter (*C. pepo*)	+	. . .	. . .
Sweet potato (*Ipomoea batatas*)	+	. . .	. . .
Tomato (*Lycopersicon esculentum*)	+	. . .	. . .
Turnip (no tops) (*Brassica rapa*)	. . .	. . .	+

[a] Sus. = very susceptible and injured by one light freeze; Mod. = moderately susceptible and may recover from one to two light freezings; Tol. = moderately tolerant and may recover from several light freezings.

toxin) and trichothecene primarily in moldy corn or moldy small grains. Information on the identification of *Fusarium* spp. and their mycotoxins is available from the Fusarium Center, Department of Plant Pathology, 211 Buckhout Laboratory, The Pennsylvania State University, University Park, PA 16802.
6. *Myrothecium* sp.: trichothecene in feed.
7. *Penicillium rubrum*: ochratoxin and rubratoxin in grains. This species also produces citreoviridin, citrinin, luteoskyrin, cyclochlorotine, tremorgenic toxins, penicillic

acid, and patulin in the same commodities discussed under *Aspergillus* species.

8. *Stachybotrys* sp.: stachybotryotoxin in fodder, hay, straw, feed, bedding, and wheat; and trichothecene in feed.

9. *Trichoderma viride*: trichothecene in feed.

Using the Field Diagnosis for Laboratory Confirmation

After the cause of the disease has been placed into a general category, the scope of the investigation is considerably narrowed. A specific cause is often apparent at this time. In any event, appropriate samples are taken or sent to the clinic laboratory for confirmation of the field diagnosis.

Unfortunately, there is an occasional overlapping of disease signatures caused by agents in different groups. Certain parts of the field disease signature may be transitory, obscured, or not yet developed at the time of the visit. Another problem is the likelihood that two or more diseases caused by agents in different groups may be present, even on the same plant part. Despite the exceptions and problems, diagnosing in the field is the logical first step and has much to offer in the successful diagnosis of the specific cause of a problem, which must be completed in the laboratory. Laboratory methods for confirming a field diagnosis are presented in the next chapter.

References

Bennett, W. F., ed. 1993. Nutrient Deficiencies and Toxicities in Crop Plants. American Phytopathological Society, St. Paul, MN.

Bonanno, A. R., and Monaco, T. J. 1986. Weed Attack Manual. North Carolina State University, Raleigh.

Brooks, M. C. 1951. Effect of black walnut trees and their products on other vegetation. Bull. W. Va. Agric. Exp. Stn. 347:1-31.

Camper, N. D., ed. 1986. Research Methods in Weed Science. 3rd ed. Southern Weed Science Society, Champaign, IL.

Cook, M. T. 1921. Wilting caused by walnut trees. Phytopathology 11:346.

Eagle, D. J., Caverly, D. J., and Holly, K. 1981. Diagnosis of Herbicide Damage to Crops. Chemical Publishing, New York.

Hardenburg, R. E., Watada, A. E., and Wang, C. Y. 1986. The commercial storage of fruits, vegetables, and florist and nursery stocks. U.S. Dep. Agric. Agric. Handb. 66.

Henley, R. W. 1981. Diagnosing plant disorders. Pages 327-350 in: Foliage Plant Production. N. Jasper, ed. Prentice-Hall, Englewood Cliffs, NJ.

Horst, R. K. 1990. Westcott's Plant Disease Handbook. 5th ed. Van Nostrand Reinhold, New York.

Humberg, N. E., ed. 1989. Herbicide Handbook. 6th ed. Weed Science Society of America, Champaign, IL.

Jones, R. K. 1978. Plant disease development calendar. N.C. State Univ. Agric. Ext. Serv. AG-135.

Kader, A. A, ed. 1992. Postharvest technology in horticultural crops. Univ. Calif. Div. Agric. Nat. Resour. Publ. 3311.

Klingman, G. C., and Ashton, F. M. 1982. Weed Science: Principles and Practices. 2nd ed. John Wiley & Sons, New York.

LaCasse, N. L., and Treshow, M. 1976. Diagnosing Vegetation Injury Caused by Air Pollution. Applied Science Association, Air Pollution Training Institute, Publ. MD-20. Environmental Protection Agency, Research Triangle Park, NC.

Lloyd, R. A., and Liptons, W. J. 1978. Handling, Transportation, and Storage of Fruits and Vegetables. 2nd ed. Vol. 1, Melons and Vegetables. AVI, Westport, CT.

Lockerman, R. H., Putnam, A. R., Rice, R. P., and Meggitt, W. F. 1975. Diagnosis and prevention of herbicide injury. Mich. State Univ. Coop. Ext. Serv. Bull. E-809.

Maloy, O. C. 1981. Diagnosis of plant diseases. Wash. State Univ. Coop. Ext. Serv. Bull. EB 0751.

Massey, A. B. 1925. Antagonism of the walnuts (*Juglans nigra* L. and *J. cinerea* L.) in certain plant associations. Phytopathology 15:773-784.

McGregor, B. M. 1987. Manual de transportes de productos tropicales. U.S. Dep. Agric. Ofic. Transp. Man. Agric. 668.

Meister, R. T., ed. 1996 (revised each year). Farm Chemicals Handbook. Meister, Willoughby, OH.

Moje, W. 1966. Organic soil toxins. Pages 533-751 in: Diagnostic Criteria for Plants and Soils. H. D. Chapman, ed. Quality Printing, Abilene, TX.

Moline, H. E. 1984. Postharvest pathology of fruits and vegetables: Postharvest losses in perishable crops. Univ. Calif. Agric. Exp. Stn. Bull. 1914.

Monaco, T. J. 1984. Principles and Methods in Weed Science: A Laboratory Manual. North Carolina State University, Raleigh.

Monaco, T. J., Bonanno, A. R., and Baron, J. J. 1986. Herbicide injury: Diagnosis, causes, prevention and remedial action. In: Southern Weed Science Society Research Manual. N. D. Camper, ed. The Society, Champaign, IL.

Ostry, M. E., and Nicholls, T. H. 1978. How to Identify and Control Noninfectious Diseases of Trees. North Central Forest Experiment Station, St. Paul, MN.

Piedrahita, O. 1984. Black walnut toxicity. Ont. Minist. Agric. Food Publ. Factsheet AGDEX 246/80, order no. 84-050.

Shurtleff, M. C. 1966. How to Control Plant Diseases in Home and Garden. 2nd ed. Iowa State University Press, Ames.

Shurtleff, M. C. 1989. The weather and plant diseases. Univ. Ill. Rep. Plant Dis. 1003.

Shurtleff, M. C., Kirby, H. W., Eastburn, D. M., and White, D. G. 1990. Mycotoxins and mycotoxicoses. Univ. Ill. Rep. Plant Dis. 1105.

Skroch, W. A., and Sheets, T. J., eds. 1977. Herbicide injury symptoms and diagnosis. N.C. State Univ. Agric. Ext. Serv. AG-85.

Snowdon, A. L. 1990. A Color Atlas of Post-harvest Diseases and Disorders of Fruits and Vegetables. Vol. 1, General Introduction and Fruits. CRC Press, Boca Raton, FL.

Snowdon, A. L. 1992. A Color Atlas of Post-harvest Diseases and Disorders of Fruits and Vegetables. Vol. 2, Vegetables. CRC Press, Boca Raton, FL.

Streibig, J. C., and Kudak, P. 1993. Herbicide Bioassays. CRC Press, Boca Raton, FL.

Swain, R. W. 1985. Plant physiological disorders. Minist. Agric. Fish. Food (G.B.) Ref. Book 223. (available in the U.S. from Bernam, Lanham, MD)

TransFresh Corporation. 1988. Fresh Produce Mixer and Loading Guide. TransFresh, Salinas, CA.

Wyllie, T. D., and Morehouse, L. G., eds. 1977–1978. Mycotoxic Fungi, Mycotoxins, Mycotoxicoses: An Encyclopedic Handbook. Marcel Dekker, New York.

Appendix 4.1. University-related plant disease and soil testing laboratories in the United States and Canada[a,b]

Country State or province	Soil testing	Plant problem diagnosis
United States		
Alabama	Soil Testing Laboratory Auburn University 118 Funchess Hall Auburn, AL 36849-5624	Plant Disease Clinic 102 Extension Hall Department of Plant Pathology Auburn University Auburn, AL 36849-5624
Alaska	Soil Testing Laboratory Agricultural Experiment Station University of Alaska 533 E. Firewood Palmer, AK 99645	Department of Plant Pathology University of Alaska Agricultural Forestry Experiment Station Fairbanks, AK 99775-0080
Arizona	Soil, Water, and Plant Tissue Testing Laboratory Department of Soils, Water, and Engineering University of Arizona Tucson, AZ 85721	Plant Disease Clinic Department of Plant Pathology University of Arizona Tucson, AZ 85721
Arkansas	Soil Testing and Research Laboratory University of Arkansas P.O. Drawer 767 Marianna, AR 72360	Plant Disease Clinic Lonoke Agricultural Center P.O. Drawer D; Hwy. 70 E Lonoke, AR 72086
California	No soil testing service is offered by a public agency	Contact your local county farm advisor or extension specialist at your nearest university
Colorado	Soil, Water, and Plant Testing Laboratory A319 NESB Colorado State University Fort Collins, CO 80523	Plant Diagnostic Clinic Department of Plant Pathology Colorado State University Fort Collins, CO 80523
Connecticut	Soil Testing Laboratory Plant Science Department University of Connecticut Storrs, CT 06268	Consumer Horticultural Center University of Connecticut Storrs, CT 06269-4087 Connecticut Agricultural Experiment Station Huntington Avenue East Haven, CT 06512
Delaware	Soil Testing Laboratory Department of Plant and Soil Science University of Delaware Newark, DE 19711	Extension Plant Pathologist 136 Townsend Hall University of Delaware Newark, DE 19717-1303
Florida	Soil Testing Laboratory University of Florida Gainesville, FL 32611	Nematode Assay Laboratory Entomology and Nematology Department Building 78, Mowry Road Gainesville, FL 32611-0830 Plant Disease Clinic P.O. Box 110830 Building 78, Mowry Road Gainesville, FL 32611-0830 Plant Diagnostic Clinic North Florida Research Center Route 3, Box 4370 Quincy, FL 32351 Plant Diagnostic Clinic SW Florida Research and Education Center P.O. Drawer 5127 Immokalee, FL 33934 Regional Plant Disease Laboratory 18905 SW 280 Street Homestead, FL 33031-3314
Georgia	Soil Testing and Plant Analysis Laboratory University of Georgia 2400 College Station Road Athens, GA 30602	Extension Plant Disease Clinic 4-Towers Building University of Georgia Athens, GA 30602

(continued on next page)

Appendix 4.1. (*continued*)

Country State or province	Soil testing	Plant problem diagnosis
United States (*continued*)		
Hawaii	Soil Testing Laboratory Agricultural Diagnostic Service Center 1910 East-West Road Sherman Hall 134 Honolulu, HI 96822	Plant Disease Clinic Agricultural Diagnostic Service Center 1910 East-West Road Sherman Hall 112 Honolulu, HI 96822
Idaho	Department of Plant and Soil Science College of Agriculture Moscow, ID 83843	Extension Plant Pathologist University of Idaho Research and Extension Center 3793 N, 3600 E. Kimberly, ID 83341 Extension Plant Pathologist Research and Extension Center Parma, ID 83660
Illinois	No soil testing service is offered by a public agency	(May–October) Plant Clinic 1401 W. St. Mary's Road Urbana, IL 61801 (November–April) N-533 Turner Hall 1102 S. Goodwin Avenue University of Illinois Urbana, IL 61801
Indiana	No soil testing service is offered by a public agency; contact the Plant and Pest Diagnostic Laboratory for a partial listing of private soil testing laboratories	Plant and Pest Diagnostic Laboratory Department of Botany and Plant Pathology 1155 Lilly Hall Purdue University West Lafayette, IN 47907
Iowa	Soil Testing Laboratory G501 Agronomy Iowa State University Ames, IA 50011	Plant Disease Clinic Department of Plant Pathology 323 Bessey Hall Iowa State University Ames, IA 50011
Kansas	Soil Testing Laboratory Agronomy Department Kansas State University Manhattan, KS 66506	Plant Disease Diagnostic Laboratory Department of Plant Pathology Throckmorton Hall Kansas State University Manhattan, KS 66506-5502
Kentucky	Soil Testing Laboratory 103 Regulatory Services Building University of Kentucky Lexington, KY 40506	(serving western Kentucky) Plant Disease Diagnostic Laboratory Research and Education Center P.O. Box 469 Highway 91 South Princeton, KY 42445 (serving central and eastern Kentucky) Plant Disease Diagnostic Laboratory Department of Plant Pathology University of Kentucky Lexington, KY 40546-0091
Louisiana	Soil Testing Laboratory Department of Agronomy Louisiana State University Baton Rouge, LA 70803	Plant Disease Diagnostic Clinic 20 H. D. Wilson Building Louisiana State University Baton Rouge, LA 70803
Maine	Maine Soil Testing Service 5722 Deering Hall University of Maine Orono, ME 04469-5722	Pest Management Office Cooperative Extension 491 College Avenue University of Maine Orono, ME 04473-1295
Maryland	Soil Testing Laboratory Agronomy Department University of Maryland College Park, MD 20742	Plant Diagnostic Laboratory Department of Plant Biology The University of Maryland College Park, MD 20742

(*continued on next page*)

Appendix 4.1. (*continued*)

Country State or province	Soil testing	Plant problem diagnosis
United States (*continued*)		
Massachusetts	Soil Testing West Experiment Station University of Massachusetts Amherst, MA 01003	No disease diagnostic service is offered by a public agency to homeowners; commercial samples are handled by individual extension specialists at the university
Michigan	MSU Soil and Plant Nutrient Laboratory A81 Plant and Soil Sciences East Lansing, MI 48824	Plant Diagnostic Clinic Department of Botany and Plant Pathology Michigan State University East Lansing, MI 48824
Minnesota	Research and Soil Testing Laboratories 135 Crops Research Building University of Minnesota 1903 Hendon Avenue St. Paul, MN 55108	(for homeowners) Dial U Clinic 145 Alderman Hall 1970 Folwell Avenue University of Minnesota St. Paul, MN 55108 (for commercial growers) Plant Disease Clinic Department of Plant Pathology 495 Borlaug Hall 1991 Upper Buford Circle University of Minnesota St. Paul, MN 55108
Mississippi	Soil Testing Laboratory Cooperative Extension Service Mississippi State University Mississippi State, MS 39762	Plant Pathology Laboratory Room 9 Bost Extension Center Box 9655 Mississippi Cooperative Extension Service Mississippi State, MS 39762
Missouri	Soil Testing Laboratory Room 23 Mumford Hall University of Missouri Columbia, MO 65211	Plant Science Unit/Diagnostic Clinics Room 45, Agriculture Building University of Missouri Columbia, MO 65211
Montana	No soil testing service is offered by a public agency	Plant Disease Clinic 525 Leon Johnson Hall Department of Plant Pathology Montana State University Bozeman, MT 59717
Nebraska	Soil Testing Laboratory Department of Agronomy University of Nebraska Lincoln, NE 68583	Plant and Pest Diagnostic Clinic Department of Plant Pathology 448 Plant Sciences University of Nebraska Lincoln, NE 68583-0722
Nevada	No soil testing service is offered by a public agency	Bureau of Plant Industry Division of Agriculture 350 Capital Hill Avenue Reno, NV 89502
New Hampshire	Analytical Services Laboratory Nesmith Hall University of New Hampshire Durham, NH 03824	Plant Diagnostic Laboratory 322 Nesmith Hall Plant Biology Department University of New Hampshire Durham, NH 03824
New Jersey	Soil Testing Laboratory Rutgers University P.O. Box 902 Milltown, NJ 08850	Plant Diagnostic Laboratory Rutgers University P.O. Box 550 Milltown, NJ 08850
New Mexico	SWAT Laboratory Agronomy and Horticulture New Mexico State University Box 30003 Las Cruces, NM 88003	Extension Plant Pathologist Box 3AE Plant Sciences Cooperative Extension Service New Mexico State University Las Cruces, NM 88003

(*continued on next page*)

Appendix 4.1. (*continued*)

Country State or province	Soil testing	Plant problem diagnosis
United States (*continued*)		
New York	CNA Labs SCAS Department Bradfield Hall Cornell University Ithaca, NY 14853	(homeowners and commercial) Insect and Plant Disease Diagnostic Laboratory 334 Plant Science Building Cornell University Ithaca, NY 14853
		(commercial ornamental samples only) Long Island Horticultural Research Laboratory 30599 Sound Avenue Riverhead, NY 11901
North Carolina	Soil Testing Laboratory Agronomic Division North Carolina Department of Agriculture 4300 Reedy Creek Road Raleigh, NC 27611	Plant Disease and Insect Clinic Department of Plant Pathology Box 7616 North Carolina State University Raleigh, NC 27695-7616
North Dakota	Soil Testing Laboratory Soil Science Department North Dakota State University Fargo, ND 58105	Plant Diagnostic Clinic Department of Plant Pathology Box 5012 North Dakota State University Fargo, ND 58105
Ohio	R.E.A.L. OARDC Wooster, OH 44691	Plant and Pest Diagnostic Clinic Department of Plant Pathology 2021 Coffey Road The Ohio State University Columbus, OH 43210
Oklahoma	SWFAL Agronomy Department 048 Agriculture Hall Oklahoma State University Stillwater, OK 74078	Plant Disease Diagnostic Laboratory Department of Plant Pathology 110 Noble Research Center Oklahoma State University Stillwater, OK 74078
Oregon	Soil Testing Laboratory Oregon State University Corvallis, OR 97331	Plant Disease Clinic Extension Plant Pathology Cordley Hall 1089 Oregon State University Corvallis, OR 97331-2903
Pennsylvania	Agricultural Analytical Service Laboratory College of Agricultural Sciences The Pennsylvania State University University Park, PA 16802	Plant Disease Clinic 220 Buckhout Laboratory The Pennsylvania State University University Park, PA 16802
Rhode Island	Soil Testing Laboratory University of Rhode Island Kingston, RI 02881	Homeowner Clinic Cooperative Extension Education Center University of Rhode Island Kingston, RI 02881
South Carolina	Soil Testing Laboratory Agricultural Service Laboratory Clemson University Clemson, SC 29634	Plant Problem Clinic Cherry Road Clemson University Clemson, SC 29634-0377
South Dakota	Soil Testing Laboratory Plant Science Department Box 2207-A South Dakota State University Brookings, SD 57007	Plant Disease Clinic Plant Science Department Box 2109 South Dakota State University Brookings, SD 57007
Tennessee	Soil and Forage Testing Laboratory University of Tennessee Knoxville, TN 37901	Plant and Pest Diagnostic Clinic P.O. Box 1071 University of Tennessee Knoxville, TN 37901-1071
Texas	Soil Testing Laboratory Room 220, Soil and Crop Sciences Texas A&M University College Station, TX 77843	Texas Plant Disease Diagnostic Laboratory Room 101, L. F. Peterson Building Texas A&M University College Station, TX 77843-2132

(*continued on next page*)

Appendix 4.1. (*continued*)

Country State or province	Soil testing	Plant problem diagnosis
United States (*continued*)		
Utah	Soil Testing Laboratory Department of Plants, Soils and Biometeorology Agricultural Science Building Utah State University Logan, UT 84322-4830	Plant Pest Diagnostic Laboratory Department of Biology Utah State University Logan, UT 84322-5305
Vermont	Soil Testing Laboratory Department of Plant and Soil Science Hills Building University of Vermont Burlington, VT 05405-0086	Plant Diagnostic Laboratory Department of Plant and Soil Science Hills Building University of Vermont Burlington, VT 05405-0086
Virginia	Virginia Tech Soil Testing Lab 145 Smyth Hall P.O. Box 10664 Blacksburg, VA 24062-0664	Plant Disease Clinic Department of Plant Pathology, Physiology, and Weed Science VPI and SU Blacksburg, VA 24061

Washington	No public agency provides soil testing; contact your
	   county extension agent for a listing of local labora-
	   tories
		(serving eastern Washington)
		Plant Diagnostic Clinic
		WSU-Prosser-IAREC
		Rt. 2, Box 2953-A
		Prosser, WA 99350-9687
		
		(serving western Washington)
		Plant Diagnostic Clinic
		WSU-Puyallup Research and Extension Center
		7612 Pioneer Way East
		Puyallup, WA 98371-4998

West Virginia	Soil Testing Laboratory 1090 Agricultural Sciences Building West Virginia University Morgantown, WV 26506	Plant Disease Diagnostic Clinic 401 Brooks Hall West Virginia University Morgantown, WV 26506
Wisconsin	Soil and Plant Analysis Laboratory 511 Mineral Point Road University of Wisconsin Madison, WI 53705	Plant Disease Clinic Department of Plant Pathology 1630 Linden Drive University of Wisconsin Madison, WI 53706
Wyoming	Soil Testing Laboratory Plant Science Department Box 3354 University of Wyoming Laramie, WY 82071	Plant Disease Clinic Department of Plant, Soil and Insect Sciences P.O. Box 3354 University of Wyoming Laramie, WY 82071-3354
Canada		
Alberta	Soil and Crop Diagnostic Center University of Alberta #905 OS Longman Building 6909 116 Street Edmonton, Alberta	Brooks Diagnostics LTD[c] Plant Diagnostic Laboratory Crop Diversification Centre-South Brooks, Alberta T1R 1E6 Regional Crop Laboratory Alberta Agriculture Box 10 Olds, Alberta T0M 1PO Regional Crop Laboratory Alberta Agriculture Provincial Building Box 7777 Fairview, Alberta TOH 1LO Alberta Environmental Centre Bag 4000 Vegreville, Alberta T0B 4L0

(*continued on next page*)

Appendix 4.1. (*continued*)

Country State or province	Soil testing	Plant problem diagnosis
Canada (*continued*)		
British Columbia	Soil Testing Unit British Columbia Department of Agriculture 1873 Spall Road Kelowna, BC VIY 4R2	Plant Diagnostic Laboratory British Columbia Ministry of Agriculture— Agriculture and Fisheries Abbotsford Agriculture Centre 1767 Angus Campbell Road Abbotsford, BC V3G 2M3
Manitoba	Department of Soil Science University of Manitoba Winnipeg, Manitoba R3T 2N2	Crop Diagnostic Centre 201-545 University Cres. Agricultural Service Complex Winnipeg, Manitoba R3T 5S6
New Brunswick	Agricultural Soils Laboratory New Brunswick Department of Agriculture P.O. Box 6000 Frederickton, New Brunswick E3B 5H1	Pest Diagnostic Laboratory New Brunswick Department of Agriculture P.O. Box 6000 Frederickton, New Brunswick E3B 5H1
Nova Scotia	Soil and Crops Branch Nova Scotia Agricultural College Truro, Nova Scotia B2N 5E3	Plant Diagnostic Laboratory Nova Scotia Department of Agriculture and Marketing Kentville Research Station Kentville, Nova Scotia B4N 1J5 Plant Diagnostic Laboratory Department of Biology Nova Scotia Agricultural College Box 550 Truro, Nova Scotia B2N 5E3
Ontario	No provincial soil testing service is offered	Pest Diagnostic Clinic Agriculture and Food Service Centre P.O. Box 3650 95 Stone Road W., Zone 2 Guelph, Ontario N1H 8J7
Prince Edward Island	Soil and Feed Testing Laboratory Department of Agriculture Box 1600 Charlottetown, Prince Edward Island C1A 7N3	Plant Pathologist Plant Health Services Department of Agriculture, Fisheries and Forestry Box 1600 Charlottetown, Prince Edward Island C1A 7N3
Quebec	Canadian Industries Limited Soil Laboratory Beloeil Works McMasterville, Quebec	Laboratoire de Diagnostic Le Service de Recherche en Phytotechnie de Quebec 2700, rue Einstein Ste. Foy, Quebec G1P 3W8

[a] Compiled by G. Ruhl, Department of Botany and Plant Pathology, Purdue University. The addresses are current as of September 1995.

[b] Many states have diagnostic laboratories associated with their state departments of agriculture or with the U.S. Department of Agriculture (USDA) research facilities. Practically all countries have state and/or federal diagnostic laboratories that offer a variety of services. These services may include soil testing, plant identifications, and plant disease, insect, and other pest diagnoses. There are also hundreds of consultants who operate private plant pest diagnostic laboratories. The services offered vary greatly from one firm to another. Some specialize only in tree or lawn problems; others cater to commercial fruit, vegetable, nursery, or greenhouse growers. The number of private laboratories has increased rapidly in recent years as many public institutions have reduced their services because of financial restraints. County, state, or federal agencies may have a partial listing of private soil testing and plant disease or other plant problem laboratories.

[c] Brooks Diagnostics LTD, a private laboratory, has taken over the testing for Alberta Special Crops and Horticultural Research Centre.

Appendix 4.2. Some problem plant species associations or planting sequences[a]

Toxic plant	Affected plants	Comments
Alfalfa (*Medicago sativa*)	Alfalfa (*Medicago sativa*) Barley (*Hordeum vulgare*) Cotton (*Gossypium* spp.) Wheat (*Triticum* spp.)	Toxic excretions from roots Growth reduced
Apple (*Malus* spp.)	Apple (*Malus* spp.) Pear (*Pyrus communis*)	Nematodes; toxic excretions from roots; growth reduced
Apricot (*Prunus armeniaca*)	Cherry, sour (*Prunus cerasus*)	Growth depressed
Barley (*Hordeum vulgare*)	Barley (*Hordeum vulgare*) Clover, red (*Trifolium pratense*) Quackgrass (*Agropyron repens*)	Growth depressed
Beans (*Phaseolus* spp.)	Pea (*Pisum sativum*)	Dried plant material reduces germination and growth
Bermuda grass (*Cynodon dactylon*)	Coffee (*Coffea* spp.)	Growth inhibited with root leachates
Bindweed, field (*Convolvulus arvensis*)	Alfalfa (*Medicago sativa*) Flax (*Linum usitatissimum*) Oat (*Avena sativa*) Wheat (*Triticum* spp.)	Germination and growth reduced
Bluegrasses (*Poa* spp.)	Tomato (*Lycopersicon esculentum*)	Growth depressed
Annual (*P. annua*)	Apple (*Malus* spp.)	Reduced growth
Kentucky (*P. pratensis*)	Bluegrass, Canada (*Poa compressa*) Redtop (*Agrostis gigantea*)	Growth depressed
Broad bean (*Vicia faba*)	Broad bean (*Vicia faba*)	Growth reduced
Brome grass, smooth (*Bromus inermis*)	Brome grass, smooth (*Bromus inermis*)	Dried roots, reduced growth
Butternut (*Juglans cinerea*)	Many plant species. (See Appendix 4.3)	Wilting and reduced growth
Cabbage (*Brassica oleracea* var. *capitata*)	Spinach (*Spinacia oleracea*)	Growth depressed
Catnip (*Nepeta cataria*)	Pea (*Pisum sativum*) Wheat (*Triticum* spp.)	Dried plant material inhibits germination and growth
Cherry		
Sour (*Prunus cerasus*)	Cherry, sour (*Prunus cerasus*) Wheat (*Triticum* spp.)	Growth depressed
Mahaleb (*P. mahaleb*)	Cherry, Mahaleb (*Prunus mahaleb*)	Nematodes (*Pratylenchus* spp.) cause reduced growth
Mazzard (*P. avium*)	Cherry, mazzard (*Prunus avium*)	Nematodes (*Pratylenchus penetrans*) cause reduced growth
Chickweed, common (*Stellaria media*)	Barley (*Hordeum vulgare*)	Yield reduced as much as 80%
Chrysanthemum spp. (Chrysanthemum varieties)	Chrysanthemum varieties Oat (*Avena sativa*)	Stunting and death Toxic effects reduced after leaching with water or 1% H_2SO_4
Citrus fruits (*Citrus* spp.)	Citrus fruits (*Citrus* spp.)	Growth of young trees suppressed Problem reduced after leaching
Clovers (*Trifolium* spp.)	Apple (*Malus* spp.) Clovers (*Trifolium* spp.) Mustards (*Brassica* spp.) Tobacco (*Nicotiana tabacum*) Tomato (*Lycopersicon esculentum*)	Growth reduced 23–94%
Red (*T. pratense*)	Barley (*Hordeum vulgare*) Clover, red (*Trifolium pratense*) Clover, white (*Trifolium repens*) Radish (*Raphanus sativus*) Rye grass, Italian (*Lolium multiflorum*)	Growth reduced
Coffee (*Coffea* spp.)	Coffee (*Coffea* spp.) Shade tree (*Inga* spp.)	Growth depressed Swollen root tips
Corn or maize (*Zea mays*)	Beet (*Beta vulgaris*) Maize (*Zea mays*) Lettuce (*Lactuca sativa*) Onion (*Allium cepa*) Pea (*Pisum sativum*) Tobacco (*Nicotiana tabacum*) Tomato (*Lycopersicon esculentum*) Wheat (*Triticum* spp.)	Decomposition of maize produces a toxin that injures roots

(continued on next page)

Appendix 4.2. (*continued*)

Toxic plant	Affected plants	Comments
Cotton (*Gossypium* spp.)	Cotton (*Gossypium* spp.) Cow pea (*Vigna unguiculata*) Wheat (*Triticum* spp.)	Depressed early growth
Dogwood (*Cornus* spp.)	Wheat (*Triticum* spp.)	Growth reduced
Fescue		
Meadow (*Festuca pratensis*)	Apple (*Malus* spp.) Fescue, meadow (*Festuca pratensis*)	Germination and growth reduced
Red (*F. rubra*)	Rape (*Brassica napus*) Wild oat (*Avena fatua*)	Root exudates reduce germination and growth
Flax (*Linum usitatissimum*)	Flax (*Linum usitatissimum*)	Growth depressed
Flower-of-an-hour (*Hibiscus trionum*)	Pea (*Pisum sativum*) Wheat (*Triticum* spp.)	Germination and growth inhibited
Garlic, field (*Allium vineale*)	Oat (*Avena* spp.)	Growth reduced
Hawkweed (*Hieracium piloselloides,* *H. umbellatum*, and *H. vulgatum*)	Various plants	Germination and seedling growth inhibited
Heath (*Erica carnea*)	Various annuals	Growth reduced
Locust, black (*Robinia pseudoacacia*)	Barley (*Hordeum vulgare*)	Reduced growth Death
Mallow (*Malva* spp.)	Pea (*Pisum sativum*) Wheat (*Triticum* spp.)	Germination and growth inhibited
Mangel (*Beta vulgaris* subsp. *vulgaris*)	Mangel (*Beta vulgaris* subsp. *vulgaris*)	Growth reduced
Maple (*Acer* spp.)	Wheat (*Triticum* spp.)	Growth reduced
Marsh-elder (*Iva xanthifolia*)	Pea (*Pisum sativum*) Wheat (*Triticum* spp.)	Germination and growth inhibited
Mayweed (*Anthemis cotula*)	Barley (*Hordeum vulgare*)	Growth reduced
Milkweed, common (*Asclepias syriaca*)	Pea (*Pisum sativum*) Wheat (*Triticum* spp.)	Germination and growth inhibited
Mountain-ash, European (*Sorbus aucuparia*)	Mountain-ash, European (*Sorbus aucuparia*)	Nematodes (*Pratylenchus* spp.); toxic excretions by roots
Mustards (*Brassica* spp.)	Kale or collards (*Brassica oleracea* var. *acephala*) Peach (*Prunus persica*)	Yield and growth reduced
Nut sedge (*Cyperus* spp.)	Alfalfa (*Medicago sativa*)	Growth reduced
Oak (*Quercus* spp.)	Wheat (*Triticum* spp.)	Reduced growth
Oat (*Avena sativa*)	Alfalfa (*Medicago sativa*) Barley (*Hordeum vulgare*) Oat (*Avena sativa*) Pea (*Pisum sativum*) Peach (*Prunus persica*) Wheat (*Triticum* spp.)	Germination and growth reduced
Orchard grass (*Dactylis glomerata*)	Alfalfa (*Medicago sativa*) Bluegrass, Kentucky (*Poa pratensis*) Clover, white (*Trifolium repens*) Clovers (*Trifolium* spp.) Orchard grass (*Dactylis glomerata*)	Root leachates cause reduced growth
Paspalum, sour (*Paspalum conjugatum*)	Coffee (*Coffea* spp.)	Root leachates inhibit growth
Peach (*Prunus persica*)	Peach (*Prunus persica*)	Peach replant problem
Pigweed (*Amaranthus spinosus*)	Coffee (*Coffea* spp.)	Growth inhibited
Pine (*Pinus* spp.)	Wheat (*Triticum* spp.)	Growth inhibited
Plantain (*Plantago* spp.)	Pea (*Pisum sativum*) Wheat (*Triticum* spp.)	Germination and growth inhibited
Plum, Myrobalan (*Prunus cerasifera*)	Pear (*Pyrus communis*)	Growth depressed
Potato (*Solanum tuberosum*)	Peach (*Prunus persica*)	Reduced growth
Purslane (*Portulaca oleracea*)	Pea (*Pisum sativum*) Wheat (*Triticum* spp.)	Germination and growth reduced

(*continued on next page*)

Appendix 4.2. (*continued*)

Toxic plant	Affected plants	Comments
Quackgrass (*Agropyron repens*)	Alfalfa (*Medicago sativa*) Barley (*Hordeum vulgare*) Clover, red (*Trifolium pratense*) Clover, white (*Trifolium repens*) Dandelion (*Taraxacum* spp.) Flax (*Linum usitatissimum*) Oat (*Avena sativa*) Pea (*Pisum sativum*) Rape (*Brassica napus*) Wheat (*Triticum* spp.)	Germination and growth reduced
Quince (*Cydonia oblonga*)	Quince (*Cydonia oblonga*)	Reduced growth
Ragweed, giant (*Ambrosia trifida*)	Pea (*Pisum sativum*) Wheat (*Triticum* spp.)	Germination and growth reduced
Rape (*Brassica napus*)	Clover, red (*Trifolium pratense*) Peach (*Prunus persica*)	Reduced growth
Redtop (*Agrostis gigantea*)	Bluegrass, Kentucky (*Poa pratensis*) Timothy (*Phleum pratense*)	Growth reduced
Rice (*Oryza sativa*)	Barnyard grass (*Echinochloa crus-galli*)	Reduced growth
Rosemary (*Rosmarinus officinalis*)	Various annuals	Growth reduced
Rye, common (*Secale cereale*)	Clover, red (*Trifolium pratense*) Maize (*Zea mays*) Rye (*Secale cereale*) Tobacco (*Nicotiana tabacum*) Wheat (*Triticum* spp.) Wild oat (*Avena fatua*)	Reduced growth
Rye grass		
Italian (*Lolium multiflorum*)	Clover, red (*Trifolium pratense*)	Reduced growth
Perennial (*L. perenne*)	Apple (*Malus* spp.) Bent grass, colonial (*Agrostis capillaris*) Bluegrass, Kentucky (*Poa pratensis*) Timothy (*Phleum pratense*)	Yield and growth reduced
Sage, common (*Salvia officinalis*)	Coffee (*Coffea* spp.)	Reduced growth
Sesame (*Sesamum indicum*)	Cabbage (*Brassica oleracea* var. *capitata*)	Withering and death
Solomon's-seal, two-leaved (*Maianthemum canadense*)	Pea (*Pisum sativum*) Wheat (*Triticum* spp.)	Germination and growth inhibited
Sorghum (*Sorghum bicolor*)	Sesame (*Sesamum indicum*) Wheat (*Triticum* spp.)	Root leachate depresses growth
Soybean (*Glycine max*)	Pea (*Pisum sativum*) Wheat (*Triticum* spp.)	Germination and growth reduced
Spurge, leafy (*Euphorbia esula*)	Pea (*Pisum sativum*) Wheat (*Triticum* spp.)	Water extract of leaves inhibits germination and growth
Spurry, corn (*Spergula arvensis*)	Barley (*Hordeum vulgare*)	Growth slightly reduced
Squashes (*Cucurbita* spp.)	Pea (*Pisum sativum*) Wheat (*Triticum* spp.)	Germination and growth reduced
Sunflower		
Showy (*Helianthus laetiflorus*)	Sunflower, showy (*Helianthus laetiflorus*)	Stunting
Western (*H. occidentalis*)	Sunflower, western (*H. occidentalis*)	Toxin in the roots Stunting
Sweet-clover, white (*Melilotus alba*)	Maize (*Zea mays*)	Germination and growth inhibited
Sweet potato (*Ipomoea batatas*)	Cow pea (*Vigna unguiculata*) Kidney bean (*Phaseolus vulgaris*) Sweet potato (*Ipomoea batatas*)	Growth reduced
Thistle		
Canada (*Cirsium arvense*)	Alfalfa (*Medicago sativa*) Buckwheat (*Fagopyrum* spp.) Flax (*Linum usitatissimum*) Oat (*Avena sativa*) Wheat (*Triticum* spp.)	Germination and growth reduced

(*continued on next page*)

Appendix 4.2. (*continued*)

Toxic plant	Affected plants	Comments
Thistle (*continued*)		
Field (*C. discolor*)	Pea (*Pisum sativum*) Wheat (*Triticum* spp.)	Germination and growth reduced
Timothy, common (*Phleum pratense*)	Apple (*Malus* spp.) Barley (*Hordeum vulgare*) Radish (*Raphanus sativus*) Tobacco (*Nicotiana tabacum*) Wheat (*Triticum* spp.)	Growth reduced
Tobacco (*Nicotiana tabacum*)	Tobacco (*Nicotiana tabacum*)	Growth reduced
Tomato (*Lycopersicon esculentum*)	Pea (*Pisum sativum*) Peach (*Prunus persica*) Tomato (*Lycopersicon esculentum*) Wheat (*Triticum* spp.)	Germination and growth reduced
Tulip-tree (*Liriodendron tulipifera*)	Wheat (*Triticum* spp.)	Growth reduced
Velvet grass (*Holcus lanatus*)	Barley (*Hordeum vulgare*)	Growth depression
Walnut, black (*Juglans nigra*)	Many species (see Appendix 4.3)	Growth reduced, wilt
Wheat (*Triticum* spp.)	Barley (*Hordeum vulgare*) Pea (*Pisum sativum*) Wheat (*Triticum* spp.)	Germination and growth inhibited
Wheat grass, slender (*Agropyron trachycaulum*)	Wheat (*Triticum* spp.)	Growth reduced
Wormwood, common (*Artemisia absinthium*)	Fennel (*Foeniculum vulgare*) Lovage (*Levisticum officinale*)	Germination and growth reduced

[a] Adapted from Moje, W., 1966, Organic soil toxins, pages 533-751 in: Diagnostic Criteria for Plants and Soils, H. D. Chapman, ed., Quality Printing, Abilene, TX.

Appendix 4.3. Susceptibility of some plant species to juglone from roots of black walnut and related species

Scientific name	Common name	Sensitivity[a]	Scientific name	Common name	Sensitivity[a]
Acer spp.	Maples	−	*Petroselinum crispum*	Parsley	−
Agrostis gigantea	Redtop	−	*Phaseolus*		
Allium cepa	Onion	−	*lunatus*	Lima bean	+/−
Anemone spp.	Anemones	−	*vulgaris*	Common bean	−
Arisaema triphyllum	Jack-in-the-pulpit	−	*Philadelphus* spp.	Mock-orange	−
Asparagus officinalis	Common asparagus	+	*Phleum* spp.	Timothy	−
Athyrium felix-femina	Northern lady fern	−	*Phlox* spp.	Phlox	−
Beta vulgaris	Beet, garden and sugar	−	*Pinus*		
Brassica oleracea			*mugo*	Swiss mountain or mugo pine	+
var. *capitata*	Cabbage	+/−	*resinosa*	Red pine	+
Capsicum annuum	Peppers	+/−	*strobus*	Eastern white pine	+
Carya spp.	Hickories	−	*Pisum sativum*	Garden pea	++
Cercis spp.	Redbud	−	*Poa* spp.	Bluegrasses	−
Chrysanthemum spp.	Chrysanthemums	+	*Podophyllum peltatum*	May-apple	−
Crocus sativus	Saffron crocus	+/−	*Polygonatum* spp.	Solomon's-seal	−
Cyclamen persicum	Florist's cyclamen	−	*Polystichum* spp.	Christmas fern	−
Dactylis glomerata	Orchard grass	−	*Potentilla fruticosa*	Bush cinquefoil	+
Danthonia spicata	Poverty oat grass	+/−	*Primula* spp.	Primroses	−
Daucus carota subsp. *sativus*	Carrot	−	*Prunus*		
Epimedium spp.	Epimedium, barrenwort	−	*avium*	Sweet cherry	+
Erythronium albidum	White dog-tooth-violet	−	*cerasus*	Sour cherry	+
Euonymus spp.	Burning-bush	−	*persica*	Peach	+/−
Forsythia spp.	Forsythias	−	*Pyrus communis*	Common pear	+
Gentiana			*Quercus* spp.	Oaks	−
asclepiadea	Gentian	−	*Ranunculus ficaria*	Pilewort	−
septemfida	Gentian	−	*Rhododendron* spp.	Azaleas, rhododendrons	+
Helleborus spp.	Green hellebore	−	*Rhus radicans*	Poison-ivy	−
Heuchera spp.	Heuchera, alumroot	−	*Rosa* spp.	Wild or native roses	−
Hosta spp.	Plantain lilies	−	*Rubus* spp.	Blackberries, raspberries	+
Hydrangea macrophylla	House or French hydrangea	+	*idaeus*	Red raspberry	+/−
Impatiens spp.	Impatiens	−	*occidentalis*	Black raspberry	−
Iris spp.	Irises	−	*Rudbeckia* spp.	Rudbeckia	−
Juniperus virginiana	Eastern red-cedar	−	*Salvia* spp.	Sages	−
Kalmia latifolia	Mountain-laurel	+	*Solanum* spp.	Nightshades	−
Ligustrum spp.	Privets	+	*tuberosum*	Potato	++
Lilium spp.	Lilies	−	*Syringa* spp.	Lilacs	+/−
Lonicera spp.	Honeysuckles	−	*vulgaris*	Common lilac	+/−
Lycopersicon esculentum	Tomato	++	*Thalictrum* spp.	Meadow-rues	−
Magnolia spp.	Magnolias	+/−	*Tricyrtis hirta*	Toad lily	−
Malus spp.	Crabapples	+/−	*Trifolium repens*	White clover	−
sylvestris	Apple	+	*Trillium* spp.	Trilliums	−
Matteuccia struthiopteris	Ostrich fern	−	*Tulipa* spp.	Tulips	−
Medicago sativa	Alfalfa	++	*Uvularia* spp.	Bellwort, wild oats	−
Myosotis alpestris	Forget-me-not	−	*Vaccinium* sp.	Blueberry	+
Narcissus spp.	Narcissus	−	*Viburnum* sp.	Viburnum	+/−
Ophiopogon spp.	Lily-turf	−	*Vinca minor*	Common periwinkle	−
Paeonia officinalis	Peony	+/−	*Viola tricolor*	Pansy	−
Parthenocissus quinquefolia	Virginia creeper	−	*Vitis* spp.	Grapes	−
Pastinaca sativa	Parsnip	−	*Zea mays*	Maize or corn	−

[a] ++ = Very sensitive and often observed; + = sensitive; +/− = variable sensitivity; and − = not sensitive and growth may be enhanced.

CHAPTER 5

Diagnosing in the Clinic

Diagnosing in the clinic laboratory, especially during the initial phases, is an extension of field diagnosing (Chapters 3 and 4) during which plants can be examined more meticulously. Many of the procedures are the same.

After the sample has been received and properly checked into the clinic laboratory (see the following pages), the diagnostician must formulate one or more tentative diagnoses that will subsequently be verified or rejected by appropriate means discussed throughout this book. As stated in earlier chapters, tentative diagnoses and probable causes are arrived at by comparing the field signature and plant symptoms with descriptions in bulletins, compendia, host indexes, and books. During the preliminary laboratory examination, which usually includes routine soil pH and soluble salts determinations in addition to a detailed examination of plants, the tentative diagnosis may be modified. At this point, the diagnosis may be completed and the appropriate persons notified. In other cases, the diagnosis may be referred to an appropriate laboratory or specialist familiar with herbicide, air pollution, soil, and genetic- or insect-induced injuries. Plant clinics are expected to diagnose diseases caused by bacteria, fungi, nematodes, viruses, and related pathogens.

This chapter discusses preliminary and routine procedures used in clinic laboratories. Procedures used for diagnosing diseases caused by abiotic agents were discussed in Chapter 4.

As stated in Chapter 2, the plant disease clinic laboratory can diagnose diseases only on the specimens received and with the help of the field information enclosed with them. This is often a serious limitation for diagnosing a field problem because even properly gathered specimens and complete field information are often inadequate. However, in conjunction with a proper field investigation and tentative diagnosis provided by an observant consultant, county agent, or grower, the clinic diagnosis is very useful in determining the actual field problem. Although many plant diagnostic clinics provide control recommendations, they should be given only by qualified field personnel familiar with grower operations, quarantines, pesticide labels, and legal implications.

Handling Plants and Clinic Routine

The handling of specimens received by the clinic varies with the type of specimen. A system must be set up to prevent further deterioration of the specimens after they have been received and to expedite rapid examination, diagnosis, and reply. If, after a preliminary examination, it is apparent that a period of more than 3–5 days is necessary to complete the diagnosis, the grower should be notified right away that the specimen has been received in satisfactory condition and that the diagnosis will require additional time (in days or weeks) to complete. Some clinics use a pro forma postcard (Fig. 5.1) to notify the grower.

When specimens are received, every effort should be made to open the packages immediately and place the specimens in such a location as to prevent further deterioration and to prepare them for examination. If the clinic facility is overloaded, it may be necessary to refrigerate new specimens. This should be done immediately. Many specimens deteriorate rapidly when packaged because of compression, anaerobic conditions, desiccation, excessive moisture, or respiratory heat buildup (see Chapter 3).

The following system may be helpful in developing a suitable clinic routine for receiving and handling specimens:

1. Open the package, and check the field information for the grower's name and address; it may be present only on the exterior mailing package. Discard excess packaging material. Occasionally, it is necessary to repackage the specimens for easier handling in the clinic and to limit drying out.

2. Log in the sample by giving it a clinic number (optional), and attach the field information sheet to the specimens. Often, growers do not have the plant clinic form (Fig. 3.2 and Appendix to Chapter 2) and submit the field information in a letter or on a piece of paper. In these cases, the clinic form is filled out as completely as possible from whatever information is available.

3. Group the samples on counters or table tops according to

the person or specialist who will examine them. Alternatively, they can be grouped by crop type, urgency of diagnosis, or stability of the specimen. Samples for nematode and virus assays or insect, plant, weed, or other identification requests are sent to the appropriate person, and the grower is notified (Fig. 5.1).

4. Initiate standard and indicated clinic tests.
   a. Samples with soil
      (1) Determine the soil pH.
      (2) Determine the soil soluble salts.
      (3) Assay for nematodes.
      (4) Assay for pathogens (e.g., *Verticillium*).
   b. Foliage
      (1) Place in a moist chamber.
      (2) Examine under microscopes.
      (3) Assay for viruses.
      (4) Culture or other assay.
   c. Seedlings
      (1) Place in a moist chamber.
      (2) Examine under microscopes.
      (3) Culture or other assay.
   d. Fleshy fruits, vegetables, etc.
      (1) Place in a moist chamber.
      (2) Examine under microscopes.
      (3) Culture or other assay.
   e. Potted plants
      (1) Keep the soil moist or place in a large plastic bag.
      (2) Clip off appropriate parts and handle as in 4b and d.

5. If the specimens cannot be examined within a reasonable period of time (2 h), they should be packaged to prevent drying and refrigerated. Cold-sensitive plants, such as tuberous roots of sweet potatoes and many ornamentals, should be placed at 10–15°C (50–60°F). See Table 4.13 for a listing of temperature-sensitive plants.

Plant Examination and Techniques

The packages in which the specimens are shipped often function as moist chambers in which bacteria, fungi, insects, mites, and other organisms may continue to develop. When the packages are being opened and specimens removed, check carefully for mold, insects, webbing, and odors. Remove the plants carefully from the packages to preserve these signs and to avoid spilling soil on foliage or mechanically injuring the plants while laying them out on the table or counter.

After the plant is neatly laid out, it should be examined

```
                 PLANT DISEASE CLINIC
          (---Clinic address and phone number---)

Dear Grower:_________________________________, Date _______.

Your ______________________________ specimen was received:

___  in good condition. You should hear from us in about ___ days
     because ____________________________________________.

___  in good condition, however, to complete the diagnosis
     please send________________________________________.

___  but is inadequate because ________________________.
     Please send a fresh specimen.

___  without adequate field information.  Please send a fresh
     sample with complete field information.

___  in good condition and referred to ________________.


CC:_______________________  ____________________________
                                  Clinician and date
```

Fig. 5.1. Pro forma postcard used to notify the grower that there will be delays in completing the diagnosis, that additional specimens are required, that the specimen has been rejected, or that the specimen has been forwarded to an appropriate specialist.

systematically. A frequent error of inexperienced diagnosticians is examine symptomatic tissue immediately. They risk not seeing subtle, but important, symptoms and signs elsewhere on the plant. The overall quality and extent of root, stem, foliage, flower, and fruit development should be noted. See Plates 2D and P and 3S.

Roots

Roots are examined for small lesions, fruiting structures, necrosis of small rootlets, swellings, mold (fungal mycelium), and sclerotia. Under the dissecting microscope, root hairs, fungal strands (mycelium) and fruiting structures, small galls and swellings, and partially embedded parasitic nematodes can be seen. Often, it is desirable to gently wash the roots under tap water to remove adhered soil and to examine them again under the dissecting microscope. Small sections of roots can be placed in opened, plastic culture (petri) plates with water for better observation. Root galls and swellings should be carefully dissected to check for endoparasitic nematodes. Cross and longitudinal cuts should be made on the larger roots to observe discoloration in the cortex and vascular systems (Plates 1H; and 5I, J, and P).

Stems

Stems are examined superficially for rots, lesions, fruiting structures, galls, cracking, insect holes, webbing, frass, mycelium, sclerotia, and slimes. Succulent stems should be pinched in several areas to assess the firmness of underlying tissue and to detect the presence of large, hard sclerotia in the pith of rotted stems. Cross and longitudinal cuts should be made at the soil line and at regular intervals up the stem to observe the cortex, vascular system, and pith for discoloration, tissue soundness, and the presence of mycelium, sclerotia, insects (webbing, adult and immature forms, and pupal cases), and "clouds" of bacteria. See Plates 1A–C, L, N, and O; and 3B, D, and S.

Foliage

Foliage is examined on both surfaces for color, color patterns, and morphological aberrations such as galling, atypical venation and leaf margins, and unusual size and shape. The surfaces of leaves should be examined for spray deposits, webbing, leaf miner injury, insect eggs, insects, and mites. Symptomatic areas are examined carefully under the dissecting microscope for the same things and also for fruiting bodies of pathogenic fungi (Plates 2D, L–P, R, and S; 3A, E, F, L, and O–R; 4; 5F, L–O, and Q; and 6A–I, L, R, and S) or bacterial ooze.

Preparing clear-tape mounts. Clear-tape mounts are excellent for examining small objects on the surface of a leaf (the phylloplane). The mounts are prepared by pressing the sticky surface of clear (not frosted) plastic tape on the upper and lower surfaces of several leaf spots. A drop of water is placed on a microscope slide, and the tape is affixed to the slide. Usually it is best to overlap the tape 1 cm around the ends of the slide to ensure proper adhesion of the tape to a dry surface (Fig. 5.2). The optical properties of tape are inferior to glass coverslips, but this tape is usually satisfactory for immediate use at magnifications up to 400×. For very poor tapes, the mount can be viewed upside down at low power. Fungal spores, mycelium, trichomes, insect eggs, pollen grains, spray deposits, mites, and many other items can then be readily seen. This method is rapid and well suited for many fungi that cause leaf spots. Often, the relationships of the spores and sporophore and other fungal and host structures are preserved for viewing. Spores of pathogenic fungi can be verified by comparing them with illustrations in bulletins and texts. The method is also suitable for viewing the surfaces of stems, flowers, and fruits and fungal colonies in culture.

Extracting foliar and stem nematodes. If foliar or stem nematodes are suspected as causes of necrosis, 1-cm^2 sections of necrotic tissue are placed in open, plastic petri dishes with water sufficient to cover the bottoms of the dishes. The tissue is then teased apart into three to six segments. If nematodes are present, large numbers (more than 15) will soon be evident in the water through the dissecting microscope. This procedure is also used to detect the pinewood nematode in trunks of trees. In this case, slivers, increment borings, or small blocks of discolored wood are placed in the dishes, left overnight, and then examined.

Observing bacterial streaming (ooze). Bacterial leaf spots, wilts, blights, and rots can be diagnosed by observing bacterial streaming (ooze) from infected tissue through the microscope. Individual leaf spots are cut in half with a sharp razor blade, excised with surrounding healthy tissue, placed in a drop of water on a microscope slide, and covered with a coverslip (Fig. 5.3). If pathogenic bacteria are present, bacterial ooze will often emanate from the edge of the cut lesion. This procedure may fail if the specimen has dried out or if the bacteria are killed by the application of bactericides or as a result of competition from saprophytes in badly rotted tissues. The presence of a few bacteria scattered throughout necrotic tissue is quite common and should not be confused with bacterial ooze.

Similarly, bacterial ooze can be seen from cut vessels of thin sections from infected vascular tissue. In rotted tissue, bacterial ooze may not be as pronounced as in tissue with leaf spots and wilts. However, many bacteria will be present at the leading edge of the infection and in apparently healthy tissue.

The relationship of foliar symptoms to the cause of disease must be carefully assessed because many injuries and infections elsewhere on the plant are manifested on the foliage (Chapter 4). It is important to notice whether symptoms are on old or young leaves. Also notice whether leaves have fallen. Leaf drop is caused by the formation of the abscission layer induced by plant hormones often released as a result of injury or infection elsewhere on the plant. The plant may also have been exposed to growth regulators from outside sources such as herbicide drift and air pollution. The location and demarcation of discolored or necrotic tissue, in relationship to the leaf margins and venation, is also indicative of the causal agent and should be noted. For example, the demarcation between necrotic and healthy tissue is sharp for many leaf spots caused by insects, desiccants, bacteria, and some fungi but is generally gradual for many leaf spots caused by fungi, viruses, air pollutants, and nutritional problems.

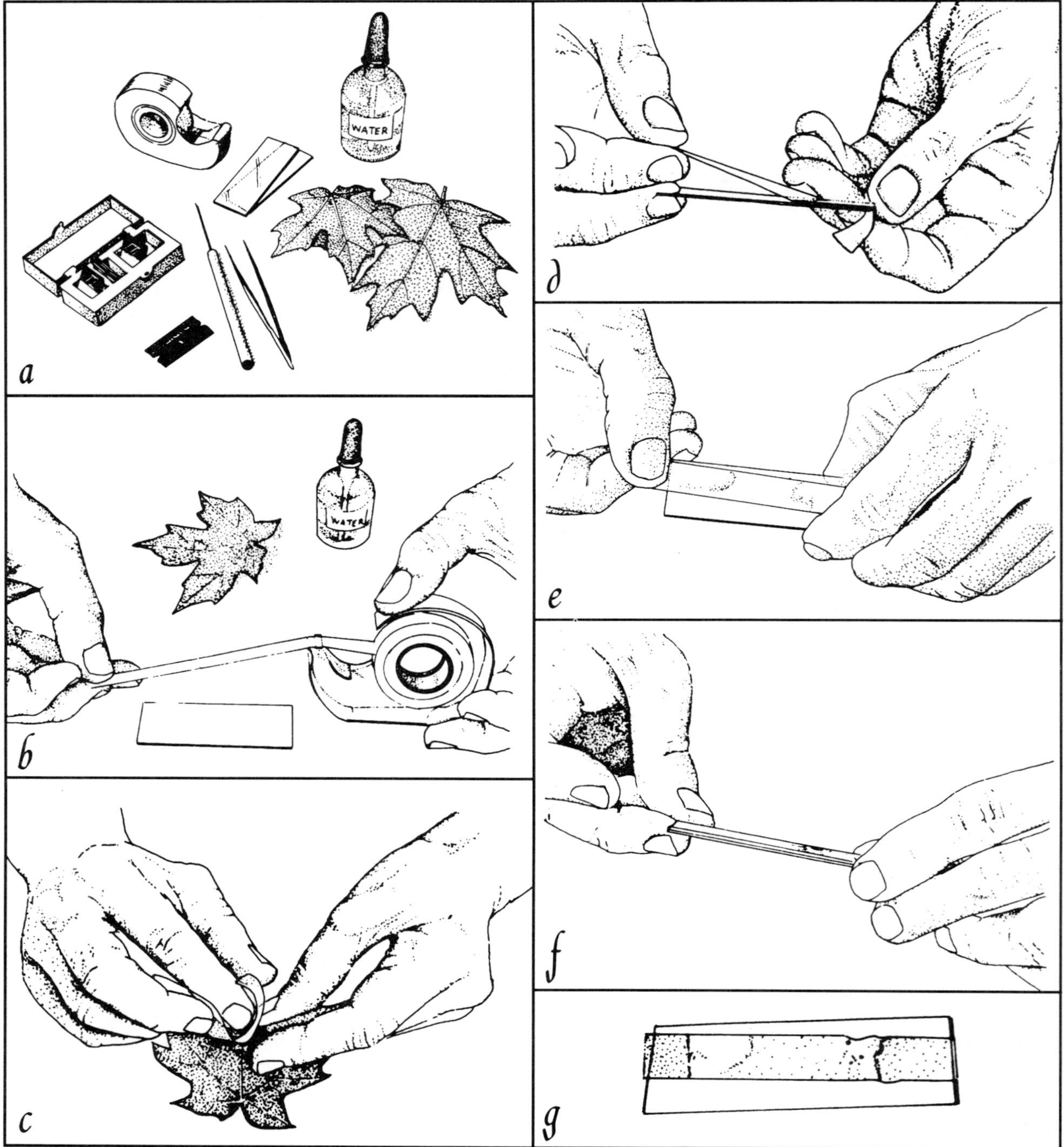

Fig. 5.2. Preparing a clear, sticky tape mount from a leaf spot for viewing through a compound microscope at 100× to 400×. **a,** Materials needed: transparent sticky tape, glass microscope slides, dropper bottle, leaf specimens, dissecting forceps, dissecting needle, razor blade, and coverslips; **b,** the tape is cut 1.3–1.5 times longer than the glass slide; **c,** the center of the tape is pressed against the upper and lower surfaces of one or more leaf spots; **d,** 1 cm of tape is stuck on the underside of the glass slide, and a drop of water or stain is placed on top of the slide; **e and f,** the tape is slowly and smoothly drawn across the slide. and the end is stuck to the opposite bottom edge of the slide; **g,** the mount is ready for viewing.

Flowers and Fruits

Flowers and fruit surfaces are examined in the same manner as the foliage. However, the interior of fruits also must be examined by making longitudinal and cross cuts through the midsections (Fig. 5.4). The cut surface of each section is examined for firmness, rot, discoloration, and seed development. Fruits (e.g., tomato) are often deformed if the seed set is poor because of a lack of pollination or because of temperature extremes during the fertilization process. It is important to check for signs of insect activity such as frass, pupal cases, webbing, and entry or exit points; often these holes are small and covered by remnants of the floral bracts. Fruit rots caused by soft rot bacteria and fungi often follow insect injuries (Plates 2L and M; 3N, P, and Q; 5C; and 6J, K, M, and N).

Enhancing the Presence of Fungi

Moist Chambers

In many cases, it is desirable to incubate portions of plants with discolored or necrotic tissue in a moist chamber to induce certain fungi (e.g., *Botrytis* spp., *Cylindrocladium crotalariae*, *Leptosphaerulina* spp., *Pythium* spp., *Rhizoctonia* spp., *Sclerotinia* [*Monilinia*] spp., and *Sclerotium rolfsii*) to form external mycelium, to sporulate, or to form sclerotia. The incubated plant tissues are observed daily for

fungal growth; woody tissues may require incubation for several days or weeks. Fungal growth is then examined through the dissecting microscope, or sticky tape mounts are prepared and examined through the compound microscope at 100× and 400× as previously described.

Moist chambers are easily prepared by placing moist (not wet) paper toweling with the specimen in a plastic bag (Fig. 5.5a and b). The plastic bag should *not* be tightly packed or sealed because an anaerobic atmosphere around the specimen promotes bacterial soft rot and should be avoided. Small moist chambers can be made by moistening round filter paper or appropriately cut paper toweling and placing in it culture (petri) dishes (Fig. 5.5a and b). These make excellent moist chambers for sections of leaves and small twigs and roots. Large petri dishes and refrigerator containers are available for larger specimens (Fig. 5.5a). Specimens in petri dish moist chambers can be viewed under the dissecting microscope without disturbing the specimen, which happens when moist chambers are made from plastic bags. Normally, moist chambers are incubated

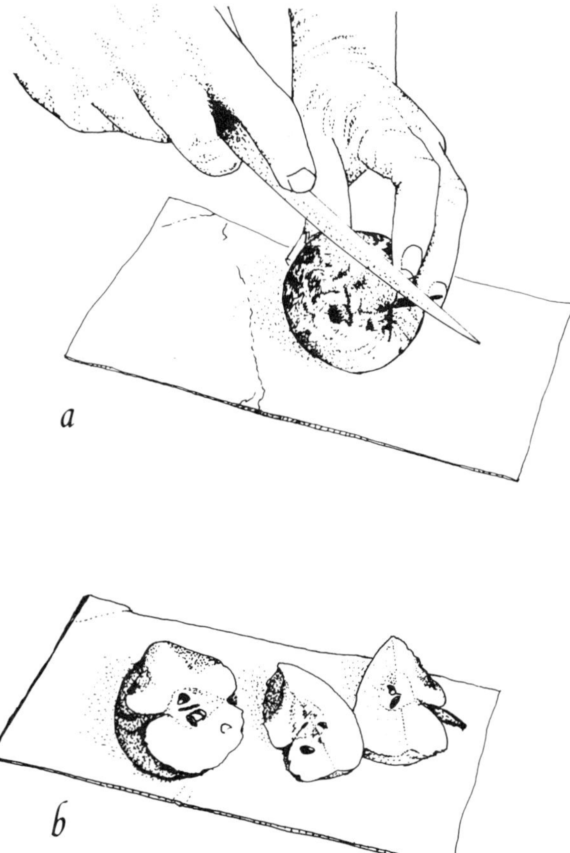

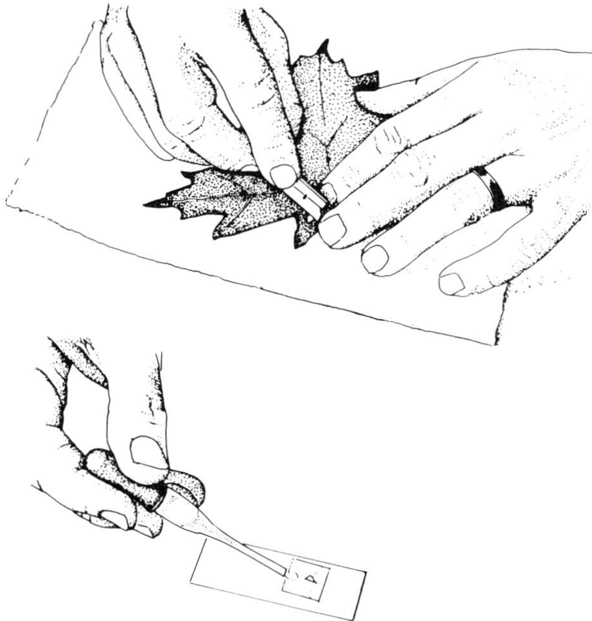

Fig. 5.3. Examining a leaf spot for bacterial ooze. Above, three straight cuts are made around the leaf spot, and the triangular specimen is freed from surrounding tissue. Below, the specimen is transferred to a drop of water. A microscope coverslip is carefully placed over the drop to avoid air bubbles, and the mount is ready for viewing.

Fig. 5.4. Cutting a fruit to examine the interior for depth of rots, seed set, and insect activity. Notice the use of a cardboard cutting board to protect the table surface. **a,** The initial cut is made longitudinally through the axis; **b,** the second cut is a cross section through the midsection of one of the halves. In this example, the seed set is satisfactory and the rot started on the surface of the apple.

Fig. 5.5. a, Moist chambers in which symptomatic plant parts are incubated for 2–10 days at room temperature to promote growth and sporulation of fungi (clockwise from upper right): refrigerator container, plastic bag, petri dish, large petri dish, and specimen container. Notice that the contents of the moist chambers have been identified and dated. **b,** Moistening paper toweling to prepare a plastic bag moist chamber for tomato foliage with leaf spots. **c,** Petri dish with moist paper toweling and small stems and leaves. **d,** Feeder roots of a wilted holly bush placed on carrot disks and set in a petri dish moist chamber. The fungus *Thielaviopsis basicola* will grow and sporulate if the disease is black rot. **e,** Cross sections of potato stems with Verticillium wilt symptoms in a water agar petri dish moist chamber. Moist paper toweling may be used instead of the agar. The fungus will grow, sporulate, and possibly form small, black sclerotia if it is the cause of the wilt.

at room temperature with diurnal lighting. If plants have been sprayed with a fungicide, fungal growth and sporulation may be inhibited. Usually, it is desirable to wash tissues, especially roots and other tissues covered with soil, in gently running tap water prior to setting up moist chambers. With larger tissues or tissues with internal infection, it may be desirable to sterilize the surfaces by immersing the specimens in a 5.25% sodium hypochlorite bleach (e.g., Clorox), diluted 1:5 with tap water, for a few seconds to a minute or longer and then rinsing in sterile water.

A modification of the moist chamber technique is well suited for enhancing the presence of some species of *Phytophthora* and *Pythium* in infected tissue for direct microscopic observation. Small (e.g., 2- × 3-mm) sections are cut from the leading edge of infected tissue, washed in distilled or deionized water, briefly surface sterilized in a fresh 1:20 dilution of liquid bleach, and placed in petri dishes with distilled or deionized water. Many of these water molds will grow into the water and often form chlamydospores or sporangia and release zoospores overnight.

Baiting

Another method commonly used to promote growth of fungi in infected tissues is baiting. For example, roots infected with the black root rot fungus (*Thielaviopsis basicola*) are washed, surface sterilized for 5 min in 70% alcohol, minced, and incubated on carrot slices in a petri dish moist chamber. The fungus readily grows and sporulates on the carrot slices (Fig. 5.5d).

Similarly, apple fruits are used to promote growth of *Phytophthora* spp. from infected roots. Surface-sterilized and washed roots are inserted into the flesh of the apple and covered with tape. In a few days, the fungus will grow into the apple tissue and can be readily observed in thin sections under the compound microscope. Potato tubers have been used to promote the growth of the late blight fungus (*Phytophthora infestans*) in a similar manner; the fungus will sporulate in a moist chamber and can be seen with the dissecting microscope.

Care and Use of Microscopes

The basic components of professional light microscopes have changed little in 75 years. In sequence, the components are light source, field condenser, field diaphragm or stops, filters, substage diaphragm, substage condenser, specimen, objective lens, and eyepiece (Fig. 5.6). Proper alignment of these components and cleanliness of lenses are essential for good resolution. The sequence and means for making these adjustments are given in the instruction booklet supplied with the instrument and should be done monthly.

Instructions supplied with the instrument for adjusting, caring for, and cleaning the microscope should be carefully followed. In clinic laboratories where microscopes are exposed to considerable dust and grit and used by many persons, cleaning and adjusting should be done weekly or at least monthly. Removal of dust and grit from the microscope stage and lenses is done at least daily. There are definite "dos and don'ts" for the care of both dissecting and compound microscopes:

1. Never wipe a dry lens with anything. Exposed surfaces of lenses in eyepieces and, to a lesser extent, objective and condenser lenses become dirty with dust and oily materials that interfere with adequate viewing. To clean the lens surfaces, dust with a camel's-hair brush and an aspirator (an ear syringe works well) or use a new Q-tip that has been dipped in alcohol. Then very gently wipe the lens with lens paper that has been moistened with lens-cleaning fluid. If these are not available, a soft, clean, lint-free cloth moistened with distilled water may be used (do *not* use saliva). Lens paper or cloth must be stored in envelopes to avoid contamination with dust. The use of lens paper or cloth impregnated with fine clay particles will ruin a lens.
2. The microscopes should be covered when not in use. Plastic bags or other airtight covers should not be used in areas of high humidity because of the risk of condensation on the lenses. Molds may grow on surfaces and between cemented lenses when temperature and relative humidity are higher than 26.7°C (80°F) and 80%, respectively. In areas of high humidity, microscopes should be stored with a small heat source such as a 25W light bulb or a noncorrosive desiccant.
3. Immediately remove dirt, grit, and liquids that fall on microscope surfaces, as described above.
4. Never use the stage platform of the dissecting microscope as a cutting or dissecting surface. This should be done on heavy cardboard, composition board, or wooden slats or in a plastic petri dish. The use of these cutting boards will prolong the usefulness of the microscope stage platform and hand instruments.

Routine Use of a Microscope

For routine observations, the following steps usually suffice for adjusting a compound microscope:

1. Adjust the light intensity of the bulb for a white color. High settings result in short bulb life, but resolution is higher.
2. Raise the substage condenser to the highest position. If the microscope is properly adjusted, the substage condenser lens will not protrude above the floor of the stage platform. (*Never* permit the substage condenser lens to protrude above the floor of the stage platform because it will be scratched by the glass slide.)
3. Turn the turret holding the objective lenses so that the 10× lens (or lower power) is in position. Place the glass slide with the specimen and coverslip on the platform,

and reduce the light intensity by closing the substage condenser. Next, turn the coarse focusing knob to the lowest setting while observing the slide and objective lens to make sure that the lens does *not* touch the coverslip on the slide. Now look through the microscope. Focus the microscope by slowing turning the coarse focusing knob so that the objective lens moves away from the glass slide. With most microscopes, all objective lenses will now be in focus, and the higher power lenses (usually 44× and 100×) will not touch the coverslip. This is not always true and is not the case if by chance the 10× objective was focused on the lower portion of a thick specimen or an unusually thick coverslip was used.

4. Open the substage diaphragm to increase brightness just to the point that objects in the field of view begin to white out. The substage diaphragm is repeatedly adjusted between this setting and the nearly closed position to obtain the best viewing. Smaller openings increase depth of field, but the resolution is decreased.

5. Adjust the field diaphragm while looking at the specimen so that the illuminated circle on the mount and the field of view are the same size. This must be done each time the objective lens is changed. If the circle of illumination is larger than the field of view, resolution deteriorates because of light scattering.

6. When viewing a specimen, first systematically examine the entire mount from the top left to the lower right with the low power objective. Interesting areas are then reexamined under the low power objective and often under the 40× or higher objective. While viewing through the 40× objective, continuously change the focal plane by turning the fine focusing knob with the left hand. In this manner, the third dimension of objects can be appreciated. The focal planes of the 40× and especially the 100× objectives are very narrow, and only a small segment of the specimen can be seen at one focus setting. When a higher magnification objective is used, it is necessary to increase the light intensity with the substage diaphragm and to reduce the diameter of the field of illumination with the field diaphragm. These adjustments are reversed when changing to lower power objectives.

7. During the day, when the microscope is used frequently

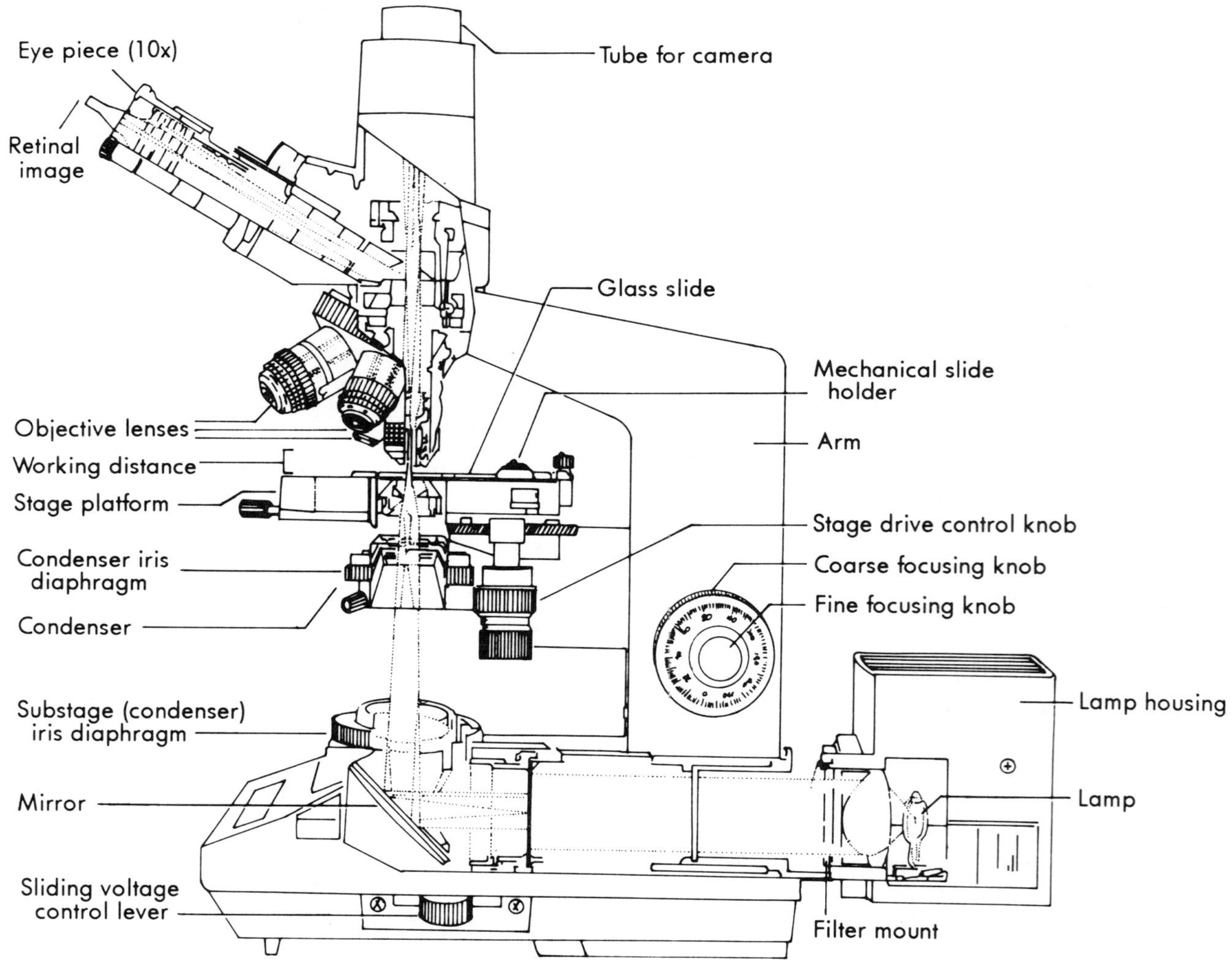

Fig. 5.6. Basic compound microscope used in many plant diagnostic clinics. Research binocular microscopes with phase contrast, dark field ultraviolet fluorescence, and other features are desirable in most clinics.

but intermittently, it is best to leave the illuminator on at a low-voltage setting. The life of the bulb is shortened if it is left on at a high setting or if it is frequently turned on and off.

Kohler Illumination

Kohler illumination is necessary to obtain the maximum resolving power of research microscopes. On microscopes with built-in illuminators, some of the adjustments are easily made. For a basic microscope equipped with a mirror and an external illuminator, the procedure is as follows:

1. Focus and center the illuminator condenser lens on the leaves of the substage diaphragm. A small, auxiliary, hand-held mirror is helpful in viewing the substage diaphragm.
2. Focus the microscope on a specimen.
3. Focus the substage condenser of the microscope until the diaphragm of the lamp (field diaphragm or stop) is in sharp focus.
4. Make sure the light from the illuminator is centered on the substage mirror.
5. Adjust the field diaphragm until only the field of view, as seen through the microscope, is illuminated.
6. Remove the microscope eyepiece and adjust the substage diaphragm until the back of the objective lens is just filled with light. The image of the lamp can be seen on the back of the objective lens.
7. The lighting intensity is adjusted for ease of viewing with neutral, clear filters at the light source. It may also be adjusted by regulating the voltage to the lamp. Light intensity is not adjusted with the substage diaphragm.

Special Effects

Special effects can easily be achieved with most research microscopes, and they are useful in some situations. Although they are normally not used in plant clinics, their use should be encouraged.

1. Birefringent objects such as cellulose, starch grains, raphides, fibrosin bodies, and some viral inclusion bodies are brilliantly illuminated against a dark background in polarized light and can be readily identified in fresh mounts. A polarizing filter is placed under the substage condenser and another on top of the eyepiece. One filter is slowly rotated until the background field becomes dark. Polarizing filters can be cut from polarizing sunglasses.
2. Very small objects, such as flagella and bacteria, can be seen in dark fields in the same manner that dust particles can be seen in a ray of sunshine in a dark room. Dark fields are made by placing an opaque disk of proper diameter in the rack directly under the condensing lens. Immersion oil is placed between the top of the condensing lens and the bottom of the glass slide. The substage and field diaphragms are then fully opened. Objects appear brilliant against a black background.
3. Colored specimens often can be seen better in colored light. Filters of complementary colors increase contrast (colors opposite on the color wheel; for primary and secondary colors they are yellow/violet, green/red, and blue/orange). Complementary colors can be approximated by staring at a colored square (2 cm^2) on a white sheet of paper for 30 s and then at a dot nearby on the paper. A light tint of the complementary hue will mysteriously appear as a square around the dot. Filters of nearly the same color as the specimen reveal detail. The human eye is most sensitive to green light (0.55 μm), and some microscopists find this color most agreeable. Many microscope lenses minimize color aberrations best at this wavelength. The resolving power of the microscope is enhanced with light of shorter wavelengths such as blue and violet. These light-quality adjustments are made by placing appropriately colored, clear filters in the rack under the condensing lens or at the light source.

In a plant disease clinic, cultures of fungi growing in a clear medium are often viewed directly in culture by removing the culture dish cover, placing a coverslip on the surface of the colony, and then placing the dish on the platform. The field and substage condenser diaphragms must be fully opened and the substage condenser fully lowered.

When the microscope is being focused on objects in a culture dish, great care must be taken to limit observations to the surface or topmost areas in the medium; otherwise the objective lens will be rammed through the glass coverslip. Plastic coverslips are recommended for these mounts. It is possible to view the culture from the underside by placing the petri dish upside down on the stage platform. However, only the 10× (or lower) objective lens may be used. The working distance between the high-power objective lenses and the specimen is only 0.73 and 0.12 mm for most 44× and 97× objective lenses, respectively. The thickness of the glass of the culture dish is much greater.

Fungi growing out of small tissues in petri dish moist chambers can be viewed with the 10× (or lower) objective lens without the use of coverslips. Focus carefully to avoid pressing the objective lens into the specimen. This procedure is used to identify *Verticillium* and *Fusarium* spp. growing out of thin cross sections of plant stems with wilt.

Injury Caused by Soil pH and Soluble Salts

The pH and soluble salts level should be routinely determined on soil accompanying plant specimens received in the clinic. These can be determined rapidly with relatively inexpensive instruments by technicians with minimal training. The information is invaluable in assessing causes of damping-off, poor growth, stem dieback, root necrosis, root pruning, foliar chlorosis and necrosis, and premature leaf drop, especially in cases where pathogens are not apparent. Salt injury is generally more severe during hot, dry weather. In many cases, diseases induced by soilborne pathogens are more severe when soil pH and soluble salts are outside the normal ranges. The percentage of diagnoses in clinics based on these determinations makes the routine effort worthwhile (see Chapter 1).

Selecting the Soil Sample

Selection of soil to sample is critical to diagnosing pH and soluble salts problems; therefore, good judgment and techniques must be used. Usually, these readings should be taken from both soils with symptomatic and normal-appearing plants and compared.

Soil samples should be taken where most of the feeder roots are located. This usually excludes the top 1–2 cm of soil and that between the rows. Determine where and when fertilizer was placed in relationship to the area sampled when interpreting pH and soluble salts readings. Fertilizer is often banded or placed on one side of the plant, and readings from soil samples taken near the band may not be representative: pH may be depressed by as much as 0.5–1.0 unit in soils with unusually high (>50) soluble salts. Readings on samples from the other side may be in the normal range.

Recent weather conditions, rainfall, drought, and irrigation activities should be noted because the effects of soluble salts are greatly influenced by soil moisture. Soluble salts also may migrate with water movement and become concentrated where moisture is lost by evaporation or transpiration.

Preparing a Soil Extract

1. Mix the soil sample well; remove rocks and plant residues.
2. Take a minimum of 50 cc of soil (measured after tapping three times to settle the soil).
3. Pour the 50 cc of soil into 100 ml of distilled or deionized water (1:2 ratio). Do *not* pour the water into the soil container and fill to the 150-ml mark; the soluble salts readings will be lower because of a higher dilution.
4. Stir for 1 min and again 30 min later. Let the soil settle to the bottom. (Readings taken after 2 min may be reasonably useful but will not be as accurate.)
5. Strain the supernatant (the liquid above the settled soil) into a clean container. A fine-mesh tea strainer (approximately 18/cm) or cheesecloth may be used to strain the solution. Avoid pouring the silt or sand accumulated at the bottom.

This solution is used for both the pH and soluble salts determinations.

An inherent discrepancy in the results obtained by this method is caused by the variable soil moisture content in field samples. This can be avoided by air drying the soil prior to preparing the soil extract. Otherwise, for wet soils, the 1:2 soil-to-water ratio should be changed to 1:1.80 for sandy soils and 1:1.50 (or lower) for clay, organic soils, and artificial soil media with high water-holding capacities.

Determining pH

The procedure outlined below is for the LaMotte model 1906 instrument. With other instruments (e.g., the Fisher model 600), procedures may differ slightly.

1. Turn the instrument on; check the battery; and then turn the instrument off.
2. Remove the protective cap from the electrode.
3. Decontaminate the electrode by shaking it to remove adhered water, and then wash it with a stream of distilled or deionized water from the dispensing wash bottle. Remove excess water with a clean paper towel or tissue. Rinse again.
4. Place the electrode in a pH 7.0 buffer. Set the temperature dial at the temperature of the solution, usually 25°C. Calibrate the instrument to read 7.0 (or the same as the standard buffer if a different buffer is used).
5. Remove the electrode and rinse (see step 3).
6. Place the electrode in a fresh pH 4.0 buffer to check the instrument at a lower reading. If it is not accurate, follow the procedures of the manufacturer to adjust the instrument internally. If readings are sluggish, the electrode probably needs to be replaced; in some cases the electrode tip can be cleaned by following the manufacturer's instructions.

   If fresh, commercially prepared buffers are not available, solutions of approximate pH levels can be prepared from grocery store items to roughly calibrate or check the pH meter: white vinegar (distilled, 5% acetic acid) = pH 2.9; a saturated solution of baking soda (sodium bicarbonate) = pH 8.0; a mixture of 70 ml of white vinegar and 30 ml of the saturated baking soda solution = pH 5.1; and black coffee = pH 5.1 ± 0.1. General Appendix J lists a number of common commercial buffers and others that can easily be prepared.
7. Remove the electrode, rinse it as in step 3, and place it in the deionized or distilled water bottle provided with the instrument. The instrument is now ready for determining the pH of soil extracts.
8. Rinse the electrode once between samples (see step 3).
9. Store the rinsed electrode in distilled water, or cover it with the moistened cap provided with the instrument.
10. The instrument should be turned off when not in use because the battery drains rapidly.
11. Precautions:

    a. Always use a fresh buffer standard (especially a pH 4.0 buffer).

b. Never store a dirty or contaminated electrode.

c. Never let the electrode dry out. It is best to keep it in distilled water or in a 1:1 solution of pH 7.0 buffer and 0.1 M KCl.

d. Never use the electrode as a stirring rod.

e. Never push the electrode into soil sediment.

When interpreting pH readings, consider the following factors:

1. pH readings may be temporarily depressed in soils with high levels of soluble, neutral salts, such as soils recently fertilized. The magnitude of the error may be as much as one pH unit but is usually 0.25–0.50 unit. The depression resulting from pH-induced neutral salts can be avoided by using a 0.01 M $CaCl_2$ solution for preparing the soil extract.

2. Symptoms caused by nutritional problems are usually induced as soil pH increasingly deviates from a "reasonable" range of about 5.3 to 6.5 or 7.0 for many plants because the availability of some nutrients is pH dependent (Fig. 5.7, Chapter 4, and Plate 6A–Q). Because soils in different areas often vary significantly in mineral content, the clinician is urged to consult with an experienced soil scientist to establish the tolerance range for each soil type and plant species. For example, plants growing in Piedmont soils in the Carolinas practically never exhibit manganese deficiency (Plate 6F–H), even at a high soil pH, because the soils are very high in manganese. Appendix 5.1 lists a number of plant species that tolerate extreme soil pH levels.

3. Plants may develop toxicity symptoms in many soils at a pH below 5.5, depending on the soil type. For example, many soils in the Piedmont and Coastal Plain in the Carolinas are high in aluminum. At pH 5.5 and below, aluminum toxicity causes root necrosis and root pruning, especially on the feeder roots, of many plants. Similarly,

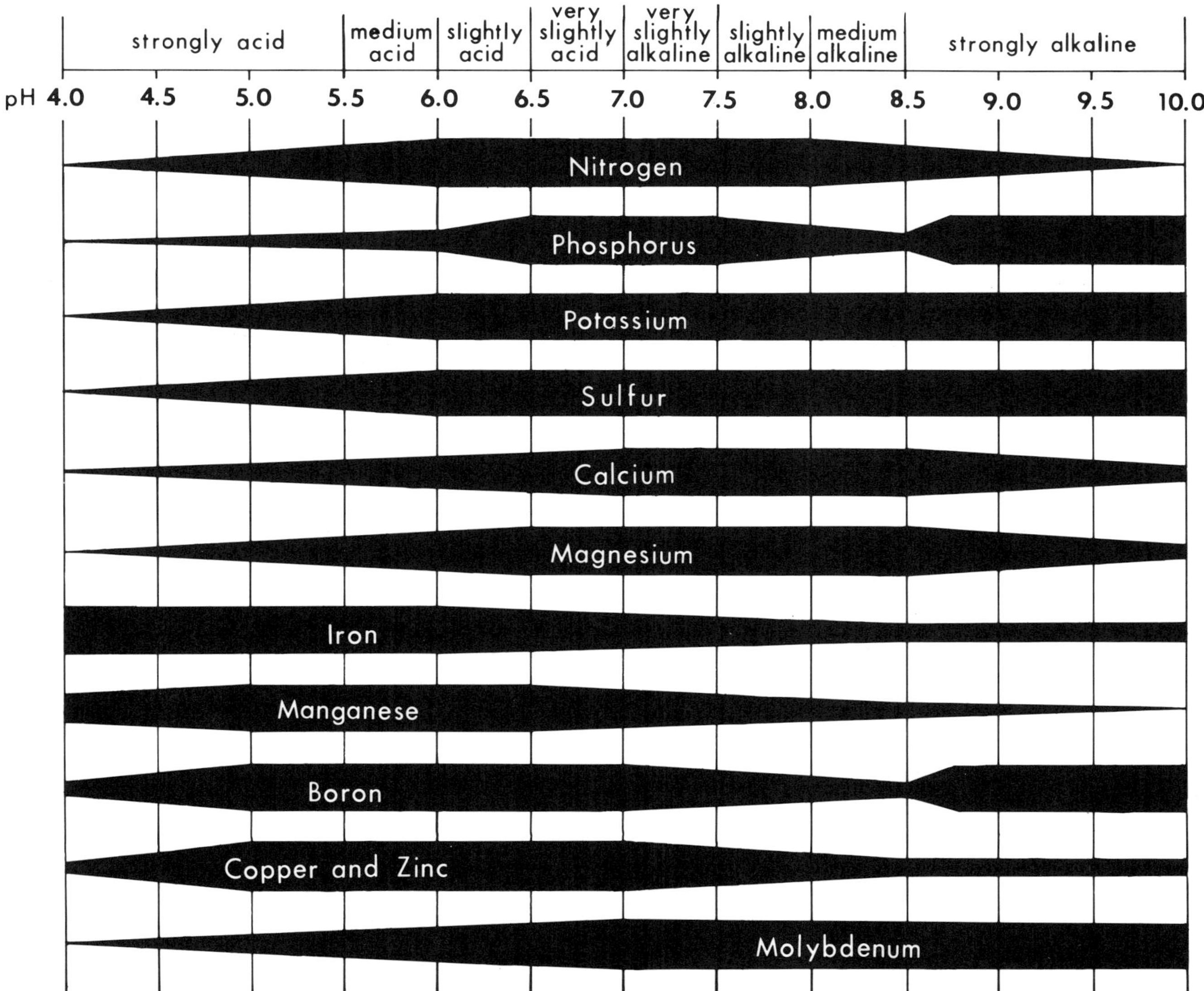

Fig. 5.7. Relationship between soil pH and availability of soil nutrients. (Redrawn from Troug, E., 1948, Lime in relationship to availability of plant nutrients, Soil Sci. 65:1-7)

Table 5.1. Soilborne diseases and soil pH favorable for disease development

Host	Disease	Pathogen or symbiont	pH[a]
Many crops	Black root rot	*Thielaviopsis basicola*	>6.2
	Cotton or Texas root rot	*Phymatotrichopsis omnivora*	>7.3
	Damping-off, root rot	*Pythium* spp.	<5.5
	Fusarium wilt, crown rot	*Fusarium* spp.	<6.0
	Root rot	*Phytophthora cinnamomi*	<6.0
	Sclerotinia blight	*Sclerotinia* spp.	>6.0
	Southern blight	*Sclerotium rolfsii*	<6.5
	Verticillium wilt	*Verticillium* spp.	>6.0
Crucifers	Clubroot	*Plasmodiophora brassicae*	<6.2–7.0
Legumes	Poor nodulation	*Rhizobium* spp.	<4.8
Potato	Common scab	*Streptomyces scabies*	>5.1
Sweet potato	Root rot (pox)	*Streptomyces ipomoea*	>5.2
Tobacco	Black shank	*Phytophthora nicotianae*	>5.5
Wheat	Take-all	*Gaeumannomyces graminis*	>6.5

[a] Beyond these approximate pH values, disease usually becomes increasingly more severe. See Engelhard, A. W., ed., 1989, Soilborne Plant Pathogens: Management of Diseases with Macro- and Microelements, American Phytopathological Society, St. Paul, MN.

manganese toxicity (Plate 6I) is a problem in certain soils in the Piedmont of Alabama, the Carolinas, Georgia, and Virginia and in the Louisiana delta region.

4. Aluminum and manganese toxicities are common in tropical oxisols and ultisols at a soil pH below 5.5. The problem is eliminated by liming, but a "chemical hardpan" remains below the tillage depth to which the lime is incorporated. Plants growing in these soils have shallow root systems and are subject to drought stress, even when ample moisture is available in the deeper soil horizons.

Many other examples undoubtedly exist. The clinician is urged to consult with local soil scientists and become familiar with soil pH problems in the area.

The soil pH can have a significant influence on some soilborne diseases (Table 5.1). In general, pH extremes and other soil stresses predispose plants to root rots caused by species of *Colletotrichum, Fusarium, Pythium,* and *Rhizoctonia* and other fungi.

Determining Soluble Salts (SS)

The metric (SI) unit for electrical conductance (ECe) is the siemens per meter (Sm^{-1}) and is equal to 1 mho per meter (mho m^{-1}). On the average, 1 mmho cm^{-1} is equal to 640 ppm or approximately one teaspoon of table salt in 2.5 gallons of distilled water. Ten large drops (0.5 ml) of a saturated solution of table salt (NaCl) at room temperature in a measured cup (236.6 ml) of deionized water will give an approximate reading of 125 SS.

The "upper safe value" for plant growth is arbitrary. In many cases, the value chosen represents the level at which growth or yield is reduced 50%. However, symptoms may appear at much lower salts levels, especially in young seedlings and new transplants, if growing conditions are unusually dry or toxic ions are present.

The relationship of soil soluble salts values, based on the preferred electrical conductivity (Sm^{-1}) method, to values obtained by the method usually used in plant disease clinics

of one volume of field soil to two volumes of distilled water varies depending on salt-saturated soil water, initial soil moisture, sample volume, bulk density, water-holding capacity, and soil or growing medium composition. An approximate conversion factor is 7.0. ECe values reported in the literature and derived soluble salts values are given in Appendix 5.2.

The procedure outlined below is for the LaMotte Dissolved Salts Meter (model 1918), which reads μmho units (10^{-6} mho); procedures may differ slightly with other instruments (e.g., Beckman model Sd-B-15).

1. Turn the instrument on, check the battery, and adjust the calibration if needed according to instructions provided with the instrument. Turn the instrument off. The standard is prepared by dissolving 0.742 g of KCl in 1,000 ml of water to give a reading of 140 SS (1,400 μmho). Deionized or distilled water should have a reading of 0.0 but typically reads 0.4 SS (4 μmho).

2. Wash the sensing element by directing a stream of distilled or deionized water on the outer surface and directly into the two side ports and bottom ports; shake to remove excess water; and rinse again.

3. Place the sensing element in distilled or deionized water.

4. Set the range knob to the highest setting (1–10,000), and then place the sensing element in the filtered soil extract deeply enough to cover the two upper ports. Tap the submerged sensing element at an angle to dislodge trapped air bubbles from the upper ports.

5. Turn the instrument on. Turn the range knob stepwise from 10,000 to lower settings so that the needle reads somewhere in the middle 75% of the arc. Make sure the sensing element is not touching the bottom or sides of the container and that the upper ports are submerged. Read the instrument, and record the data.

6. Remove the sensing element, wash as described in step 2, and store as described in step 3.

7. The total soluble salts determination is based on electrical conductivity (ECe) measurements. Terms in use include millimhos (mmho), micromhos (μmho), milli-

siemens (mS), and microsiemens (μS). The following relationships may be helpful:

10^{-3} mho
  = 1.0 mmho
  = 1.0 mS
  = 640–700 ppm
10^{-5} mho
  = 1.0 SS index
  = 6.4–7.00 ppm (one teaspoon of table salt in 2.5 gallons of water)
10^{-6} mho
  = 1.0 μmho
  = 1.0 μS
  = 0.640–0.700 ppm

To determine soluble salts (= 10^{-5} mho or 10^{-5} S), multiply the instrument value (in μmho or μS) by 0.10.

Some instruments read millimhos or millisiemens, and these values are multiplied by 100 to obtain soluble salts values.

Table 5.2. Salt index of common fertilizers[a]

Material and analysis, %	Salt index[b]	Partial salt index[c]
Ammonia, anhydrous, 82.2	47.1	0.572
Ammonium nitrate, 35.0	104.7	2.990
Ammonium nitrate-lime (Cal-Nitro), 20.5	61.1	2.982
Ammonium sulfate, 21.2	69.0	3.253
Calcium carbonate (limestone), 56.0	4.7	0.083
Calcium nitrate, 11.9	52.5	4.409
Calcium sulfate (gypsum), 32.6	8.1	0.247
Cottonseed meal, 6.9	3.5	0.501
Diammonium phosphate, 53.8	34.2	1.614,[d] 0.637[e]
Dolomite (calcium and magnesium carbonates), 20.0	0.8	0.042
Epsom salts, 16.4	44.0	2.687
Kainit, 12.5	105.9	8.475
Manure salts, 20.0	112.7	5.636
Monoammonium phosphate, 61.7	29.9	2.453,[d] 0.485[e]
Monocalcium phosphate, 56.3	15.4	0.274
Nitrogen solution, 37.0	77.8	2.104
Organic ammoniates, 5.0	3.5	0.702
Potassium chloride, 50.0	109.4	2.189
Potassium chloride, 60.0	116.3	1.936
Potassium nitrate, 46.6	73.6	5.336,[d] 1.580[f]
Potassium sulfate, 54.0	46.1	0.853
Sodium chloride, 53.0	153.8	2.899
Sodium nitrate, 16.5	100.0	6.060
Sulfate of potash-magnesia, 21.9	43.2	1.971
Superphosphate, 20.0	7.8	0.390
Superphosphate, 45.0	10.1	0.224
Urea, 46.6	75.4	1.618

[a] Adapted from Rader, L. F., White, L. M., and Whittaker, C. W., 1943, The salt index—A measure of the effect of fertilizers on the concentration of the soil solution, Soil Sci. 55:201-218.
[b] Increase in osmotic pressure of the soil solution produced by a material incorporated into the soil over that produced by the same weight of sodium nitrate × 100. Sodium nitrate is the reference compound and is assigned a value of 100. Materials with high salt indexes are more apt to cause salt injury than those with low indexes.
[c] Salt index per unit (20 lb) of the plant nutrient supplied. Two figures indicate two nutrients in the fertilizer.
[d] Nitrogen.
[e] P_2O_5.
[f] K_2O.

Several factors must be taken into account when interpreting soluble salts readings in connection with possible plant injury:

1. Recent weather patterns and irrigation. Plants can tolerate much higher levels of soluble salts if the soil is continuously wet and the transpiration rate is low (under cool, cloudy conditions).
2. The amount, kind, and placement of fertilizer. Fertilizers differ greatly in salts values. For example, sodium nitrate has a salt index of 100 and a ratio of 6.060 per unit of plant food, while urea has a salt index of 75.4 and a ratio of 1.618 per unit of plant food (Table 5.2). In soils with salt problems, needed nutrients should be supplied from fertilizers with low partial salt indexes per unit of plant nutrient (Table 5.2).
3. Soil type. Plants can tolerate high levels of soluble salts better in soils with a high clay or organic matter content because of the higher water-holding capacities (Table 5.3) than in sandy soils with low water-holding capacities.
4. Migration of dissolved fertilizers and salts in the soil water. During dry weather, salts tend to move upward and concentrate on the soil surface. During wet weather, salts tend to move downward in the leaching water. Root and plant damage may occur if the salt concentration becomes high, even temporarily, in the root zone.
5. The presence of toxic ions such as boron (Plate 6Q).
6. Plant species and, in some cases, the variety or cultivar (Table 5.4 and Appendix 5.2). Seed germination problems may not be closely related to the values given in the tables.

The general procedure used by many plant clinics is to look for extreme soil soluble salts values. Readings below 14 SS usually indicate an actual or developing nutritional

Table 5.3. Relationship between soil type or growing medium and irrigation water hazards and the upper safe limits of soluble salts readings for susceptible plants

	Soluble salts upper safe value (10^{-5} mho/cm)[a]	Comments
Soil type[b]		
Sandy	20–40	. . .
Loamy	30–50	. . .
Clay	40–60	. . .
Organic	40–80	. . .
Greenhouse potting medium	100–250	. . .
Hydroponic solution	187–300	Direct reading of solution
Irrigation water hazard		
Low (excellent)	10–25	Direct reading of
Medium (good)	25–75	water; to obtain
High (usable)	75–225	ppm multiply by
Very high (doubtful)	225–300	7 (approximate)
Not usable	>300	

[a] Many soluble salts conductivity meters read electrical conductivity in millimhos/cm (10^{-3} mho or mmho/cm); others read micromhos/cm (10^{-6} mho or μmho/cm). Thus, multiply mmho readings by 100 and μmho by 0.10 to obtain soluble salts values.
[b] Readings taken from extracts of one volume of air-dried field soil to two volumes of distilled water.

Table 5.4. Effects of soil soluble salts on most crops

Upper safe value of conductivity		Plant response	
Soluble salts $(10^{-5}$ mho)[a]	ECe mmho[b]	Class	Comment
0–14	0–2.0	None	Negligible effect: very low fertility probable
14–32	2.0–4.6	VS	Very sensitive: plants may be injured
32–66	4.6–9.4	S	Sensitive: plants may be injured
66–105	9.4–15.0	MT	Moderately tolerant: plants may be injured
105–147	15.0–21.0	T	Tolerant: plants may be injured
above 147	above 21.0	U	Unsuitable for crops, especially when soil is dry

[a] Many soluble salts conductivity meters read electrical conductivity in millimhos/cm (10^{-3} mho or mmho/cm); others read micromhos/cm (10^{-6} mho or µmho/cm). Thus, multiply mmho readings by 100 and µmho by 0.10 to obtain soluble salts values.

[b] Most reports of crop response to soil soluble salts are based on the more precise electrical conductivity in mmho (abbreviated ECe × 10^3 or ECe mmho) from soil solutions determined by the saturated paste extract method. While stirring with a spatula, the diagnostician slowly adds distilled water to air-dried soil until the soil paste glistens and flows slightly when the container is tipped (this is the saturation point). After 1 h, the sample is checked; free water should not be standing on the surface, and the paste should not have thickened. The paste is then transferred to filter paper in a funnel. The soil water is removed by vacuum, and the electrical conductivity is taken.

deficiency, especially of nitrogen and to a lesser extent potassium. Readings above 150 SS clearly indicate a salts problem in mineral soils. Extreme readings are very helpful in diagnosing plant problems and are encountered frequently.

Soil soluble salts values between 14 and 147 are somewhat more difficult to interpret. The usual procedure is to first use Table 5.3 to obtain "upper safe values" for the soil type, Table 5.4 to determine plant response, and Appendix 5.2 to establish the tolerance of the plant species to soil salts.

For example, consider a reading of 30 SS from a sandy field with field beans, *Phaseolus vulgaris*. Table 5.3 gives 20 to 40 as the upper safe range in sandy soil. Table 5.4 indicates that "very sensitive" and possibly "sensitive" plants may be injured. Appendix 5.2 classifies this plant as "sensitive" and lists 25 SS as the upper safe value. A "high

soil soluble salts" problem is very probable. This would certainly be the case if field conditions had been dry. It is hoped that plant symptoms and field distribution (Chapter 4) would be consistent with this diagnosis. If the soil were clay or organic and soil moisture continuously adequate, soluble salts would not have prevented normal growth.

If the sensitivity of the plant species to soluble salts is not listed in Appendix 5.2, consider its center of origin. Often the following relationships are valid: low tolerance = epiphytes and plants from hydrophilic environments such as rain forests; high tolerance = plants from xerophilic and saline environments such as deserts and maritime and estuarine habitats.

Further modification may be made by relating recent weather patterns, irrigation, fertilizer placement, migration of salts in the soil, presence of toxic elements, and plant variety or cultivar to probable salts injury.

Appendix 5.1. Plant species that may tolerate soils that are very strongly acidic (pH < 4.5) or strongly alkaline (pH > 8.3) without serious injury[a]

Scientific name	Common name	pH range or tolerance[b]
Abies		
balsamea	Balsam fir	4.5–6.0
fraseri	Fraser fir	4.5–6.0
Acer negundo	Box elder	Alkaline tolerant
Agrostis capillaris	Rhode Island bent grass	4.5–6.5
Albizia sp.	. . .	Alkaline tolerant
Allium canadense	Wild garlic	4.5–6.0
Amelanchier sp.	Serviceberry	Acid tolerant
Anacardium occidentale	Cashew	Acid tolerant
Ananas comosus	Pineapple	Acid tolerant
Andromeda		
glaucophylla	Bog-rosemary	3.0–5.0
polifolia	Downy bog-rosemary	4.5–6.0
Andropogon		
gayanus	. . .	Acid tolerant
scoparius	Little blue stem	4.5–6.0
Anemone		
nemorosa	European wood anemone	4.0–6.5
quinquefolia	Wood anemone	4.0–5.0
Arachis hypogaea	Peanut or groundnut	Acid tolerant
Arctostaphylos sp.	Manzanita, bearberry	Acid tolerant
alpina	Alpine or black bearberry	4.5–5.0
uva-ursi	Common bearberry	4.5–6.0
Arethusa bulbosa	Dragon's-mouth or wild pink	4.5–5.5
Arnica montana	Arnica	4.5–5.5
Artocarpus heterophyllus	Jackfruit	Acid tolerant
Asclepias syriaca	Common milkweed	4.0–5.0
Asperula odorata	Sweet woodruff	4.0–8.0
Asplenium		
bradleyi	Bradley's spleenwort	4.0–7.0
ebenoides	Spleenwort	5.5–8.5
montanum	Mountain spleenwort	4.0–6.0
pinnatifidum	Lobed spleenwort	4.5–7.0
platyneuron	Spleenwort	5.0–8.5
resiliens	Little ebony spleenwort	6.0–8.5
trichomanes	Maidenhair spleenwort	5.5–8.5
Aster dumosus	Bushy aster	4.5–5.0
Athyrium		
angustum	Highland lady fern	4.5–7.0
asplenioides	Lowland lady fern	4.5–8.0
Avena sativa	Common oat	4.5–7.5
Averrhoa carambola	Carambola	Acid tolerant
Begonia socotrana	Begonia	4.5–8.0
Berberis thunbergii	Japanese barberry	Alkaline tolerant
Bertholletia excelsa	Brazil nut	Acid tolerant
Beta vulgaris	Sugar beet	5.5–8.5
Betula		
lenta	Black, sweet, or cherry birch	4.5–7.0
pendula	European white birch	4.5–6.0
Brachiaria		
decumbens	Signal grass	Acid tolerant
humidicola	Signal grass	Acid tolerant
Brachypodium pinnatum	Japanese false brome grass or tar grass	6.5–8.5
Cajanus cajan	Pigeon pea	Acid tolerant
Calendula officinalis	Pot-marigold	5.0–8.5
Calla palustris	Water-arum or wild calla	4.0–5.5
Calluna vulgaris	Heather	4.5–6.5
Calopogonium mucunoides	Calopo	Acid tolerant
Camellia japonica	Common camellia	4.5–5.5
Camptosorus rhizophyllus	Walking fern	5.5–8.5
Capsicum annuum	Bell pepper, paprika pepper, or pimento	7.0–8.5
Carex		
globularis	Sedge	4.0–5.0
panicea	Sedge	4.0–8.0
vulgaris	Sedge	4.0–8.0
Cassiope hypnoides	Moss bush cassiope	4.5–5.5
Casuarina sp.	Casuarina	Alkaline tolerant
Ceanothus americanus	New Jersey tea	4.5–6.0
Celastrus scandens	Bittersweet	4.5–6.0

(*continued on next page*)

Appendix 5.1. (*continued*)

Scientific name	Common name	pH range or tolerance[b]
Celtis sp.	Hackberry	Alkaline tolerant
Centrosema pubescens	Butterfly pea	Acid tolerant
Cercocarpus sp.	Mountain-mahogany	Alkaline tolerant
Chamaecyparis thyoides	White cedar	4.5–6.0
Chamaedaphne calyculata	Leather-leaf or cassandra	3.0–6.0
Chara vulgaris	Stinking chara	7.0–8.5
Chimaphila umbellata	Pipsissewa	4.5–6.0
Chrysanthemum alpinum	Daisy	4.5–7.0
Citrus		
aurantifolia	Lime	Acid tolerant
limon	Lemon	6.0–8.3
maxima	Pummelo or shaddock	6.0–8.3
× *paradisi*	Grapefruit	Acid tolerant
sinensis	Sweet orange	6.0–8.3
Clethra alnifolia	Sweet pepper-bush	4.5–6.0
Clintonia borealis	Blue-bead or corn lily	4.0–5.0
Cocos nucifera	Coconut palm	Acid tolerant
Coffea arabica	Common or arabica coffee	Acid tolerant
Convallaria majalis	Lily-of-the-valley	4.5–6.0
Coptis groenlandica	Canker-root or golden-thread	4.5–5.0
Corallorrhiza		
maculata	Large coralroot	4.5–6.5
odontorhiza	Small coralroot	4.5–7.0
Corema conradii	Broom crowberry or poverty grass	4.0–5.0
Coreopsis sp.	Tickseed or coreopsis	Acid tolerant
Cornus canadensis	Bunchberry or dwarf cornel	4.5–5.5
Cosmos bipinnatus	Common cosmos	4.5–8.0
Cyclamen persicum	Florist's cyclamen	4.5–7.5
Cyperus cylindricus	Pine-barren cyperus	4.5–5.0
Cypripedium acaule	Moccasin-flower or pink lady's-slipper	4.0–5.0
Cystopteris		
bulbifera	Bulblet or berry bladder fern	5.0–8.5
fragilis	Brittle or fragile fern	5.0–8.5
Cytisus sp.	Broom	Acid tolerant
Dalbergia nigra	Jacarandá	Acid tolerant
Darlingtonia californica	California pitcher-plant	4.0–5.5
Daucus carota	Wild carrot	4.5–6.5
Dennstaedtia punctilobula	Hay-scented fern	4.0–7.0
Deschampsia flexuosa	Crinkled hair grass	4.0–5.0
Desmodium		
gyroides	. . .	Acid tolerant
heterophyllum	. . .	Acid tolerant
ovalifolium	. . .	Acid tolerant
Deutzia sp.	Deutzia	Alkaline tolerant
Diapensia lapponica	Diapensia	4.5–5.0
Digitaria decumbens	Pangola grass	Acid tolerant
Dionaea muscipula	Venus's-flytrap	4.5–7.0
Distichlis spicata	Seashore salt grass	6.5–8.5
Drosera longifolia	Sundew	4.0–6.0
Elaeagnus angustifolia	Russian olive	Alkaline tolerant
Elaeis guineensis	African oil palm	Acid tolerant
Empetrum nigrum	Black crowberry	4.5–5.5
Epigaea repens	Trailing arbutus	4.5–7.0
Equisetum		
arvense	Field horsetail	4.5–6.0
sylvaticum	Wood horsetail	4.5–6.0
Erica carnea	Spring heath or snow-heather	4.5–6.0
Eucalyptus grandiflora	Eucalyptus	Acid tolerant
Eupatorium album	White thoroughwort	4.5–5.0
Fagopyrum esculentum	Common buckwheat	4.5–7.0
Festuca		
ovina	Sheep fescue	4.5–6.5
pratensis	Meadow fescue	4.5–7.0
Forsythia sp.	Forsythia	Alkaline tolerant
Fothergilla gardenii	Dwarf or witch alder	5.0–6.0
Fraxinus velutina	Velvet ash	Alkaline tolerant
Galactia striata	Milk-pea	Acid tolerant
Galax aphylla	Galax	4.5–6.0
Galium saxatile	Rock bedstraw	4.0–6.0

(*continued on next page*)

Appendix 5.1. (*continued*)

Scientific name	Common name	pH range or tolerance[b]
Gardenia sp.	Gardenia	Acid tolerant
Gaultheria		
hispidula	Creeping snowberry or moxie-plum	4.5–5.7
procumbens	Checkerberry or wintergreen	4.5–7.0
Gaylussacia		
baccata	Black huckleberry	4.0–5.0
brachycera	Box huckleberry	4.0–6.0
dumosa	Dwarf huckleberry	4.5–5.0
frondosa	Dangleberry	4.5–6.0
Gentiana acaulis	Gentian	4.5–6.0
Geum aleppicum	Avens	4.5–5.0
Gmelina arborea	Gmelina	Acid tolerant
Guilielma gasipaes	Pejibaye, gachipaes, or peach palm	Acid tolerant
Gypsophila sp.	Baby's-breath	Acid tolerant
Habenaria		
blephariglottis	White fringed or snowy orchid	3.5–5.0
ciliaris	Yellow fringed orchid	4.0–5.0
clavellata	Small green wood or green woodland orchid	4.5–5.0
cristata	Crested yellow or golden fringed orchid	4.0–5.0
flava	Tubercled orchid	4.5–5.0
Helianthemum canadense	Frost-wort or frostweed	4.5–5.0
Helonias bullata	Swamp pink	4.0–6.0
Hevea brasiliensis	Rubber-tree	Acid tolerant
Hibiscus syriacus	Rose-of-Sharon or shrub althea	Alkaline tolerant
Hydrangea macrophylla	House or French hydrangea	4.0–7.5
Hydrocotyle umbellata	Water pennywort	4.5–6.5
Hyparrhenia rufa	Jaragua grass	Acid tolerant
Hypericum gentianoides	Pine-weed	4.0–5.0
Hypnum sp.	Hypnum moss	3.5–5.0
Hypoxis hirsuta	Golden star-grass	4.0–5.0
Ilex		
aquifolium	English holly	4.0–5.5
glabra	Inkberry	4.5–5.0
Iris		
cristata	Dwarf-crested iris	4.5–7.0
prismatica	Slender blue flag iris	4.5–7.0
verna	Violet or vernal iris	4.0–5.0
Juncus aristulatus	Grass-leaf rush	4.5–5.0
Kalmia		
angustifolia	Lambkill or sheep-laurel	4.5–6.0
latifolia	Mountain-laurel	4.5–6.0
polifolia	Bog kalmia	4.5–6.0
Kerria spp.	. . .	Alkaline tolerant
Kummerowia stipulacea	Korean lespedeza	4.0–6.5
Lagerstroemia indica	Crape-myrtle	Acid tolerant
Larix		
decidua	European larch	4.5–6.5
laricina	Tamarack	4.5–6.5
Ledum groenlandicum	Labrador tea	3.0–5.0
Leiophyllum buxifolium	Sand-myrtle	4.0–6.0
Lespedeza striata	Japanese lespedeza	4.0–6.5
Leucothoe axillaris var. *editorum*	Drooping laurel	4.5–6.0
Liatris spicata	Blazing-star	4.5–5.0
Lindera benzoin	Spicebush	4.5–6.0
Linnaea borealis var. *americana*	American twinflower	4.0–5.0
Linum floridanum	Southern wild flax	4.5–5.0
Liparis loeselii	Loesels twayblade	4.5–6.5
Lithodora fruticosa	Gromwell	4.5–5.5
Logodium palmatum	Climbing fern	4.0–5.0
Loiseleuria procumbens	Alpine or trailing azalea	4.5–5.0
Lonicera fragrantissima	Fragrant honeysuckle	Alkaline tolerant
Lupinus sp.	Lupine	Acid tolerant
Lycopersicon esculentum	Tomato	4.5–7.5
Lycopodium		
annotinum	Stiff or bristly club-moss	4.5–5.0
clavatum	Common club-moss or wolf's-claws	4.5–6.0
complanatum	Ground-cedar or Christmas-green	4.0–6.0
depressum	Pine-barren club-moss	4.5–5.0
obscurum var. *dendroideum*	Flat-branch ground-pine	4.5–7.5
selago	Club-moss	4.5–5.0

(*continued on next page*)

Appendix 5.1. (*continued*)

Scientific name	Common name	pH range or tolerance[b]
Lyonia		
ligustrina	Male-blueberry	4.7–6.0
mariana	Stagger-bush	4.0–6.0
Lysimachia quadrifolia	Whorled loosestrife	4.0–5.0
Magnolia sp.	Magnolia	Acid tolerant
virginiana	Sweet-bay magnolia or sweet-bay	4.0–5.0
Maianthemum canadense	Two-leaved Solomon's-seal	4.5–7.0
Malaxis unifolia	Green adder's-mouth	4.5–5.5
Malus sargentii	Sargent crabapple	Alkaline tolerant
Mangifera indica	Mango	Acid tolerant
Manihot esculenta	Cassava	Acid tolerant
Mariscus mariscoides	Twig rush	4.5–5.0
Medeola virginica	Indian cucumber-root	4.0–7.0
Medicago lupulina	Black medic or yellow-trefoil	4.5–6.5
Melampyrum lineare	Cow-wheat	4.0–5.0
Melica uniflora	One-flowered melic	4.0–7.5
Melinis minutiflora	Molasses grass	Acid tolerant
Menziesia pilosa	Minnie-bush or Allegheny menziesia	4.5–8.0
Mercurialis perennis	Mercury	4.5–8.0
Molinia caerulea	Moor grass	4.0–5.0
Moneses uniflora	Wood-nymph or one-flowered wintergreen	4.5–6.0
Musa		
× *paradisiaca*	Plantain cooking banana	Acid tolerant
× *sapientum*	Banana	Acid tolerant
Myrica cerifera	Wax myrtle or waxberry	4.0–6.0
Nardus stricta	White bent grass	4.5–6.5
Nicotiana tabacum	Tobacco	4.5–7.5
Onoclea sensibilis	Sensitive fern	4.5–8.0
Ophioglossum vulgatum	Adder's-tongue	4.1–6.0
Oryza sativa	Rice	Acid tolerant
Osmunda		
cinnamomea	Cinnamon fern	4.5–6.5
claytoniana	Interrupted fern	4.5–7.5
regalis	Royal fern	4.0–6.0
Oxalis acetosella	European wood-sorrel, Irish shamrock	4.0–6.0
Oxydendrum arboreum	Sourwood, sorrel-tree, or titi	4.5–6.5
Pachysandra terminalis	Japanese-spurge	4.5–5.5
Panicum		
lucidum	Panic grass	4.5–5.0
maximum	Guinea grass	Acid tolerant
meridionale	Panic grass	4.5–5.0
oricola	Panic grass	4.5–5.0
Paronychia argyrocoma	Whitlow-wort	4.0–5.0
Paspalum		
notatum	Bahia grass	Acid tolerant
plicatulum	Brown-seed paspalum	Acid tolerant
Passiflora edulis	Purple granadilla, passion-fruit, or granadilla	Acid tolerant
Paullinia cupana	Guaraná	Acid tolerant
Pellaea atropurpurea	Purple cliffbrake	5.5–8.5
Pennisetum purpureum	Napier grass	Acid tolerant
Philadelphus sp.	Mock-orange	Alkaline tolerant
Phoenix dactylifera	Date palm	Alkaline tolerant
Phyllodoce caerulea	Mountain heath or heather	4.5–5.5
Picea		
mariana	Black spruce	4.0–6.0
rubens	Red spruce	4.0–6.0
Pieris sp.	Andromeda or fetterbush	Acid tolerant
Pinus		
banksiana	Jack pine	4.5–6.0
caribaea	Cuban or Caribbean pine	Acid tolerant
echinata	Yellow or short-leaf pine	4.5–6.0
palustris	Long-leaf pine	4.5–6.0
pungens	Table mountain pine	4.5–6.0
resinosa	Red pine	4.5–6.0
rigida	Pitch pine	4.5–6.0
serotina	Pond pine	4.5–6.0
strobus	Eastern white pine	4.5–7.5
taeda	Loblolly pine	4.5–6.0
Piper nigrum	Black pepper	Acid tolerant

(*continued on next page*)

Appendix 5.1. (*continued*)

Scientific name	Common name	pH range or tolerance[b]
Pogonia ophioglossoides	Rose pogonia or snakemouth	4.0–5.0
Polygala paucifolia	Flowering-wintergreen	4.0–5.0
Polypodium		
aureum	Hare's-foot fern or golden polypody	4.0–5.0
polypodioides	Resurrection fern or polypody	4.5–6.0
Polystichum lonchitis	Northern holly fern	4.5–6.0
Populus		
fremontii	Fremont cottonwood	Alkaline tolerant
tremuloides	Quaking or trembling aspen	3.8–5.5
Potamogeton sp.	Pondweed	6.5–8.5
Potentilla		
fruticosa	Shrubby cinquefoil	Alkaline tolerant
recta	Upright or sulphur cinquefoil	4.0–5.5
tridentata	Three-toothed cinquefoil	4.5–6.0
Psidium guajava	Common guava	Acid tolerant
Pteridium aquilinum	Common bracken	4.5–6.0
Pueraria phaseoloides	Tropical kudzu	Acid tolerant
Punica granatum	Pomegranate	Acid tolerant
Pyrola		
asarifolia var. *incarnata*	Wintergreen	4.5–5.0
rotundifolia var. *americana*	One-sided wintergreen	4.5–7.0
Pyxidanthera barbulata	Flowering moss or pyxie	4.5–6.0
Quercus		
ilicifolia	Scrub oak	4.5–7.5
palustris	Pin oak	Acid tolerant
rubra	Northern red oak	4.5–7.5
Rhexia virginica	Deer grass or handsome Harry	4.5–5.5
Rhododendron		
albiflorum	Rhododendron	4.5–6.0
canescens	Hoary azalea or Florida pinxter	4.5–5.0
carolinianum	Carolina rhododendron	4.5–7.0
caucasicum	Rhododendron	4.5–6.0
ferrugineum	Rhododendron	4.5–7.0
lapponicum	Lapland rhododendron	4.5–6.0
maximum	Rose-bay rhododendron or great laurel	4.5–7.0
mucronatum	Snow azalea	4.5–6.0
nudiflorum	Pinxterbloom	4.5–5.5
obtusum var. *amoenum*	Kurume azalea	4.5–6.0
viscosum	Swamp or clammy azalea	4.0–6.0
Rhodora canadense	Rhodora	4.5–6.0
Robinia pseudoacacia	Black locust	Alkaline tolerant
Rubus		
allegheniensis	Common or Allegheny blackberry	4.5–6.0
chamaemorus	Cloudberry or salmonberry	4.5–6.0
flagellaris	Northern dewberry	4.0–6.0
occidentalis	Black raspberry	4.5–6.5
pubescens	Dwarf red raspberry	4.0–6.0
Rumex acetosa	Garden sorrel or sour dock	4.0–7.0
Saccharum officinarum	Sugarcane	Acid tolerant
Salix		
babylonica	Weeping willow	Acid tolerant
uva-ursi	Bearberry or creeping willow	4.5–5.0
Sarracenia		
drummondii or *leucophylla*	Drummond pitcher-plant	4.0–5.0
flava	Yellow or trumpet pitcher-plant	3.5–6.0
minor	Hooded pitcher-plant	4.0–6.0
psittaccina	Common pitcher-plant	4.0–5.0
purpurea	Southern or sweet pitcher-plant	3.5–6.5
rubra or *jonesii*	Candler pitcher-plant	4.0–5.0
sledgeii or *alata*	Pale pitcher-plant	3.5–5.0
Scheuchzeria palustris	Arrow grass	4.5–6.0
Schizachyrium scoparium	Little blue-stem	4.5–6.5
Schizaea pusilla	Curly-grass fern	4.0–5.0
Secale cereale	Common rye	4.5–8.0
Setaria italica	Foxtail or Italian millet	4.5–6.5
Sinapis alba	White mustard	4.5–7.5
Smilacina		
racemosa	False spikenard	4.0–7.0
trifolia	Dwarf false Solomon's-seal	4.5–5.0

(*continued on next page*)

Appendix 5.1. (*continued*)

Scientific name	Common name	pH range or tolerance[b]
Solanum tuberosum	Potato	Acid tolerant
Solidago fistulosa	Pine-barren goldenrod	4.5–5.0
Sophora japonica	Japanese pagoda-tree	Alkaline tolerant
Sorbus americana	American mountain-ash	4.5–5.5
Sorghum		
bicolor	Broom corn	4.5–6.5
	Sorghum	4.5–7.5
× *drummondii*	Sudan grass	4.5–6.5
Sphagnum sp.	Sphagnum moss	3.5–6.0
Spinacia oleracea	Spinach	5.5–8.5
Spiraea × *vanhouttei*	Bridal-wreath	Alkaline tolerant
Spiranthes		
beckii	Beck's ladies'-tresses	4.5–5.0
cernua	Common or nodding ladies'-tresses	4.0–5.0
vernalis	Early ladies'-tresses	4.5–5.0
Stellaria borealis	Chickweed	4.5–5.5
Stenanthium gramineum	Feather-fleece or feather-bells	4.0–5.0
Streptopus		
amplexifolius	White mandarin or twisted-stalk	4.5–5.0
roseus	Rose mandarin or rosy twisted-stalk	4.5–5.5
Stylosanthes		
capitata	Pencil flower	Acid tolerant
guianensis	Brazilian lucerne	Acid tolerant
scabra	. . .	Acid tolerant
viscosa	. . .	Acid tolerant
Tagetes patula	French marigold	4.5–7.5
Taraxacum officinale	Common dandelion	5.0–8.5
Thelypteris		
bootii	Boot's shield fern	4.5–8.0
cristata	Crested shield or wood fern	4.5–8.0
marginalis	Evergreen wood fern	5.0–8.5
noveboracensis	New York fern	4.0–7.0
palustris	Marsh or snuffbox fern	4.0–6.0
phegopteris	Long beech fern	4.5–8.0
simulata	Massachusetts fern	4.0–5.0
spinulosa	Spinulose shield fern	4.5–8.0
Trientalis borealis	Starflower	4.0–5.0
Trifolium agrarium	Yellow or hop clover	4.0–6.0
Trillium undulatum	Painted trillium	4.5–7.0
Tsuga caroliniana	Carolina or southern hemlock	4.5–6.0
Ulex europaeus	Gorse, furze	4.5–6.0
Vaccinium		
angustifolium var. *laevifolium*	Sweet blueberry	4.0–5.5
atrococcum	Black high-bush blueberry	4.5–6.0
corymbosum	High-bush blueberry	3.4–6.5
erythrocarpum	Blueberry	4.5–6.0
macrocarpon	American or large cranberry	4.2–6.0
myrtillus	Whortleberry or bilberry	3.5–5.0
oxycoccus	Small cranberry	4.2–6.0
uliginosum	Bog bilberry	4.5–6.0
vacillans	Low or sweet blueberry	4.5–6.0
Viburnum		
acerifolium	Maple-leaved viburnum	4.0–5.0
dilatatum	Linden's viburnum	Alkaline tolerant
Vicia villosa	Hairy or winter vetch	4.5–7.0
Vigna unguiculata	Cow pea, southern pea, or black-eyed pea	4.5–7.0
Viola palustris	Alpine marsh or European swamp violet	4.0–6.5
Washingtonia sp.	Washington palm	Alkaline tolerant
Woodwardia		
areolata	Nettle chain fern	4.0–7.5
virginica	Virginia chain fern	4.0–6.0
Xerophyllum asphodeloides	Turkey's-beard	4.0–5.0
Zantedeschia aethiopica	Garden or florist's calla	5.0–8.5
Ziziphus jujuba	Common or Chinese jujube	Alkaline tolerant
Zornia latifolia	. . .	Acid tolerant

[a] Data from Sánchez, P. A., and Salinas, J. G., 1981, Low-input technology for managing oxisols and ultisols in tropical America, Adv. Agron. 34:279-406; Smith, M. D., ed., 1984, The Ortho Problem Solver, 2nd ed., Chevron Chemical Co., San Francisco; and Spurway, C. H., 1941, Soil reaction (pH) preferences of plants, Mich. State Coll. Spec. Bull. 306.

[b] Acid tolerant = plants normally grow in soils with a pH of 4.5–5.5, and alkaline tolerant = plants normally grow in soils with a pH of 7.5–8.4. In both cases, some plant species or cultivars may show no injury symptoms at soil pH beyond these ranges. Injury is often dependent on the composition of the soil or growing medium.

Appendix 5.2. Plant species and their sensitivity to soil soluble salts and, where indicated, to airborne salt spray

Plant	Common name	Class[a]	Upper safe value[b] SS (10^{-5} mho/cm)	ECe	Source[c]
Abelia × *grandiflora*	Glossy abelia	VS	14	2.0	d
Abelmoschus esculentus	Okra	S	. . .	. . .	b,c,e
Abies balsamea	Balsam fir	S roots	. . .	. . .	f
Acer					
ginnala	Amur maple	MT foliage	. . .	. . .	k
negundo	Box elder or ash-leaved maple	MT	. . .	. . .	f
		S foliage	. . .	. . .	k
platanoides	Norway maple	MT	. . .	. . .	f
		R foliage	. . .	. . .	k
rubrum	Red maple	S	. . .	. . .	f
		MT foliage	. . .	. . .	k
saccharinum	Silver maple	S	. . .	. . .	f
		MT foliage	. . .	. . .	k
saccharum	Sugar maple	S	. . .	. . .	f
		MT foliage	. . .	. . .	k
Aesculus hippocastanum	Common horse-chestnut	R foliage	. . .	. . .	k
Agropyron					
cristatum	Fairway crested wheat grass	T	103	14.7	a,b,c,e
desertorum	Standard crested wheat grass	T	111	16.0	a,c,e
elongatum	Tall wheat grass	T	136	19.4	a,b,c,e
trachycaulum	Slender wheat grass	MT	. . .	. . .	c,e
Agrostis stolonifera	Creeping bent grass	MS	. . .	. . .	c,e
Ailanthus altissima	Tree-of-heaven	R foliage	. . .	. . .	k
Allium cepa	Onion	S	30	4.3	a,b,c,e
Alnus rugosa or *A. incana*	Speckled alder	S	. . .	. . .	f
		MT foliage	. . .	. . .	k
Alopecurus pratensis	Meadow foxtail	MS	47	6.7	a,b,c,e
Amelanchier laevis	Allegheny serviceberry	S foliage	. . .	. . .	k
Apium graveolens var. *dulce*	Celery	S	24	3.5	c,h
Arachis hypogaea	Peanut or groundnut	MS	34	4.9	a,c,e
		VS	20	. . .	g
Araucaria heterophylla	Norfolk Island pine	T	. . .	. . .	j
Arbutus unedo	Strawberry-tree	S	28	4.0	d
Arctotheca calendula	Arctotheca	T	. . .	. . .	j
Asparagus officinalis	Cultivated asparagus	T	77	11.0	h
Astragalus spp.	Milk-vetch	MT	. . .	. . .	c,h
Avena sativa	Common oat	MT	. . .	. . .	c,h
Baccharis pilularis	Coyote-bush or kidney-wort	T	. . .	. . .	j
Begonia × *rex-cultorum*	Rex begonia	S	40	. . .	g
Berberis					
aquifolium	Oregon grape	VS	14	2.0	d
thunbergii	Japanese barberry	S	. . .	. . .	f
Beta					
vulgaris	Common or garden beet	MT	67	9.6	a,b,c,e
subsp. *vulgaris*	Sugar beet	T	108	15.3	a,b,c,e
Betula					
papyrifera	Paper or canoe birch	MT	. . .	. . .	f
		MT foliage	. . .	. . .	k
pendula	European white birch	MT	. . .	. . .	f
populifolia	Gray birch	MT foliage	. . .	. . .	k
Bougainvillea spectabilis	Brazil bougainvillea	T	70	10.0	c,d
Brassica					
napus	Rape or colza	T	91	13.0	c,h
oleracea					
var. *acephala*	Kale	MT	49+	7+	c,h
var. *botrytis*	Cauliflower	MS	28+	4.0	c,h
var. *botrytis*	Broccoli	MT	57	8.2	a,b,c,e
var. *capitata*	Cabbage	MS	49	7.0	a,b,c,e
Bromus					
catharticus	Rescue grass	T	. . .	. . .	c,h
inermis	Smooth brome	MT	. . .	. . .	b,c,e,h
Buxus					
microphylla var. *japonica*	Japanese boxwood	MT	44	6.4	c,d
sempervirens	Common boxwood	S	30	. . .	f,g
Callistemon viminalis	Weeping bottlebrush	MT	44	6.4	c
		MT	56	8.0	d
		T	. . .	. . .	j

(continued on next page)

Appendix 5.2. (*continued*)

Plant	Common name	Class[a]	Upper safe value[b] SS (10^-5 mho/cm)	ECe	Source[c]
Capsicum frutescens cv. grossum	Bell or sweet pepper	MS	35	5.1	a,b,c,e
Caragana arborescens	Siberian pea-tree	R foliage	...	...	k
Carissa macrocarpa	Natal-plum	T	70	10.0	c,d
Carpinus caroliniana	American hornbeam or blue-beech	S	...	...	f
Carthamus tinctorius	Safflower	MT	69	9.9	a,b,c,e
Carya spp.	Hickories	S	...	...	f
ovata	Shagbark hickory	T foliage	...	...	k
Catalpa speciosa	Northern catalpa	MT foliage	...	...	k
Chaenomeles japonica	Flowering quince	S foliage	...	...	k
Chamaerops humilis	European fan palm or dwarf hair palm	MT	56	8.0	d
Chloris gayana	Rhodes's grass	MS	...	...	c,e,h
Chrysanthemum × morifolium	Florist's chrysanthemum	T	75	...	g
Citrus					
aurantiifolia	Lime	MT	...	...	c
limon	Lemon	S	34	4.8	a,c,e,h
× *paradisi*	Grapefruit	S	34	4.9	a,c,e,h
reticulata	Tangerine	MT	...	...	c
sinensis	Sweet orange	S	34	4.8	a,c,e,h
Coprosma repens	Mirror-plant	MT	...	...	j
Cordyline indivisa	Blue dracaena	MT	56	8.0	d
Cornus					
racemosa	Gray dogwood	VS foliage	...	...	k
sericea	Red osier dogwood	VS foliage	...	...	k
Cortaderia selloana	Pampas grass	T	...	...	j
Cotoneaster congesta	Pyrenees cotoneaster	VS	14	2.0	d
Crataegus spp.	Hawthorns	S foliage	...	...	k
Cucumis					
melo var. *reticulatus*	Muskmelon	S	64	9.1	a,c,h
sativus	Cucumber	MS	44	6.3	a,c,e
Cucurbita maxima	Winter squash	MS	...	...	c,h
Cydonia oblonga	Common quince	MT foliage	...	...	k
Cynodon dactylon	Bermuda grass	T	103	14.7	a,b,c,e
		S	30	...	g
Dactylis glomerata	Orchard grass, cock's-foot	MT	67	9.6	a,b,c,e
Daucus carota subsp. *sativus*	Carrot	S	32	4.6	a,b,c,e
Delosperma alba	White ice-plant	T	70	10.0	d
Dieffenbachia spp.	Dieffenbachia	T	100		l
Distichlis stricta	Desert salt grass	T	...	...	c
Dodonaea viscosa var. *arborescens*	Dodonaea	S	50	7.1	c
		MT	42	6.0	d
Dracaena endivisa	Dracaena	MT	62	8.9	c
Drosanthemum hispidum	Rosea iceplant	T	70	10.0	d
Elaeagnus					
angustifolia	Russian olive or oleaster	T	...	...	f
		T foliage	...	...	k
commutata	Silverberry	S	49	7.0	c
pungens	Thorny elaeagnus	MT	42	6.0	d
Elymus					
angustus	Altai wild rye	T	...	...	c,e
canadensis	Canada wild rye	MT	...	...	c,h
junceus	Russian wild rye	T	...	...	c,e
smithii	Western wheat grass	MT	...	...	c
triticoides	Beardless wild rye	MT	77	11.0	b,c,e
Eragrostis spp.	Love grasses	MS	56	7.9	a,c,e
Eremochloa ophiuroides	Centipede grass	VS	20	...	g
Euonymus					
alatus	Winged spindle-tree	S	...	...	f
		T foliage	...	...	k
japonica	Evergreen euonymus	MT	64	9.1	c
		MT	56	8.0	d
Fagus					
grandifolia	American beech	S	...	...	f
		VS foliage	...	...	k
sylvatica	European beech	S	...	...	f
Feijoa sellowiana	Pineapple guava or feijoa	VS	18	2.6	c
		VS	14	2.0	d

(*continued on page 119*)

Color Plates

PLATE 1. FIELD DISTRIBUTION PATTERNS

1A. Yellows or wilt of celery caused by *Fusarium oxysporum* f. sp. *apii*. (Courtesy A. O. Paulus)

1B. Wilt of sweet potato caused by *Fusarium oxysporum* f. sp. *batatas*. (Courtesy A. O. Paulus)

1C. Corm rot and shoot blight of gladiolus caused by *Fusarium oxysporum* f. sp. *gladioli*. (Courtesy University of Illinois)

1D. Brittle root of horseradish caused by *Spiroplasma citri*. (Courtesy University of Illinois)

1E. Walnut wilt of tomato. (Courtesy University of Illinois)

1F. Damage to corn (maize) caused by the sting nematode, *Belonolaimus* sp. (Courtesy D. W. Dickson)

1G. Snap bean with severe root rot caused by *Pythium* sp. (Courtesy A. O. Paulus)

1H. Damage to sugar beet caused by the cyst nematode *Heterodera schachtii*. (Courtesy BASF)

1I. Marigold aster yellows, caused by a mycoplasma. (Courtesy R. K. Jones)

1J. Zoysia patch in a golf fairway caused by *Gaeumannomyces incrustans*. (Courtesy University of Illinois)

1K. Seedling blight of white pine caused by *Cylindrocladium* sp. (Courtesy North Carolina State University)

1L. A "hot spot" of onion white rot, caused by *Sclerotium cepivorum*. (Courtesy University of Illinois)

1M. Paraquat injury (plants in the wheel tracks survived). (Courtesy L. E. Bode)

1N. Tomato plants susceptible and resistant to wilt caused by *Fusarium oxysporum* f. sp. *lycopersici*. (Courtesy University of Illinois)

1O. Wilt of potato caused by *Verticillium dahliae*. (Courtesy A. O. Paulus)

1P. Severe spring freeze damage to *Taxus* sp. (Courtesy D. F. Schoeneweiss)

2A. Leafhopper "burn" injury to potato. (Courtesy R. C. Rowe)

2B. Flea beetle injury to potato tubers. (Courtesy University of Wisconsin)

2C. Cloudy spot of tomato caused by stinkbug feeding. (Courtesy Purdue University)

2D. Red spider mite damage to snap bean leaves. (Courtesy BASF)

2E. Bird damage to grain sorghum. (Courtesy L. E. Claflin)

2F. Dodder (*Cuscuta* sp.) on onions. (Courtesy University of Illinois)

2G. Witch-weed (*Striga asiatica*) attacking corn (maize). (Courtesy R. E. Eplee)

2H. Broomrape (*Orobanche* sp.) in *Cineraria*. (Courtesy British Ministry of Agriculture)

2I. True or leafy mistletoe (*Phoradendron serotinum*, syn. *P. flavescens*) on walnut. (Left, courtesy University of Illinois and right, courtesy Illinois Natural History Survey)

2J. Dwarf mistletoe (*Arceuthobium* sp.) on white fir (*Abies concolor*). (Courtesy Purdue University)

2K. Zinnia aster yellows, caused by a mycoplasma. (Courtesy University of Illinois)

2L. Pea anthracnose, caused by *Colletotrichum gloeosporioides*. (Courtesy D. J. Hagedorn)

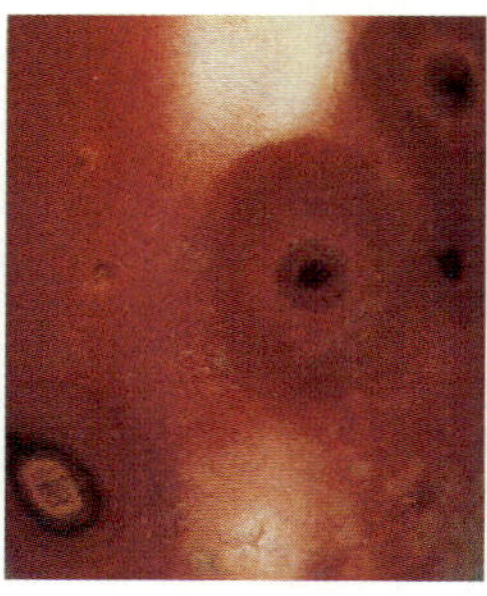

2M. Tomato anthracnose ripe rot, caused by *Colletotrichum coccodes*. (Courtesy Purdue University)

2N. Watermelon anthracnose, caused by *Colletotrichum orbiculare*. (Courtesy Purdue University)

2O. Leaf blister of red oak, caused by *Taphrina caerulescens*. (Courtesy Purdue University)

2P. Blight of *Calceolaria* blooms caused by *Botrytis cinerea*. (Courtesy T. H. Bowyer)

2Q. Carrot motley dwarf, caused by carrot mottle and carrot red leaf viruses. (Courtesy A. O. Paulus)

2R. Angular leaf spot of cucumber caused by *Pseudomonas syringae* pv. *lachrymans*. (Courtesy Purdue University)

2S. Damping-off of petunia seedlings. (Courtesy North Carolina State University)

3A. Downy mildew on onion, caused by *Peronospora destructor*. (Courtesy A. O. Paulus)

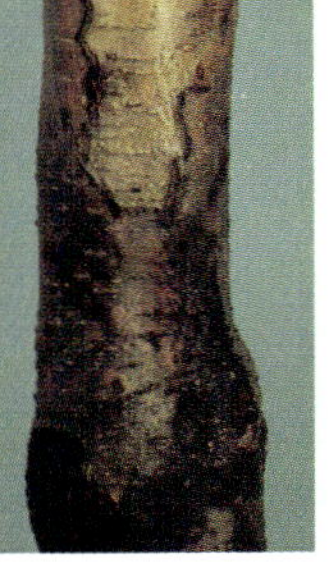

3B. Apple collar rot caused by *Phytophthora cactorum*. (Courtesy Purdue University)

3C. Elm yellows or phloem necrosis, caused by a mycoplasmalike organism. (Courtesy Illinois Natural History Survey)

3D. Stem blight of *Vinca minor* caused by *Ascochyta phaseolorum*. (Courtesy P. C. Pecknold)

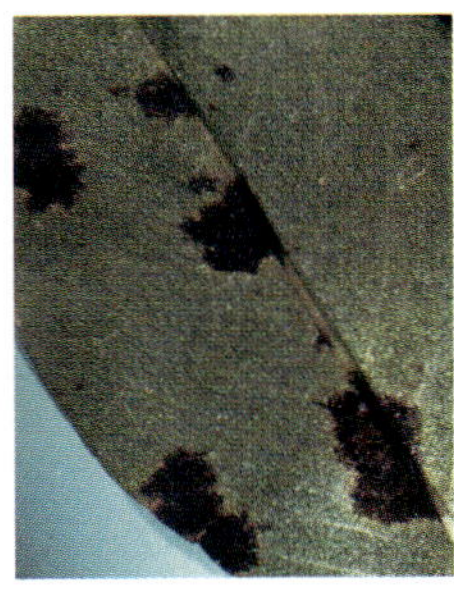

3E. Late leaf spot of peanut caused by *Phaeoisariopsis personata*. (Courtesy E. Dutky)

3F. Brown needle spot of pine, caused by *Mycosphaerella dearnessii*. (Courtesy Purdue University)

3G. Pepper leaf infected with tobacco etch virus. (Courtesy T. H. Bowyer)

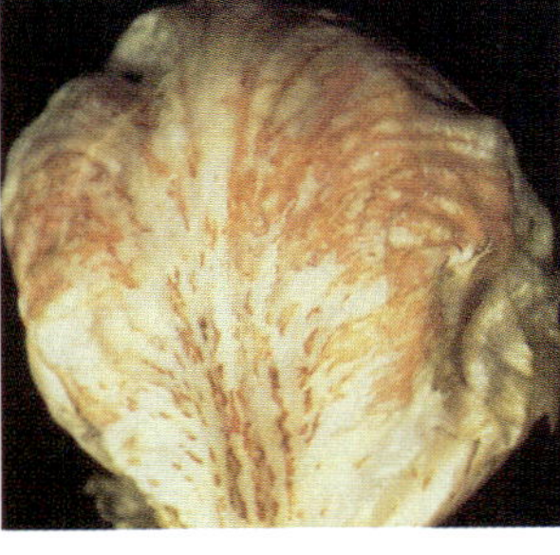

3H. Lettuce infected with tomato spotted wilt virus. (Courtesy USDA)

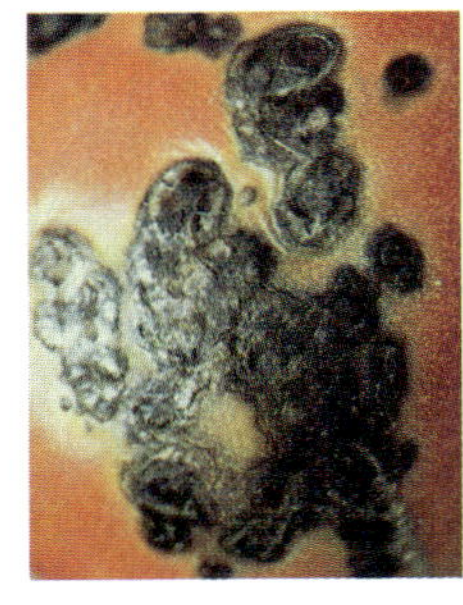

3I. Bacterial spot of tomato caused by *Xanthomonas campestris* pv. *vesicatoria*. (Courtesy Purdue University)

3J. Root rot of Fraser fir (*Abies fraseri*) caused by *Phytophthora cinnamomi*. (Courtesy North Carolina State University)

3K. Root and butt rot of *Quercus virginiana* caused by *Ganoderma applanatum*. (Courtesy T. H. Bowyer)

3L. Rosette or witches'-broom of rose (cause unknown). (Courtesy F. J. Crowe)

3M. Sugar beet yellows, caused by the beet yellows virus. (Courtesy BASF)

3N. Pumpkin infected with a ringspot virus. (Courtesy University of Illinois)

3O. White rust on spinach caused by *Albugo occidentalis*. (Courtesy A. O. Paulus)

3P. Sooty blotch, caused by *Gloeodes pomigena,* and flyspeck, caused by *Schizothyrium perexiguum,* on apple. (Courtesy University of Illinois)

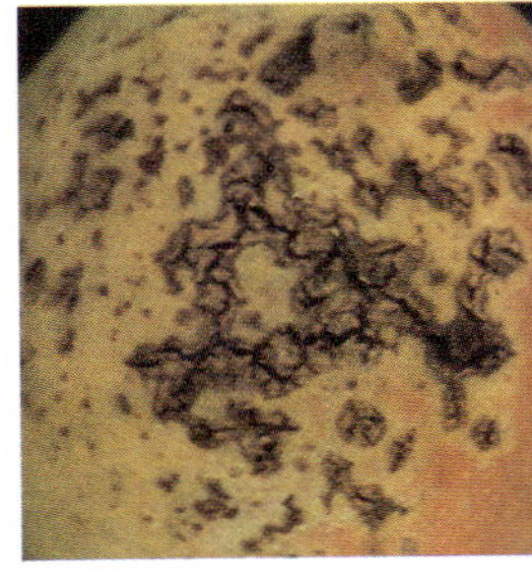

3Q. Bacterial spot of peach caused by *Xanthomonas campestris* pv. *pruni.* (Courtesy University of Illinois)

3R. Powdery mildew on snap bean caused by *Erysiphe polygoni*. (Courtesy D. J. Hagedorn)

3S. A slime mold fruiting on cucumber. (Courtesy T. H. Bowyer)

PLATE 4. AIR POLLUTION INJURIES

4A. Ambient ozone injury to butternut squash. (Courtesy A. L. Heagle)

4B. Ambient ozone injury to cantaloupe. (Courtesy University of Illinois)

4C. Ambient ozone injury to pole bean cultivar Kentucky Wonder. (Courtesy A. L. Heagle)

4D. Ozone injury to lettuce cultivar Buttercrunch. (Courtesy A. L. Heagle)

4E. Ambient ozone injury to poison ivy. (Courtesy A. L. Heagle)

4F. Ambient ozone injury to okra cultivar Emerald. (Courtesy A. L. Heagle)

4G. Ozone injury to Bel-W3 flue-cured tobacco. (Courtesy A. L. Heagle)

4H. Ambient ozone injury to peanut. (Courtesy A. L. Heagle)

4I. Ambient ozone injury to white pine (*Pinus strobus*). (Courtesy North Carolina State University)

4J. Ambient ozone injury to Ohio buckeye (*Aesculus glabra*). (Courtesy A. L. Heagle)

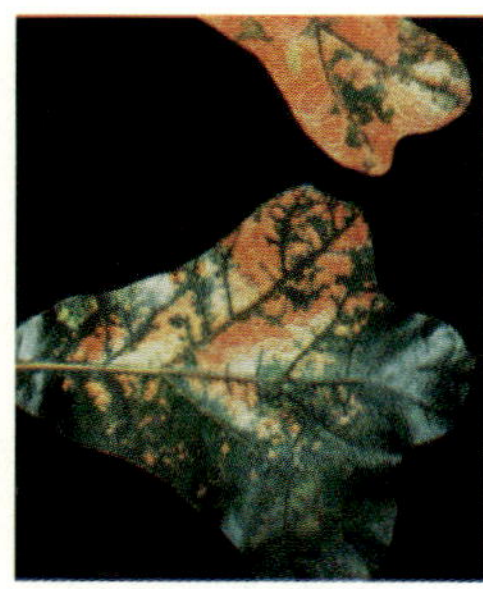

4K. Sulfur dioxide injury to blackjack oak (*Quercus marilandica*). (Courtesy A. L. Heagle)

4L. Sulfur dioxide injury to blackberry. (Courtesy A. L. Heagle)

4M. Sulfur dioxide injury to evening primrose. (Courtesy A. L. Heagle)

4N. Sulfur dioxide injury to squash. (Courtesy A. L. Heagle)

4O. Sulfur dioxide injury to a single-cross hybrid of corn. (Courtesy A. L. Heagle)

4P. Severe chlorine injury to rhododendron. (Courtesy E. Dutky)

4Q. Chlorine injury to field or dent corn. (Courtesy University of Illinois)

4R. Fluoride injury to field corn. (Courtesy University of Illinois)

4S. Fluoride injury to poplar (*Populus* sp.). (Courtesy University of Illinois)

4T. Fluoride injury to lilac. (Courtesy University of Illinois)

PLATE 5. CHEMICAL INJURIES

5A. 2,4-D injury to pin oak (*Quercus palustris*). (Courtesy Illinois Natural History Survey)

5B. Uneven ripening of grapes caused by 2,4-D drift injury. (Courtesy University of Illinois)

5C. 2,4-D drift injury to tomato fruit (left) and stem (right). (Left, courtesy University of Illinois and right, courtesy Purdue University)

5D. Accidental 2,4,5-T spray injury to white pine (*Pinus strobus*). (Courtesy University of Illinois)

5E. Bradford's pear (*Pyrus calleryana*) injured by a soil application of bromacil. (Courtesy E. Dutky)

5F. Boxwood (*Buxus* sp.) injured by a soil application of simazine. (Courtesy Maryland Department of Agriculture)

5G. Atrazine carryover injury to soybeans. (Courtesy E. L. Knake)

5H. Atrazine carryover injury to wheat headlands area. (Courtesy University of Illinois)

5I. Triazine carryover injury to field corn seedlings. (Courtesy University of Illinois)

5J. Profluralin (Tolban) herbicide carryover injury to field corn seedlings. (Courtesy University of Illinois)

5K. Trifluralin (Treflan) spray injury to a creeping bent grass putting green. (Courtesy University of Illinois)

5L. Dalapon soil carryover injury to citrus. (Courtesy J. B. Sinclair)

5M. Nicotine sulfate fog injury to a chrysanthemum bloom. (Courtesy Purdue University)

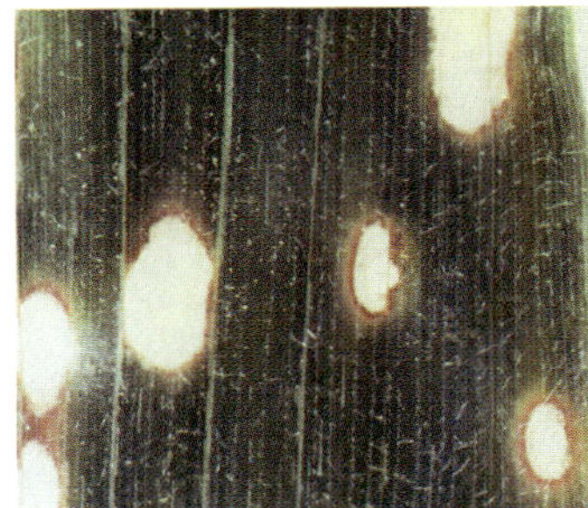

5N. Paraquat drift injury to a field corn leaf. (Courtesy University of Illinois)

5O. Glyphosate (Roundup) soil contamination injury to cantaloupe seedlings. (Courtesy E. Dutky)

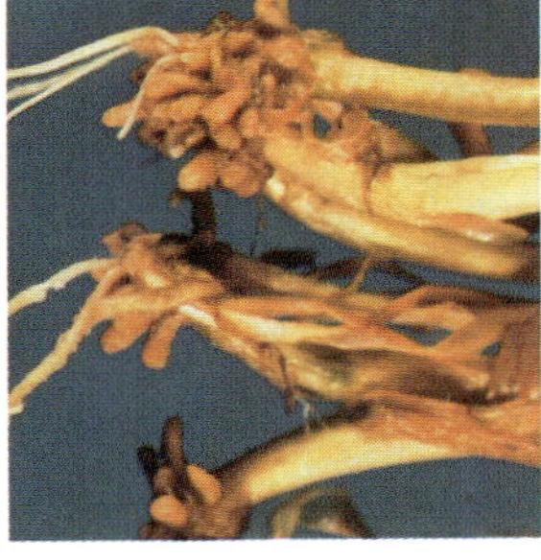

5P. Oryzalin (Surflan) carryover injury to wheat seedlings. (Courtesy E. Dutky)

5Q. Soluble salts injury to lettuce. (Courtesy J. Hartman)

PLATE 6. NUTRIENT DEFICIENCY AND TOXICITY SYMPTOMS AND GENETIC DISORDERS

6A. Nitrogen deficiency in cucumber fruit. (Courtesy J. W. Courter)

6B. Phosphorus deficiency in field corn. (Courtesy C. T. Schiller)

6C. Iron deficiency in blueberry. (Courtesy University of Illinois)

6D. Chlorosis of raspberry leaves caused by iron deficiency. (Courtesy University of Illinois)

6E. Zinc deficiency in *Vinca minor*. (Courtesy University of Illinois)

6F. Gray speck of oat caused by manganese deficiency. (Courtesy J. A. Browning)

6G. Manganese deficiency in peach leaves. (Courtesy University of Illinois)

6H. Marsh spot caused by manganese deficiency in a pea shoot (left) and seeds (right). (Courtesy D. J. Hagedorn)

6I. Manganese toxicity in tomato. (Courtesy University of Illinois)

6J. Blossom-end rot of squash fruit caused by calcium deficiency. (Courtesy Purdue University)

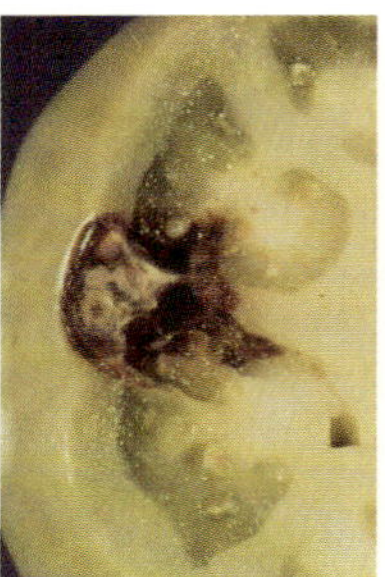

6K. Internal blossom-end rot of tomato caused by calcium deficiency. (Courtesy Purdue University)

6L. Physiological leaf roll of tomato caused in part by calcium deficiency. (Courtesy University of Illinois)

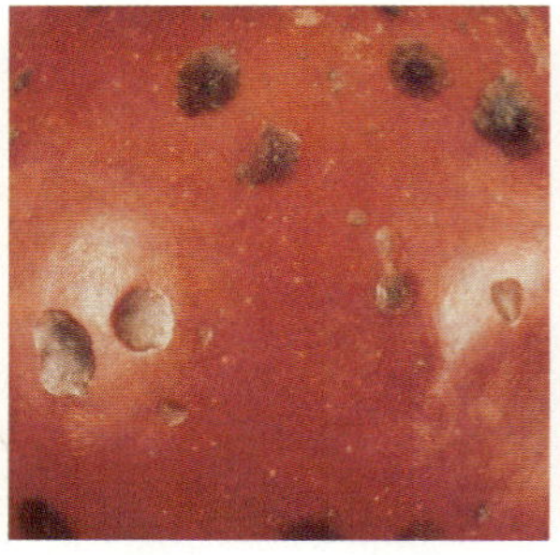

6M. Jonathan spot on apple caused in part by calcium deficiency. (Courtesy Purdue University)

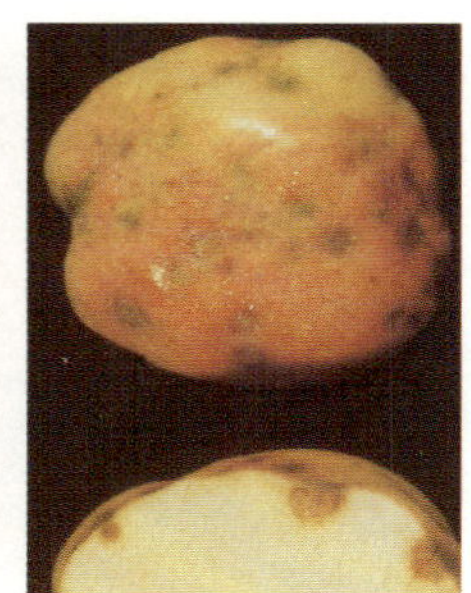

6N. Bitter pit of apple caused in part by calcium deficiency. (Courtesy Illinois Natural History Survey)

6O. Boron deficiency in cauliflower. (Courtesy A. F. Sherf)

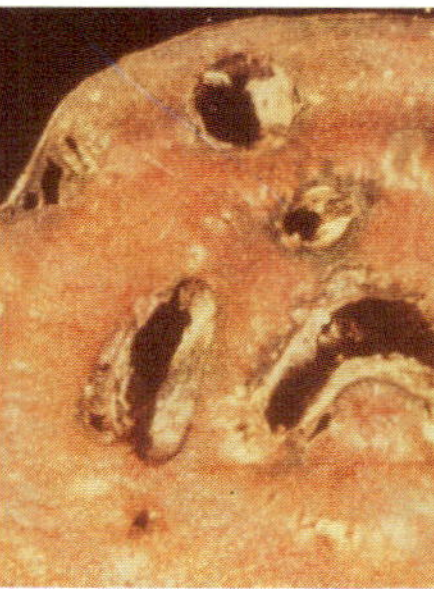

6P. Boron deficiency (internal black spot) in garden beet root. (Courtesy University of Wisconsin)

6Q. Boron toxicity in snap bean. (Courtesy D. J. Hagedorn)

6R. Strawberry June yellows, an inherited disorder (cause unknown). (Courtesy J. P. Fulton)

6S. Genetic variegation in field corn. (Courtesy University of Illinois)

Appendix 5.2. (*continued*)

Plant	Common name	Class[a]	Upper safe value[b]		Source[c]
			SS (10⁻⁵ mho/cm)	ECe	

Plant	Common name	Class[a]	SS (10^{-5} mho/cm)	ECe	Source[c]
Festuca pratensis	Meadow fescue	MT	93	13.3	a,c,e,h
		S	30	. . .	g
Ficus carica	Common fig	MT	59	8.4	a,c,h
Forsythia × intermedia	Border forsythia	MT	. . .	. . .	f
		MT foliage	. . .	. . .	k
Fragaria chiloensis	Chilean strawberry	VS	18	2.5	a,c,e,g
Fraxinus					
americana	White ash	MT	. . .	. . .	f
		T foliage	. . .	. . .	k
pennsylvanica	Green or red ash	MT	. . .	. . .	f
		MT foliage	. . .	. . .	k
Gazania spp.	Gazanias	T	. . .	. . .	j
Gleditsia triacanthos	Honey locust	T	. . .	. . .	f
		R foliage	. . .	. . .	k
Glycine max	Soybean	MT	52	7.5	a,b,c,e
		VS	20	. . .	g
Gossypium hirsutum	Upland cotton	T	121	17.3	a,b,c,e
Hedera canariensis	Algerian ivy	S	24	3.4	c
		S	28	4.0	d
Helianthus tuberosus	Jerusalem artichoke	T	. . .	. . .	k
Hibiscus rosa-sinensis	Chinese hibiscus or rose-of-China	S	32	4.5	c,d
Hippophae rhamnoides	Sea-buckthorn	R foliage	. . .	. . .	k
Hordeum vulgare	Common barley, forage	MT	91	13.0	a,b,c,e
	Common barley, grain	T	126	18.0	a,b,c,e
Ilex					
cornuta	Chinese holly	VS	24	3.4	c,d,g
crenata	Japanese holly	S	40	. . .	g
Ipomoea batatas	Sweet potato	S	42	6.0	a,b,c,e
Juglans					
nigra	Black walnut	S	. . .	. . .	f
		MT foliage	. . .	. . .	k
regia	Persian or English walnut	S	34	4.8	a
		MT foliage	. . .	. . .	k
Juniperus					
chinensis cv. pfitzeriana	Pfitzer juniper	T	. . .	. . .	f
		T foliage	. . .	. . .	k
communis	Common juniper	S	45	6.4	c,d
		VS	20	. . .	g
horizontalis cv. andorra	Andorra juniper	MT	. . .	. . .	f
		T foliage	. . .	. . .	k
virginiana	Red-cedar	MT roots	. . .	. . .	f
		T, MT foliage	. . .	. . .	f,k
Kolkwitzia amabilis	Beauty-bush	S foliage	. . .	. . .	k
Lactuca sativa	Lettuce	S	36	5.1	a,b,c,e
Lampranthus sp.	Trailing ice-plant	T	. . .	. . .	j
productus	Purple ice-plant	T	70	10.0	d
Lantana camara	Lantana or yellow-sage	S	36	5.1	c,d
Larix spp.	Larches	S	. . .	. . .	f
Leucophyllum frutescens	Cenizo or barometer-bush	T	70	10.0	d
Ligustrum spp.	Privets	MT foliage	. . .	. . .	k
amurense	Amur privet	MT	. . .	. . .	f
lucidum	Glossy or wax-leaf privet	MT	46	6.6	c,d
Linum usitatissimum	Common flax	S	41	5.9	a,b,c,d
Lolium perenne	Perennial rye grass	MT	85	12.2	a,b,c,e
Lonicera spp.	Honeysuckles	T foliage	. . .	. . .	k
japonica	Japanese honeysuckle	MT	. . .	. . .	f
tatarica	Tatarian honeysuckle	T	. . .	. . .	f
Lotus					
corniculatus	Bird's-foot trefoil	MT	70	10.0	a,b,c,e
uliginosus	Big trefoil	S	35	4.9	a,c,e,h
Lycium barbarum	Matrimony-vine	T	. . .	. . .	f
Lycopersicon esculentum	Tomato	S	53	7.6	a,b,c,e
Malus spp.	Apples and crabapples	S foliage	. . .	. . .	k
baccata	Siberian crabapple	MT	. . .	. . .	f
sylvestris	Apple or crabapple	S	34	4.8	a,c,e,g
Medicago sativa	Alfalfa or lucerne	MS	62	8.8	a,b,c,e

(*continued on next page*)

Appendix 5.2. (*continued*)

Plant	Common name	Class[a]	SS $(10^{-5}$ mho/cm)	ECe	Source[c]
			Upper safe value[b]		
Melilotus spp.	Sweet-clovers	MS	. . .	. . .	c,h
Mesembryanthemum croceum	Croceum ice-plant or fig marigold	T	70	10.0	d
Morus spp.	Mulberries	T	. . .	. . .	f
		S foliage	. . .	. . .	k
Nandina domestica	Heavenly-bamboo or sacred-bamboo	S	34	4.8	c,d
Nerium oleander	Oleander	S	64	9.1	c
		MT	56	8.0	d
Nicotiana tabacum	Tobacco	S	40	. . .	g
Olea europaea	Common olive	MT	59	8.4	a,c,e,h
Oryza sativa	Rice or upland rice	MS	50	7.2	a,b,c,e
Paspalum dilatatum	Dallis grass	MT	. . .	. . .	c,h
Pelargonium × hortorum	Zonal or florist's geranium	T	75	. . .	g
Persea americana	Avocado	S	26	3.7	a,c,e,h
Phalaris					
arundinacea	Reed canary grass	MT	. . .	. . .	c,e,h
stenoptera	Harding's grass	MT	78	11.2	a,b,c,e
Phaseolus vulgaris	Beans, various	S	25	3.6	a,b,c,e
Philadelphus spp.	Mock-oranges	MT foliage	. . .	. . .	k
Phleum pratense	Common timothy	S	42	6.0	c,e
Phoenix dactylifera	Date palm	T	125	17.9	a,c,e,h
Photinia frazeri	Photinia	VS	14	2.0	d
Phyla nodiflora	Mat-grass or turkey-tangle	T	. . .	. . .	j
Picea					
abies	Norway spruce	S roots	. . .	. . .	f
		MT foliage	. . .	. . .	f,k
glauca	White spruce	S roots	. . .	. . .	f
		ST foliage	. . .	. . .	f,k
pungens	Colorado blue spruce	S roots	. . .	. . .	f
		RT foliage	. . .	. . .	f,k
Pinus					
banksiana	Jack pine	R foliage	. . .	. . .	k
halepensis	Aleppo pine	MT	56	8.0	d
		R foliage	. . .	. . .	k
mugo	Mugo or Swiss mountain pine	R foliage	. . .	. . .	k
nigra	Austrian pine	T,R foliage	. . .	. . .	f,k
resinosa	Red or Norway pine	S roots	. . .	. . .	f
		S foliage	. . .	. . .	k
strobus	Eastern white pine	S roots	. . .	. . .	f
		VS foliage	. . .	. . .	f,k
sylvestris	Scotch or Scot's pine	MT foliage	. . .	. . .	f
		S foliage	. . .	. . .	k
thunbergiana	Japanese black pine	MT	42	6.0	d
Pisum sativum	Garden pea	S	21+	3+	c,h
Pittosporum					
crassifolium	Karo	T	. . .	. . .	j
tobira	Japanese pittosporum	S	50	6.0	c
		S	28	4.0	d
Podocarpus macrophyllus	Yew podocarpus	VS	14	2.0	d
Populus					
alba	White poplar	T	. . .	. . .	f
deltoides	Eastern cottonwood	R foliage	. . .	. . .	k
grandidentata	Large-toothed aspen	T foliage	. . .	. . .	k
nigra var. *italica*	Lombardy poplar	S	. . .	. . .	f
		T foliage	. . .	. . .	k
tremuloides	Trembling or quaking aspen	T foliage	. . .	. . .	k
Prunus sp.	Plum	S	30	4.3	a,c,e
armeniaca	Apricot	S	27	3.8	a,c,e,h
domestica	Prune plum, common or garden plum	S	30	4.3	c,e,h
dulcis	Almond	S	30	4.3	a,c,e,h
persica	Peach	S	29	4.1	a,c,e,g
		S foliage	. . .	. . .	k
virginiana	Black chokecherry	T foliage	. . .	. . .	k
Pseudotsuga menziesii	Douglas fir	S roots	. . .	. . .	f
Puccinellia nuttalliana	Nuttall's alkali grass	T	. . .	. . .	c,h
Punica granatum	Pomegranate	MT	59	8.4	a,c,h

(*continued on next page*)

Appendix 5.2. (*continued*)

Plant	Common name	Class[a]	Upper safe value[b] SS (10^{-5} mho/cm)	ECe	Source[c]
Pyracantha					
braperi	Pyracantha	S	51	7.3	c
coccinea	Scarlet or everlasting fire-thorn	MT	. . .	. . .	f
fortuneana	Pyracantha	MT	42	6.0	d
Pyrus communis	Common pear	S	34	4.8	a,c,h
		T foliage	. . .	. . .	k
Quercus					
alba	White oak	T	. . .	. . .	f
macrocarpa	Bur oak	T	. . .	. . .	f
rubra	Northern red oak	T foliage	. . .	. . .	k
virginiana	Southern live oak	R	. . .	. . .	f
Raphanus sativus	Radish	S	35	5.0	a,c,e,h
Rhaphiolepis indica	Indian hawthorn	MT	42	6.0	d
Rhododendron spp.	Azaleas	VS	30	. . .	g
		VS[3]	75	. . .	i
	Rhododendrons	VS	30	. . .	g
Rhus					
trilobata	Skunk-bush	T	. . .	. . .	f
typhina	Staghorn sumac	R foliage	. . .	. . .	k
Ribes nigrum	European black currant	MT	. . .	. . .	f
Ricinus communis	Castor bean	MT	46	6.5	h
Robinia pseudoacacia	Black locust	T	. . .	. . .	f
		R foliage	. . .	. . .	k
Rosa spp.	Roses	VS	18	2.6	c,d
		S	40	. . .	g
multiflora	Multiflora rose	S	. . .	. . .	f
Rosmarinus					
officinalis	Rosemary	MT	56	8.0	c,d
Rubus spp.	Blackberries	VS	27	3.8	a,c,e
idaeus	Red or European red raspberry	VS	22	3.2	a,c,e
ursinus	California blackberry	VS	26	3.8	a,c,e
Saccharum officinarum	Sugarcane	MS	71	10.2	c,e
Saintpaulia ionantha	Common African violet	VS	50	. . .	g
Salix spp.	Willows	T	. . .	. . .	f
alba	White willow	S foliage	. . .	. . .	k
var. *vitellina*	Golden willow	T	. . .	. . .	f
nigra	Black willow	MT foliage	. . .	. . .	k
purpurea var. *nana*	Arctic blue willow	S	. . .	. . .	f
Sanguisorba sp.	Burnet, forage	VS	21	. . .	h
Secale cereale	Common rye, grain	MT	70	10.0	h
	Common rye, hay	MT	60	8.5	c,h
Sesbania macrocarpa	Peatree or Colorado river hemp	MT	66	9.4	a,b,c,e
Setaria italica	Italian or foxtail millet	MS	. . .	. . .	c,e
Shepherdia argentea	Silver buffalo-berry	T	. . .	. . .	f
Solanum tuberosum	Potato or white Irish potato	S	41	5.9	a,b,c,e
Sorbus aucuparia	European mountain-ash	T foliage	. . .	. . .	k
Sorghum					
bicolor	Sorghums, broom-corn, milo, kaffir	MT	77	11.0	a,b,c,e
× *drummondii*	Sudan grass	MT	101	14.4	a,c,e
Spinacia oleracea	Spinach	MS	60	8.7	a,b,c,e
Spiraea					
× *bumalda*	Bumalda spirea	S foliage	. . .	. . .	k
× *vanhouttei*	Bridal-wreath	S	. . .	. . .	f
Sporobolus airoides	Alkali sacaton	MT	. . .	. . .	c
		T	126	18.0	e
Syringa					
amurensis var. *japonica*	Japanese tree lilac	T foliage	. . .	. . .	k
vulgaris	Common lilac	T foliage	. . .	. . .	k
Syzygium paniculatum	Brush cherry	T	70	10.0	d
Tamarix pentandra					
or *T. ramosissima*	Tamarix	T	. . .	. . .	f
Taxus sp.	Yew	MT foliage	. . .	. . .	k
cuspidata	Japanese yew	T foliage	. . .	. . .	f
Thuja spp.	Arborvitaes	T roots	. . .	. . .	f
		S foliage			

(*continued on next page*)

Appendix 5.2. (*continued*)

Plant	Common name	Class[a]	Upper safe value[b]		Source[c]
			SS $(10^{-5}$ mho/cm)	ECe	
Thuja (continued)					
orientalis	Oriental arborvitae	MS	51	7.2	c
		S foliage	...	...	k
Tilia					
americana	American linden	S	...	...	f
		MT foliage	...	...	k
cordata	Little-leaf or European linden	S	...	...	f
Trachelospermum jasminoides	Star- or confederate-jasmine	VS	25	3.5	c
		VS	14	2.0	d
Trifolium spp.	Alsike, ladino, and red clovers	S	40	5.7	a,b,c,e
alexandrinum	Berseem or Egyptian clover	MT	72	10.3	a,c,e
Triticum aestivum	Common or bread wheat	MT	91	13.0	a,b,c,e
Tsuga canadensis	Canada or eastern hemlock	S roots	...	...	f
		S,VS foliage			f,k
Ulmus					
americana	American or white elm	MT	...	...	f
		MT foliage	...	...	k
pumila	Siberian or dwarf elm	T	...	...	f
		MT foliage	...	...	k
Vaccinium corymbosum	High-bush blueberry	VS	20	...	g
Viburnum spp.	Viburnums	S	32	4.6	c,f
		S	28	4.0	d
opulus	European cranberry-bush	S foliage	...	...	k
Vicia					
faba	Broad or horse bean	MS	53	7.6	a,b,c,e
sativa	Common vetch	MS	53	7.5	a,c,e
Vigna unguiculata	Cow pea, southern pea, or black-eyed pea	S	34	4.9	a,c,e
Vitis spp.	Grapes	S	47	6.7	a,c,e,h
Weigela florida cv. Eva Rathke	Weigela	MT	...	...	f
Xylosma senticosum	Shiny xylosma	S	37	5.3	c,d
Yucca filamentosa	Adam's-needle or silk-grass	T	...	...	f
Zea mays	Maize or dent corn, forage	S	60	8.6	a,c,e
	Maize or dent corn, grain	S	41	5.9	a,c,e
		S	30	...	g
	Sweet corn	S	41	5.9	a,b,c,e

[a] VS (32) = very sensitive; S (66) = sensitive; MT (105) = moderately tolerant; T (147) = tolerant; and R = resistant. In many cases, there is good agreement in publications on the sensitivity of plants to soil soluble salts. Discrepancies among some reports may be caused by different methodology, cultivars, and soil texture and the presence of toxic ions.

[b] See Table 5.3 for ranges and upper safe values of electrical conductivity (ECe) of soil extracts. Readings taken from an extract of one volume of air-dried field soil to two volumes of distilled water. Many soluble salts conductivity meters read electrical conductivity in millimhos/cm (10^{-3} mho or mmho/cm); others read micromhos/cm (10^{-6} mho or μmho/cm). Thus, multiply mmho readings by 100 and μmho by 0.10 to obtain soluble salts (SS) values.

[c] a. Ayers, R. S. 1977. Quality of water for irrigation. J. Irrig. Drain. Div. Proc. Am. Soc. Civ. Eng. 103(IR2):135-150.

b. Bernstein, L. 1964. Salt tolerance of plants. USDA Inf. Bull. 283.

c. Carter, D. L. 1982. Salinity and plant productivity. In: Handbook of Agricultural Productivity. Vol. 1, Plant Productivity. M. Rechcigl, Jr., ed. CRC Press, Boca Raton, FL.

d. Francois, L. E. 1980. Salt injury to ornamental shrubs and ground covers. USDA Home Garden Bull. 231.

e. Maas, E. V., and Hoffman, G. J. 1977. Crop salt tolerance—Current assessment. J. Irrig. Drain. Div. Proc. Am. Soc. Civ. Eng. 103(IR2):115-134.

f. Pellett, N. E. Salt tolerance of trees and shrubs. Univ. Ver. Agric. Ext. Serv. Hort. Brieflet 1212 (this entry is based only on observations of susceptibility, moderate tolerance, and tolerance of vegetation along roadways subject to salt runoff and sprays from deicing operations).

g. Plant Disease and Insect Clinic, N.C. State Univ., Raleigh.

h. Richards, L. A., ed. 1954. Diagnosis and improvement of saline and alkali soils. USDA Handb. 60.

i. Smith, G. 1978. Testing by the grower for soluble salts. Proc. Int. Plant Propag. Soc. 28:21-25.

j. Smith, M. D., ed. 1984. The Ortho Problem Solver, 2nd ed. Chevron Chemical Co., San Francisco.

k. Lumis, G. P., Hoftra, G., and Hall, R. 1977. Salt damage to roadside plants. Can. Minist. Agric. Food, Factsheet AGDEX 275/690.

l. J. Trolinger (*personal communication*).

Measurements and Conversions

Weights

1 gram (g)
  = 1,000 milligrams
  = 0.035274 ounce (oz) avoir-
    dupois (av)
  = 0.0022046 pound
  = 0.001 kilogram

1 ounce avoirdupois (oz av)
  = 0.911458 troy oz
  = 28.349527 grams
  = 0.0625 pound (lb)

1 ounce (troy)
  = 1.097 oz av
  = 31.10348 grams

1 kilogram (kg)
  = 1,000 grams
  = 35.273957 oz av
  = 2.20462 pounds avoirdupois

1 pound avoirdupois (lb av)
  = 16 oz av
  = 14.5833 troy oz
  = 453.5924 grams
  = 0.4535924 kilogram

1 ton (short, U.S.)
  = 2,000 pounds (av)
  = 2,430.56 troy pounds
  = 0.892857 long ton
  = 0.907185 metric ton
  = 907.18486 kilograms
  = 32,000 oz

1 ton (long, U.S.)
  = 2,240 pounds
  = 2,722.22 troy pounds
  = 1.120 short tons
  = 1.016 metric tons
  = 1,016.04 kilograms

1 ton (metric)
  = 2,204.6 pounds
  = 1.1023 short tons
  = 0.984 long ton
  = 1,000 kilograms

Area Measures

1 square millimeter (sq mm)
  = 0.01 sq centimeter
  = 0.000001 sq meter
  = 0.00155 sq inch

1 square centimeter (sq cm)
  = 100 sq millimeters
  = 0.155 sq inch
  = 0.001076 sq foot

1 square inch (sq in)
  = 6.451626 sq centimeters
  = 0.0069444 sq foot

1 square foot (sq ft)
  = 144 sq inches
  = 0.111111 sq yard
  = 0.0929 sq meter
  = 0.003673 sq rod

1 square yard (sq yd)
  = 9 sq feet
  = 1,296 sq inches
  = 0.83613 sq meter
  = 0.03306 sq rod

1 square meter (sq m)
  = 10.76387 sq feet
  = 1,550 sq inches
  = 1.195985 sq yards
  = 0.039537 sq rod
  = 1 million sq millimeters
  = 10,000 sq centimeters

1 square rod (sq rod)
  = 30.25 sq yards
  = 25.29295 sq meters
  = 272.25 sq feet

  = 0.00625 sq acre
  = 0.0025293 hectare

1 square acre
  = 43,560 sq feet
  = 4,840 sq yards
  = 160 sq rods
  = 4046.893 sq meters
  = 0.404687 hectare
  = 0.0015625 sq mile
  = strip 8 feet wide and 1 mile long
    (approximate)

1 hectare (ha)
  = 2.471 acres
  = 395.367 rods
  = 10,000 sq meters
  = 0.01 sq kilometer
  = 0.0039 sq mile

1 square kilometer (sq km)
  = 0.3861 sq mile
  = 247.1 acres
  = 100 hectares
  = 1 million sq meters

1 square mile (sq mi)
  = 640 acres
  = 1 section
  = 258.9998 hectares
  = 102,400 sq rods
  = 3,097,600 sq yards
  = 2.589998 sq kilometers

Cubic Measures

1 cubic centimeter (cc)
  = 0.06102 cu inch
  = 1,000 cu millimeters
  = 0.000001 cu meter

1 cubic inch (cu in.)
  = 16.38716 cubic centimeters
  = 0.0005787 cu foot
  = 0.004329 gallon (U.S.)

1 cubic foot (cu ft)
= 0.80356 bushel
= 1,728 cu inches
= 0.037037 cu yard
= 0.028317 cu meter
= 7.4805 U.S. gallons
= 6.229 British or Imperial gallons
= 28.317 liters
= 29.922 quarts (liquid)
= 25.714 quarts (dry)

1 cubic foot of water
= 62.43 pounds
(1 pound of water
= 27.68 cu inches
= 0.1198 U.S. gallon
= 0.01602 cu foot)

1 cubic foot (cu ft) of dry soil
(approximate)
= sandy (90 lb),
loamy (80 lb),
clayey (75 lb)

1 bushel (bu) dry soil (approximate)
= sandy (112 lb),
loamy (100 lb),
clayey (94 lb)

1 cubic yard (cu yd)
= 27 cu ft
= 46,656 cu inches
= 764.559 liters
= 292 U.S. gallons
= 168.176 British gallons
= 1,616 pints (liquid)
= 807.9 quarts (liquid)
= 21.694 bushels
= 0.764559 cubic meter

1 cubic meter (cu m)
= 1.30794 cu yards
= 35.3144 cu feet
= 28.3776 bushels
= 264.173 gallons
= 2,113.4 pints (liquid)
= 1,056.7 quarts (liquid)
= 61,023 cu inches

Volumes

Dry

1 quart (U.S.)*
= 67.2 cu inches
= 2 pints

*1 pint or quart dry measure is about 16%
larger than 1 pint or quart liquid measure

= 1.1012 liters
= 0.125 peck
= 0.03125 bushel
= 0.038889 cu foot

1 peck (U.S.)
= 0.25 bushel
= 2 gallons
= 8 quarts
= 16 pints
= 32 cups
= 8.80958 liters
= 537.605 cu inches

1 bushel (U.S.)
= 4 pecks
= 32 quarts
= 64 pints
= 128 cups
= 1.2445 cu feet
= 35.2383 liters
= 0.304785 barrel
= approximately 1/20 cubic yard

Liquid

1 milliliter (ml)
= 1 cubic centimeter
= 0.001 liter
= 0.061 cu inch
= 0.03815 fluid ounce

1 teaspoon (tsp)
= 5 milliliters
= 0.17 fluid ounce

1 tablespoon (Tsp)
= 3 teaspoons
= 14.8 milliliters or cubic centimeters
= 1/2 fluid ounce
= 0.902 cu inch
= 0.063 cup

1 fluid ounce (fl oz, U.S.)
= 2 tablespoons
= 0.125 cup
= 0.0625 pint
= 0.03125 quart
= 0.00781 gallon
= 29.573 milliliters
= 1.80469 cu inches
= 0.029573 liter

1 cup
= 16 tablespoons
= 8 fluid ounces

= 0.5 pint
= 236.6 cubic centimeters or
milliliters

1 pint (U.S.)*
= 16 fluid ounces
= 32 tablespoons
= 2 cups
= 0.125 gallon
= 473.167 milliliters
= 1.04 pounds of water
= 28.875 cu inches
= 0.473167 liter
= 0.01671 cu foot

1 liter (l)
= 2.1134 pints
= 1.0567 liquid quarts (U.S.)
= 0.9081 dry quart (U.S.)
= 0.264178 gallon (U.S.)
= 1,000 milliliters or cubic centimeters
= 33.8147 fluid ounces
= 61.025 cu inches
= 0.0353 cu foot
= 0.028378 bushel

1 quart (U.S.)*
= 2 pints
= 4 cups
= 32 fluid ounces
= 57.749 cu inches
= 64 tablespoons
= 0.25 gallon
= 0.94633 liter
= 0.3342 cu foot

1 gallon (U.S.)
= 4 quarts
= 8 pints
= 16 cups
= 128 fluid ounces
= 0.1337 cu foot
= 0.83268 British or Imperial
gallon
= 3,785.4 milliliters or cubic
centimeters
= 231 cu inches
= 8.337 pounds water
= 3.782 kilograms

1 barrel (U.S.)
= 31.5 gallons
= 7,056 cu inches
= 0.11924 cu meter

Linear (or Distance) Measures

1 millimeter (mm)
    = 0.1 centimeter
    = 0.01 decimeter
    = 0.001 meter
    = 1,000 micrometers
    = 0.03937 inch (about 1/25 inch)

1 centimeter (cm)
    = 10 millimeters
    = 0.01 meter
    = 0.3937 inch
    = 0.03281 foot
    = 0.010936 yard (U.S.)

1 inch (in.)
    = 25.4 millimeters
    = 2.54 centimeters
    = 0.0254 meter
    = 0.083333 foot
    = 0.027778 yard
    = 0.00505 rod

1 foot (ft)
    = 12 inches
    = 0.333 yard
    = 30.48 centimeters
    = 0.3048 meter

1 yard (yd)
    = 36 inches
    = 3 feet
    = 0.181818 rod
    = 0.9144 meter

1 meter (m)
    = 100 centimeters
    = 39.37 inches
    = 3.2808 feet
    = 1.09361 yards
    = 0.1988 rod
    = 0.001 kilometer

1 rod
    = 5.5 yards
    = 16.5 feet
    = 198 inches
    = 5.02921 meters
    = 0.003125 mile

1 kilometer (km)
    = 1,000 meters
    = 3,280.8 feet
    = 1,093.6 yards
    = 0.62137 statute mile
    = 0.53961 nautical mile

1 mile (statute)
    = 5,280 feet
    = 1,760 yards
    = 1,609.35 meters
    = 320 rods
    = 0.86836 nautical mile

Dilutions

1 part per million (ppm)
    = 1 milligram per liter or kilogram
    = 0.0001%
    = 0.013 ounce by weight in 100 gallons
    = 0.379 gram in 100 gallons
    = 1 inch in nearly 16 miles
    = 2 crystals of sugar in a pound
    = 1 ounce of salt in 62,500 pounds of sugar
    = 1 minute of time in about 2 years
    = a 1-gram needle in a 1-ton haystack
    = 1 ounce of sand in 34¼ tons of cement
    = 1 ounce of dye in 7,530 gallons of water
    = 1 pound in 500 tons
    = 1 penny of $10,000

1 part per billion (ppb)
    = 1 microgram per liter or kilogram
    = 1 inch in nearly 16,000 miles
    = 1 drop in 20,000 gallons
    = 1 ounce in 753 million gallons
    = 2 crystals of sugar in 1,000 pounds

1 percent (%)
    = 10,000 parts per million
    = 1 gram per 100 milliliters
    = 10 grams per liter
    = 1.28 ounces by weight per gallon
    = 8.336 pounds per 100 gallons

Miscellaneous Weights and Measures

1 micron or micrometer (µm)
    = 0.00039 inch

1 acre inch of water
    = 27,154 gallons
    = 624.23 gallons per 1,000 sq feet

1 pound per cubic foot
    = 0.26 gram per cubic inch

1 gram per cubic inch
    = 3.78 pounds per cubic foot

Weights per Unit Area

1 ounce per square foot
    = 2,722.5 pounds per acre

1 ounce per square yard
    = 302.5 pounds per acre

1 ounce per 100 square feet
    = 27.2 pounds per acre

1 ounce per 1,000 square feet
    = 2.72 pounds per acre

1 pound per 100 square feet
    = 435.6 pounds per acre

1 pound per 1,000 square feet
    = 43.56 pounds per acre

1 pound per acre
    = 1 ounce per 2,733 square feet (0.37 oz/1,000 sq ft)
    = 0.0104 gram per square foot
    = 1.12 kilograms per hectare

100 pounds per acre
    = 2.5 pounds per 1,000 square feet
    = 1.04 grams per square foot

5 gallons per acre
    = 1 pint per 1,000 square feet
    = 0.43 milliliter per square foot

100 gallons per acre
    = 2.5 gallons per 1,000 square feet
    = 1 quart per 100 square feet
    = 935 liters per hectare

1 quart per 100 gallons (approximate)
    = 10 milliliters per gallon

1 pound per gallon
    = 120 grams per liter

1 kilogram per 100 square meters
    = 2.05 pounds per 1,000 square feet
    = 1 kilogram per hectare
    = 89 pounds per acre

1 kilogram per hectare
    = 0.0205 pound per 1,000 square feet
    = 0.1 kilogram per 100 square meters
    = 0.90 pound per acre

Metric-English Conversion Factors

To change	To	Multiply by
Inches	Centimeters	2.54
Centimeters	Inches	0.3937
Feet	Meters	0.3048
Meters	Inches	39.37
Square inches	Square centimeters	6.452
Square centimeters	Square inches	0.155
Square yards	Square meters	0.836
Square meters	Square yards	1.196
Cubic yards	Cubic meters	0.765
Cubic meters	Cubic yards	1.308
Cubic inches	Cubic centimeters	16.387
Cubic centimeters	Cubic inches	0.061
Cubic centimeters	Fluid ounces	0.034
Fluid ounces	Cubic centimeters	29.57
Quarts	Liters	0.946
Liters	Quarts	1.057
Grams	Ounces (avoirdupois)	0.0352
Ounces (avoirdupois)	Grams	28.349
Pounds (avoirdupois)	Kilograms	0.454
Kilograms	Pounds (avoirdupois)	2.2046
Pounds (apothecary)	Kilograms	0.373
Kilograms	Pounds (apothecary)	2.205
Ounces (apothecary)	Grams (sq ft)	31.103
Grams (sq ft)	Pounds (acre)	96

Temperature Conversions

Fahrenheit to Celsius (Centigrade)

To convert from Fahrenheit to Celsius, subtract 32 from the Fahrenheit reading, multiply by 5, and divide the product by 9. Example:

$$131°F - 32 = 99$$
$$99 \times 5 = 495$$
$$495 \div 9 = 55°C$$

Celsius (Centigrade) to Fahrenheit

To convert from Celsius to Fahrenheit, multiply the Celsius reading by 9, divide the product by 5, and add 32. Example:

$$25°C \times 9 = 225$$
$$225 \div 5 = 45$$
$$45 + 32 = 77°F$$

Glossary

Å: *see* Ångstrom

a- or **an-** (prefix): not having; not; without; lacking; denotes the opposite

A horizon: surface layer of a mineral soil varying in thickness and having maximum organic matter accumulation, maximum biological activity, and/or eluviation by water of materials (e.g., iron and aluminum oxides and silicate clays)

ab- (prefix): position away from

abbr.: abbreviation

aberrant (n. aberration): atypical; deviant; straying from the normal type

abiotic (syn. inanimate): of or pertaining to the nonliving

abrupt: as if cut off transversely; truncate; without a thin, sterile margin

abscise (n. abscission): to fall off, as with leaves, flowers, fruits, or other plant parts, during a natural process involving formation of a separation layer or wall causing the plant part(s) to fall or shed

abscission layer or **zone**: layer of thin-walled cells extending across the base of the petiole or peduncle, whose breakdown separates the leaf or fruit from the stem, causing the leaf or fruit to fall; separation layer

accession number: numerical code given to a culture, plant, or specimen in a collection of microorganisms or plants

-aceae (suffix): attached to the base of a name to form the family name

achene: simple, dry, one-celled, one-seeded, indehiscent fruit with the seed attached at only one point to the ovary wall

acid soil: soil with an acid reaction (pH below 7.0); more technically, a soil having a preponderance of hydrogen ions over hydroxy ions in solution (commonly measured from a soil-water extract with a pH meter). *See* pH

acquired resistance (syn. induced resistance or acquired immunity): noninherited resistance response developed by a normally susceptible host after a predisposing treatment, such as inoculation with a bacterium, fungus, or virus or treatment with certain chemicals

acquisition access period: period of time that a vector has access to a source of inoculum

acquisition feeding period: time during which a vector feeds on an infected plant to acquire a virus for subsequent transmission, i.e., to become viruliferous

acquisition threshold period: minimum feeding time for a vector to acquire a virus or other pathogen

actinomycete: organism characterized by fine, branching filaments (usually less than 1 μm in diameter that readily break into fragments resembling bacterial cells) and classified as a gram-positive bacterium in the order Actinomycetales (except for members of Mycobacteriaceae)

acute: (of disease) as opposed to chronic, producing symptoms that are severe, of sudden onset, and rise in severity within a short period of time; *see* chronic; (of shape) sharp or tapered to a fine point; less than a right angle; (of pileus margins) tapering to a thin edge

adaptation (adj. adaptive): from an evolutionary standpoint, characteristic of a living organism that improves its chances for survival in the environment of its habitat; change brought about in the population of an organism as a result of exposure to a particular set of environmental conditions; change in structural or functional development of an individual or population enabling the organism to adjust to environmental conditions

adj.: adjective

adventitious: pertaining to plant parts, e.g., roots, tubers, flowers, and buds, produced in an unusual or irregular position or at an unusual time of development; arising sporadically

aerobe: microorganism living or active only in the presence of free molecular oxygen. *See* anaerobe

aerobic: pertaining to an environment or condition in which the concentration of oxygen is sufficient for chemical, physical, or metabolic processes

aerobic respiration: respiration in the presence of free oxygen. *See* anaerobic respiration and respiration

aflatoxins: mycotoxins (or series of toxic metabolites) produced by strains of *Aspergillus flavus* and other species of this fungus growing on peanuts (groundnuts), maize (corn), cereals, etc. that cause aflatoxicoses in animals

agar or **agar-agar**: nonnitrogenous gelatinous mixture of dried polysaccharides, some anionic, extracted from certain red algae (seaweeds, Rhodophyceae) and used for preparing semisolid culture media on which microorganisms are grown and studied (forms a gel with water below about 40°C and is used at a concentration of 1.5–2.0% to solidify media)

agaric: a member of the family Agaricaceae; a gill-bearing fungus such as a mushroom

agent of disease: organism (pathogen) or abiotic factor that causes disease

aggressiveness: relative ability of a plant pathogen to colonize and cause damage to plants. *See* virulence

agonomycetes: filamentous fungi (Hyphomycetes, Deuteromycotina) that do not form spores other than chlamydospores; may be states of members of Ascomycetes, Basidiomycetes, or Deuteromycetes. *See Mycelia sterilia*

agronomy: art and science of crop production and soil management

air-dry: state of dryness at equilibrium with the moisture content in the surrounding atmosphere

air pollutants: contaminants, often man made, in the air (those most commonly affecting plants, especially lichens, are ozone, sulfur dioxide, fluorides, and particulate deposits, e.g., smoke, heavy metals, and photochemical smog)

alate: (adj.) winged; (n.) a winged form in the life cycle of certain insects (e.g., aphids)

albino: plant or part of a plant that is white, i.e., lacking chlorophyll (albinism is usually lethal in higher plants)

-ales (suffix): attached to the base of a generic name to form the name of an order

alkaline soil: soil with a basic or "sweet" reaction above pH 7.0. *See* pH

alkaloid: nitrogenous organic compound with alkaline (basic) properties, usually poisonous, produced by certain plants and fungi

allele (adj. allelic)**:** one of two or more alternate forms of a gene derived from the same gene by mutation and occupying the same locus on a particular chromosome; a form of gene. *See* gene

allelopathy (adj. allelopathic)**:** ability of one species to inhibit or prevent the growth of another species through the excretion of toxic substance(s)

allergy: condition of hypersensitivity to a specific substance (spores of molds, e.g., *Alternaria, Aspergillus, Cladosporium,* and *Penicillium* spp., cereal rusts, smuts, lichens, etc. in the air commonly cause allergies in humans)

alluvial soil: soil recently developed from deposited material and exhibiting essentially no horizon development or modifications

alternate: pertaining to bud or leaf arrangement in which there is one bud or one leaf at a node on a stem, i.e., not opposite or whorled

alternate host: plant genus or species different from the principal host on which a pathogen, e.g., a heteroecious rust, must develop to complete its life cycle (not to be confused with secondary host or alternative host)

alternative host: one of a parasite's several hosts (unlike an alternate host, not required for completion of the developmental cycle of the parasite)

ambient temperature: air temperature at a given time and place; not radiant temperature

amendment, physical: any substance added to the soil for the purpose of altering its physical condition (e.g., sand, clay, peat moss, organic compost matter, calcined clay, lime, sulfur, gypsum, sawdust, and bark)

amino acids: organic, nitrogenous acids from which protein molecules are constructed (there are 20 common amino acids in living organisms, each with the basic formula NH_2-CHR-COOH)

amoeboid: not having a cell wall and changing in form like an amoeba; moving by temporary processes (pseudopodia) from the surface of the vegetative body

amorphous: without definite form, shape, or organized structure

amphigenous: growing on either the upper or lower surface of leaves; (of fungi) growing all around or on two sides

anaerobe (adj. anaerobic)**:** organism able to grow or occur in the absence of free (molecular) oxygen, a condition usually high in carbon dioxide

anaerobic respiration: respiration in the absence of free oxygen. *See* aerobic respiration and respiration

anamorph: fungus in its asexual or imperfect reproductive state

anatomy: science and study of the shape and structure of organisms and their parts

angiosperm: flowering plant; a plant bearing seeds that develop in an enclosed carpel

Ångstrom (Å): unit of length equal to 1/10 millimicron or nanometer (nm), 1/10,000 micron or micrometer (μm), or 1/10,000,000,000 of a meter (1×10^{-10} m) (used for measuring wavelength and to express dimensions of intercellular structures of microorganisms and viruses)

anion: negatively charged ion that, during electrolysis, is attracted to positively charged surfaces (contrasted with cation)

annual: plant that completes its life cycle from seed and dies within 1 year or less; fungus fruiting body in an active, living state for one season, occasionally surviving into the beginning of the next year

annual ring: single-year, cylindrical growth of secondary xylem added to a woody plant stem by the cambium

antagonism: phenomenon in which a microorganism produces toxic metabolic products that kill, injure, or inhibit the growth of some other microorganism(s) in close proximity; a general term for counteraction between organisms or groups of organisms; (of pesticides) the action of two or more pesticides that reduces the effectiveness of one or all of the pesticide components

anterior: at or toward the front or head, as opposed to posterior; in front

anther: saclike, pollen-bearing portion (stamen) of a flower, usually borne on a filament

anthesis: developmental stage in flowering at which anthers rupture and shed pollen; a state of full bloom

anthocyanin: one of the water-soluble flavonoid pigments that occur in the vacuole of a cell and account for many of the usually red or blue, sometimes purple, colors in leaves, flowers, and fruits

antibiosis: antagonistic association between two organisms or between one organism and a metabolic product of another organism to the detriment of one of the organisms

antibiotic: (adj.) damaging to life; (n.) a chemical compound produced by one microorganism, which in very small amounts inhibits growth or kills other living organisms

antibody: new or altered specific protein (modified serum immunoglobulin molecule) produced by an animal's B lymphocytes in response to an antigenic stimulus (foreign substance), which reacts (binds) specifically with the antigen and renders it harmless (antibodies that cause lysis are called lysins; agglutination, agglutinins; and precipitation, precipitins). *See* ELISA and serology

antidesiccant: chemical compound that minimizes water loss from a plant through transpiration

antiseptic: substance that inhibits or destroys growth and activities of bacteria and other microorganisms in living tissues

antiserum (pl. antisera): blood serum (fluid fraction of coagulated blood) of any warm-blooded animal that contains antibodies to one or more specified antigens, e.g., antiovalbumin

apex (pl. apices, adj. apical): tip or end; uppermost point; part of the root or shoot containing the apical meristem

aphid: small, sucking insect of the family Aphididae (order Homoptera) that produces honeydew and may injure plants when populations are high or when serving as a virus vector

apical: at or near the end or tip (apex)

apomixis (adj. apomictic): development of unfertilized sexual cells into spores, etc.; the asexual (vegetative) production of seedlings in the usual sexual structures of the flower but without the mingling and segregation of chromosomes (seedling characteristics are the same as those of the maternal parent)

applanate: thin; flattened horizontally

appressed (syn. adpressed): closely flattened out or pressed against a surface

apterous: pertaining to the wingless stage of the life cycle of certain insects (e.g., aphids). *See* alate

arcuate: strongly curved as a bow; arclike

arid: dry

arthropod: member of the phylum Arthropoda, which consists of animals with articulated bodies and limbs and includes insects, arachnids, and crustaceans

articulate: jointed; segmented

ascomycete (adj. ascomycetous): member of Ascomycotina, or sac fungi, one of the major divisions of fungi,

which produce sexual spores (ascospores), often eight in number, within an ascus

ascospore: sexually produced spore formed inside an ascus by "free cell formation" after the union of two nuclei

ascus (pl. asci): typically a saclike or clavate cell in which meiosis occurs and ascospores are formed (usually eight) by "free cell formation" and borne in a fungal fruit body (ascocarp) after which the spores are often forcibly discharged into the air

-ase (suffix): used in the names of enzymes, e.g., pectinase

aseptic: free of microorganisms

asexual: without sex organs, gametes, or sexual spores; pertaining to the imperfect (anamorph) or vegetative state of an organism; imperfect

asexual reproduction: any type of reproduction not involving the union of gametes (karyogamy) and meiosis and independent of sexual processes

assimilative (n. assimilation): pertaining to growth before reproduction; nonreproductive; vegetative; pertaining to the transformation of chemical energy (foods) into living protoplasm

asymmetrical: lacking symmetry about a definite axis

ATCC: American Type Culture Collection, 12301 Parklawn Drive, Rockville, MD 20852-1776 (Telex ATCCROVE 908-768; fax 301-231-5826); a large collection of bacteria, fungi (including yeasts), viruses, viroids, cells, etc. available on request by payment of a small fee

atrophy: to progressively decline in size (as an organ or an entire body); to waste away

attenuate (adj. attenuated): to weaken or decrease in virulence or pathogenicity; to thin, extend, or gradually taper; to draw out, to narrow, more or less to a point; to thin in consistency

auricle: one of a pair of clawlike appendages in grasses that occurs at the base of the leaf blade or at the apex of the leaf sheath

auto- (prefix): self-inducing; self-producing

autoclave: chamber used to sterilize with pressurized steam (a temperature of 121°C must be maintained for a minimum of 15 min to kill all pathogens)

automixis: self-fertilization by the fusion of two closely related sexual cells or nuclei

auxin: plant growth-regulating substance controlling cell elongation; a natural hormone that regulates plant growth, particularly through cell enlargement rather than cell division

available nutrient: that portion of a molecule or compound in the soil that can be readily absorbed, assimilated, and utilized by growing plants

available water: that portion of water in a soil than can be readily absorbed by plant roots; soil moisture held in the soil between field capacity (FC) and permanent wilting point (PWP) (available water = FC − PWP)

avirulent: nonpathogenic; unable to cause disease

awl-shaped: fairly long, narrow, and pointed

awn: bristlelike structure at the apex of the outer bract of some cereal and grass flowers

axenic or **gnotobiotic culture:** growth without another organism being present; germ-free, pure, uncontaminated culture

axial: belonging to, around, in the direction of, or along an axis

axil: upper angle between the petiole of a leaf (or spikelet) and the stem axis

axillary: pertaining to buds, branches, or meristems that occur in the axil of a leaf

azonate or **azonous:** uniform in color, without distinct zones, furrows, or concentric markings

B horizon: soil layer of varying thickness (usually beneath the A horizon) characterized by an accumulation of silicate clays, iron and aluminum oxides, and humus, alone or in combination, and/or a blocky or prismatic structure

bacillar or **bacilliform:** shaped like a short, blunt, thick, cylindrical rod, rounded at the ends; bacillus-shaped

bacillus: short, rod-shaped, spore-forming, obligately aerobic or facultatively anaerobic bacterium; a member of *Bacillus*, a genus of the family Bacillaceae

-bacter (suffix): used to form many generic names for bacteria

bactericide: chemical or physical agent that kills bacteria

bacteriophage (phage): virus that infects, replicates in, and causes lysis of bacterial cells

bacteriostat: chemical or physical agent that inhibits the growth of, but does not kill, bacteria

bacterium (pl. bacteria, adj. bacterial): microscopic (0.25–2.0 μm), prokaryotic, one-celled organism that lacks a nuclear membrane, usually lacks chlorophyll, exists as a parasite or saprophyte, and increases by binary fission, i.e., a division of a mother cell into two like daughter cells (bacteria that attack plants cause diseases such as crown gall, wilts, cankers, rots, leaf spots, and blights)

balanced salt solution: solution that controls the osmolarity of nutrient solutions and has a composition that maintains a balance of the requirements of the cells for which they are providing nutrients

bar: unit of pressure used to express water potential (1 atmosphere = 1.013 bar; equal to 1 million dynes/cm^2)

bare-root: pertaining to plants from which the soil surrounding the roots has been removed

bark: in woody plants, the aggregation of tissues outside the cambium

basal: related to, located at, or forming near the base of a structure

basidiomycete: fungus characterized by septate mycelium, often with clamp connections, that forms sexual spores (basidiospores) on a basidium; one of the Basidiomycotina

basidium (pl. basidia, adj. basidial): specialized cell or organ, often club shaped, in which karyogamy and meiosis occur, followed by production of externally borne, haploid basidiospores (generally four)

basipetal: toward the bottom, as in the production of structures (such as conidia) in succession from the apex toward the base, making the apical part the oldest

beetle: any insect of the order Coleoptera characterized by elytra (thickened outer wings), chewing mouthparts, and complete metamorphosis

berry: simple, fleshy fruit formed from a single ovary, which has a fleshy wall and includes one or more carpels and seeds

bi- (prefix): two; twice; twofold

biennial: plant that develops seed at the end of its second year of growth and then usually dies; a plant that completes its life cycle within 2 years

bifurcate: to divide into two branches or forks

bilateral symmetry: that condition whereby the two halves of the body of an organism (e.g., a nematode) are exact counterparts of the other, i.e., mirror images, if divided in one plane (bilaterally symmetrical organisms are characterized by ventral and dorsal sides that are marked by openings); the condition of having distinct and similar right and left sides, e.g., flowers of sweet pea or snapdragon

binucleate: containing two nuclei

bio- (prefix): life

bioassay: test involving the response of a living cell or organism to an artificial stimulus; quantitative estimation of biologically active substances by the extent of their actions, under standardized conditions, on living organisms; determination of the amount of a virus present by measuring its biological activity, e.g., infectivity of the host

biocide: compound that is toxic to all forms of life

biologic form or **race:** physiologic race

biological management or **control:** disease or pest control through counterbalance of microorganisms and other natural components of the environment; control of pests by means of living predators, parasites, disease-producing organisms, competitive microorganisms, or decomposing plant material, which reduce the population of a pathogen, plant, or insect

biology: science or study dealing with living organisms or their fossils

biotic (syn. animate): of or pertaining to living organisms

biotype: subspecies group of organisms that differ in one or a few biochemical, physiologic, or behavioral properties; a group of individuals with a like genetic makeup; a subdivision of a pathologic race; a geographical isolate or a clone of the same vector species. *See* physiologic race

birefringence: double refraction; polarization of light passing through crystals and other substances (such substances appear bright when viewed with cross-polarizing filters in a microscope)

blade: flat or expanded part of a leaf or petal

blast or **blasting:** sudden death of young buds, flowers, or fruits; failure to produce fruit or seeds

bleached: white to straw colored; used to describe areas of necrotic tissue

bleeding: plant sap and/or microbial materials (e.g., slime flux) exuding from a wound

blight: general term for sudden, severe withering and/or killing of leaves, flowers, shoots, fruit, or the entire plant (usually young, growing tissues are attacked and damage is extensive; may be coupled with the name of the host part attacked, e.g., leaf or needle blight, tip blight, shoot blight, flower blight)

blind shoot: flower bud that fails to form or develop

blotch: large, irregular, usually superficial, necrotic area on a leaf, shoot, stem, or fruit (blight affects the entire leaf; blotch affects a major portion of the leaf; and spot affects a small portion of the leaf)

bole: trunk or main stem of a tree

boot: sheath of the uppermost leaf that encloses the head of a cereal or grass plant

borer: insect or insect larva that tunnels within the stems of woody or herbaceous plants

bract: reduced leaf associated with a flower or inflorescence; modified leaf of the axil from which a flower arises

bramble: cane bush, e.g., raspberry and blackberry (*Rubus* spp.), usually with spines (fruit is an aggregate and commonly referred to as a berry)

breaking: disease symptom caused by a virus in which addition or loss of flower color creates a variegated pattern. *See* flower break

bright-field microscope or **light microscope:** microscope that uses transmitted light

broadcast: to scatter seed, fertilizers, or pesticides uniformly over the soil surface rather than in rows

broadleaf or **broad-leaved:** pertaining to any plant with a broad, flat leaf; other than a grasslike plant

brood: eggs and young animals (especially insects); a generation of an insect species

brooming: profuse branching of woody stems from a single stem position; cluster of stunted, slender twigs in crowded bunches from a stem with shortened internodes (a witches'-broom)

broomrape: whitish to yellowish, annual, leafless herb (family Orobanchaceae, the broomrape family) that grows as a parasite on the roots of another plant

brown rot: (of wood) decay resulting from selective removal of cellulose and hemicellulose, leaving a brown, amorphous residue that usually cracks into cubical blocks and consists largely of slightly modified lignin

bud sport: mutation arising in a bud and producing a genetically different shoot (includes changes caused by gene mutation, somatic reduction, chromosome deletion, or polyploidy)

bud trace: microscopic vascular connection between stem and bud

budbreak: period of time during which resting buds resume growth

budding: special type of grafting in which a single bud is used as a scion; a method of vegetative (asexual) reproduction of plants by implanting the stem bud of a desired species or cultivar onto the rootstock of another plant, often a different species or cultivar; (of fungi) the development of a spore from a small outgrowth (bud) on a parent fungal cell; a method of asexual reproduction in unicellular fungi (e.g., yeasts) or in spores; (of viruses) method of release of enveloped virus particles from the cells in which they have grown

buff: pale tan to creamy gray or yellow

buffer: mixture of an acid and a base capable of maintaining hydrogen-ion concentration and thereby avoiding rapid changes in the acidity or alkalinity of a solution

bug: (true bug) any insect of the order Hemiptera (especially suborder Heteroptera), characterized in part by piercing-sucking mouthparts, a triangular scutellum, two pairs of wings, and gradual metamorphosis (it is incorrect to apply this term to all insects or creatures that resemble bugs)

bulb: short, flattened, usually globose or disk-shaped, underground, perennial storage organ composed of concentric layers of overlapping fleshy scale leaves attached to a stem plate at the base; essentially a subterranean bud

bulblet: small, immature bulb produced on the stem in the axils of leaves, below the soil line, or at the base of an older, "mother" bulb

bulbous: bulblike, as a stipe enlarged at the base

butt rot: decay in the heartwood of the basal part of a living tree

°C: degrees Celsius (formerly degrees centigrade); unit of temperature equal to 0.01 between the boiling and freezing points of water at a standard pressure

C horizon: soil layer beneath the B layer that is relatively unaffected by biological activity and pedogenesis and is lacking in properties diagnostic of an A or B horizon

calcareous or **calciferous:** containing calcium carbonate (lime) or lime compounds; (of color) chalk white; chalky

calico: distinctive yellow leaf pattern associated with certain plants infected with a virus

callus (pl. calluses, adj. callous): tissue overgrowth (mass of large, thin-walled, undifferentiated cells) around a wound or canker and in tissue culture developing from cambium or other parenchyma cells with meristematic potential

calyx (pl. calyxes): outermost flower whorl; sepals collectively

cambium (pl. cambia): sheath of meristematic cells in stems and roots; **cork cambium:** lateral meristem that produces cork; **vascular cambium:** one- or two-celled meristematic layer between the phloem and xylem that produces both these tissues and results in secondary diameter growth (if the cambium is destroyed, as may occur in banding trees, the plant dies)

campanulate, campanuliform, or **campanaceous :** bell-shaped

cane : externally woody, internally pithy, usually flexible stem of certain plants, e.g., brambles, roses, and vines; (of nematodes) a thickening with a lack of ornamentation of the posterior cuticle

canker : definite, localized, usually dry, dead, often sunken, raised, or cracked area (necrotic lesion) on a stem, twig, branch, limb, or trunk surrounded by living tissues (may girdle affected parts, resulting in a dieback starting from the tip; usually differs from rot in having a definite line of demarcation; may be caused by pathogens or by injuries of various types); **annual canker :** canker that enlarges only once and does so within an interval shorter than the growth cycle of the plant, usually less than 1 year; **diffuse canker :** canker that enlarges without characteristic shape or noticeable callus formation at its margins; **perennial canker :** canker that enlarges during more than 1 year; **target canker :** perennial canker (e.g., that caused by a *Nectria* sp.) that includes concentric ridges of callus

canopy : expanded leafy top of a plant or plants

capillary water : water retained in the spaces among and on the surfaces of soil particles after drainage with a tension greater than 60 cm of water

capsulate : enclosed in a capsule

capsule : hyaline, gelatinous, relatively thick layer of muco-polysaccharides (and occasionally protein) that surrounds some kinds of bacterial and yeast cells; simple, dehiscent fruit of the pod type, composed of two or more carpels, that is dry at maturity, usually opening to disperse seeds in a dry condition; spore case of a moss or liverwort; (of nematodes) membrane or saclike form enclosing a structure or organ

carbohydrate : compound (foodstuff) comprised of carbon, hydrogen, and oxygen with the last two frequently in a 2:1 ratio (CH_2O); various chemical compounds of carbon, hydrogen, and oxygen, such as sugars, starches, and celluloses

carbonaceous or **carbonous :** hard, black, brittle, easily broken, and resembling charcoal or cinders

carcinogen : substance or agent capable of producing or inciting cancer

cardinal temperatures : minimum, optimum, and maximum temperatures for growth of an organism or germination of its propagules

carotenoid : one of a large class of related polyene compounds, including carotenes and xanthophylls, that are mostly yellow or red or more rarely colorless

carpel : simple pistil or a member of a compound pistil; the ovule-bearing structure of a flower in angiosperms (often regarded as a single, modified, seed-bearing leaf)

carrier : plant or animal harboring an infectious disease agent (e.g., a virus) but not showing marked symptoms (a carrier plant can be a source of infection to others; an insect contaminated externally with an infectious agent, e.g., a bacterium, fungus, nematode, or virus, sometimes is called a carrier); *see* vector; liquid or solid inert material (e.g., dust, clay, oil, or water) added to a pesticidal compound to prepare a proper formulation (provides for more uniform distribution)

caryopsis : small, one-seeded, dry, indehiscent fruit with the seed coat and a thin pericarp completely united and adhering to the seed; the "seed" (grain) or fruit of grasses (Poaceae)

casting : premature loss of abscised leaves or twigs

catalyst : any substance that accelerates a chemical reaction without entering or being used up in the reaction

catenate or **catenulate :** developing in chains or in an end-to-end series

caterpillar : "wormlike" stage (larva) of a moth or butterfly

cation : positively charged ion, in contrast to anion

catkin : type of flower cluster (spike) generally bearing either female (pistillate) or male (staminate) flowers; conelike male or female fruit of angiosperms

causal agent or **causal organism :** agent or organism (e.g., bacterium, fungus, mycoplasma, nematode, virus, or viroid) that incites a given disease or injury (e.g., a sharp drop of temperature causing frost crack on a tree or dog injury to turf)

cedar apple : popular term given to the hard, brown, rust gall produced on junipers by the cedar-apple rust fungus, *Gymnosporangium juniperi-virginianae*

cell : structural and functional unit of all plant and animal life (the living organism may have only one cell, e.g., bacteria, to hundreds of billions, e.g., a large tree; the essential features of a cell are its one or more nuclei, limited by a membrane, and living protoplasm; plant cells are surrounded by a wall)

cell sap : liquid in plant cells

cell wall : protective, resistant, but permeable polysaccharide structure secreted externally to the cell membrane in plants, bacteria, fungi, and certain other organisms; the barrier that develops between nuclei and delimits the two daughter cells during mitosis

cellulolytic : of certain enzymes able to digest cellulose

cellulose : complex polysaccharide composed of hundreds of β-glucose molecules (polymers) linked in an unbranched chain that makes up 40–55% by weight of the cell walls of plant species

centrifugal : from the center outward; around the margin

centripetal : toward the center

cereal : grass grown primarily for its edible seed

certification : of seeds, transplants, cuttings, other plant parts, or nursery stock, the production and sale, under inspection, of seeds or plant parts, the progeny of foundation, registered, or certified seed, or other plant material to maintain genetic (varietal) identity and purity and freedom from harmful diseases, insect and mite pests, and weed seeds (approved and certified by an official certifying agency)

chaff: nonseed portion of a mature grass or cereal head

chelating or **sequestering agent:** organic compound (such as EDTA) that attracts and tightly binds with specific bivalent (divalent) and trivalent metallic cations (often used to correct nutrient deficiencies, to inhibit biological interactions that require bivalent ions, to assist in virus purification, and for other purposes)

chemotaxis or **chemotropism:** movement or growth of organisms in response to chemical stimuli; a taxic or tropic reaction to a chemical, especially in relation to food material

chemotherapy: treatment of plant disease by chemicals (chemotherapeutants) absorbed and translocated internally (the chemical agent has a toxic effect directly or indirectly on the pathogen without causing undue injury to the host plant or affecting host-cell metabolism)

chestnut brown: dark reddish brown

chilling injury: direct and/or indirect injury to plants or plant parts from exposure to low, but above-freezing temperatures (as high as 9°C or 48°F)

chip bud: method of grafting a small piece of scion, bearing a single bud, onto the stock of another plant

chisel: tillage implement with one or more cultivator-type shanks to which are attached knifelike units that shatter or break up hard, compact layers, usually in the subsoil

chlorophyll (adj. chlorophyllous): green, light-sensitive pigment found chiefly in the chloroplasts of leaves and other green parts of higher plants and used by the plant with the energy of sunlight to convert water and carbon dioxide into food in a process called photosynthesis

chloroplast: specialized cytoplasmic organelle (plastid) in plant cells that contains chlorophyll and is the site of photosynthesis

chlorosis (adj. chlorotic): paling, yellowing, or whitening of normally green tissue characterized by the partial to complete destruction of chlorophyll (may be caused by a virus; lack of or unavailability of some element such as iron, manganese, zinc, nitrogen, boron, or magnesium; lack of oxygen in a waterlogged soil; alkali injury; or some other factor). *See* yellowing

chromosome: one of the self-replicating, strandlike bodies within eukaryotic cell nuclei formed from chromatin and bearing an aggregate of genes (hereditary determinants); a store of genetic information composed of DNA and associated proteins (in diploid organisms, one chromosome of each pair is inherited from each parent; in bacteria, the entire genome is contained within one double-stranded, circular DNA molecule)

chronic: pertaining to slowly developing, persistent, or recurring symptoms that appear over a long period of time; to an infection that lingers, often without symptoms; or to a condition that is of long duration. *See* acute

cineraceous or **cinereous:** ash gray; dirty white

cinnabar or **cinnabarine:** reddish orange (vermilion red)

cinnamon: light yellowish brown

circum- (prefix): all around; round about

cirrhus or **cirrus** (pl. cirrhi or cirri; syn. "spore horn"): curled, tendril-like mass of exuded spores held together by a slimy matrix as it issues from an ostiole; elaborate cephalic appendage found in certain nematodes

cladode: stem with leaflike form; on certain cacti, a joint or stem segment

clavate or **claviform:** club-shaped; narrowing in the direction of the base; (of an agaric stipe) narrowing to the apex

claypan: compact, slightly permeable layer of varying thickness and depth in the subsoil having a much higher clay content than the overlying material (usually hard when dry, plastic and sticky when wet)

clod: compact, coherent mass of soil produced artificially, usually by tillage operations, especially when performed on soils that are either too wet or too dry

clone: (n.) aggregate of individual organisms produced asexually (vegetatively) originating from one sexually produced individual (e.g., rooted cuttings) or from a mutation (any plant propagated vegetatively is considered to be a genetic duplicate of its parent); (in virology) a population of recombinant DNA molecules all carrying the same inserted sequence; (in microbiology) a colony of microorganisms containing a specific DNA fragment inserted into a vector; a population of cells or organisms of identical genotype; (v.) to use in vitro recombination techniques to insert a particular DNA sequence into a vector

CMI AAB: Commonwealth Mycological Institute/Association for Applied Biologists in the United Kingdom

co- (prefix): together, with

coalesce (adj. coalescent): to grow or join together into one body or spot; overlap; merge

coccoid: sphere-shaped

coccus (pl. cocci): spherical bacterial cell

Coelomycetes: division of the Deuteromycotina (Fungi Imperfecti) whose members produce conidia in pycnidial, pycnothyrial, acervular, cupulate, or stromatic conidiomata. *See* Hyphomycetes

coeno- (prefix): living together; e.g., multinucleate

cold frame: unheated plant bed on the ground enclosed by side walls that are usually 12–14 in. high and covered with glass or plastic (useful for growing and protecting young plants during early spring; heat comes from the sun)

coleoptile: ephemeral, nonpigmented tissue that sheaths the first true leaf in seedlings of certain monocots and protects the plumule as it grows through the soil and emerges through the surface

collar: light-colored band at the junction of the leaf blade and sheath on the outside of a grass leaf; the lower part of a herbaceous stem or trunk of a tree

collar rot: rotting of the stem at or about soil level

collateral host: secondary, alternative, or subordinate host

collenchyma (adj. collenchymatous): tissue, composed of elongated, parenchymatous living cells with unevenly thickened primary walls, in elongating, soft stems and certain other parts of plants that helps support the plants

colloid: state in which a finely divided solid remains suspended in a liquid (colloidal systems are usually more stable than emulsions or suspensions and are generally electrically charged and usually turbid); (of soil) organic and inorganic matter with very small particle size and a correspondingly large surface area per unit of mass

colonize (n. colonization): to become established in a host after infection, as a pathogen. *See* infect

colony: macroscopically visible group of individuals, e.g., bacteria and yeasts, usually of the same species, living in close association, often on a solid culture medium (in mycelial fungi, the term usually refers to a group of hyphae growing out of a single point and forming a round or globose mycelium)

color break: *see* flower break and breaking

commensalism (adj. commensal): form of symbiosis in which one organism lives with another organism without causing injury (one or neither may benefit). *See* mutualism and symbiosis

compatible: (of plants) pertaining to different types of plants that set fruit when cross-pollinated or make a successful graft union when intergrafted; (of mating types, strains, etc.) able to be cross-mated, cross-fertile; (of pesticides) pertaining to two or more compounds able to be mixed together without deleterious effects; (of disease) pertaining to an interaction between a plant and a pathogen resulting in disease. *See* incompatible

complete fertilizer: any fertilizer containing the three basic elements most commonly deficient in soil: nitrogen (N), phosphorus (P), and potassium (K)

complete flower: one that has all the male and female flower parts (sepals, petals, stamens, and pistil or pistils) all attached to a receptacle

compost: mixture of organic materials that have been piled, mixed, moistened, and allowed to decompose biologically, sometimes with added soil and mineral fertilizers

concentric: pertaining to one or more circles within one another with a common center (describes a frequent symptom of numerous diseases caused by fungi, viruses, or bacteria). *See* ring spot

conic, conical, or **conoid**: more or less cone-shaped

conidiogenous cell: any fungus cell that directly produces one or more conidia

conidioma (pl. conidiomata): any organized hyphal structure (or asexual fruit body) bearing or containing conidia, e.g., acervulus, coremium, pycnidium, sporodochium, and synnema. *See* conidiophore

conidiophore (pl. conidiophores): specialized, simple or branched hyphal cell or group of cells frequently bearing conidiogenous cells that produce conidia. *See* conidiogenous cell

conidium or **conidiospore** (pl. conidia): any asexual, non-motile spore (except intercalary chlamydospores or sporangiospores) that develops externally or is liberated from a conidiogenous cell (the term is generally restricted to members of Ascomycotina, Basidiomycotina, and Deuteromycotina)

conk: fruit body (basidiocarp, basidioma, or sporophore) usually of a polypore wood-rotting fungus (family Polyporaceae) formed on tree stumps, branches, trunks, or occasionally on lumber (usually spongy to hard, becoming large when mature, and persisting for 1 year or more)

contagious: capable of spreading from one to another

contaminant: substance or microorganism that renders an otherwise pure system unclean or unsuitable for the intended purpose (bacteria and fungi [e.g., *Alternaria, Aspergillus, Mucor, Penicillium,* and *Rhizopus* spp.] and mites are common contaminants in culture of fungi in petri dishes)

control (of plant diseases): prevention, retardation, or alleviation of disease development; disease management

controlled-atmosphere (CA) storage: process of fruit storage in which the CO_2 level is increased to 1–3% and the O_2 level is decreased to 2–3%

corky: firm but not hard; of a density similar to cork

corolla: petals, collectively, of a flower

cortex (adj. cortical): primary, soft, mostly parenchymatous tissue between the epidermis and the phloem in stems and the pericycle in roots; (of fungi) a more or less thick outer covering; periderm; the upper and lower or outer layers of a lichen thallus; (of nematodes) the outermost layer (cortical) of the cuticle

cotyledon: seed leaf, one in the monocotyledons and two in the dicotyledons; primary embryonic leaf; a food-digesting and food-storing part of an embryo

cover crop: close-growing, natural or introduced crop grown primarily to improve and maintain soil structure, add organic matter, and prevent soil erosion

crawler: newly hatched or first-stage insect (e.g., a whitefly or scale insect) that is still able to crawl

crop rotation: growing of crop plants in different locations in a systematic sequence to help control insects, diseases, and weeds, improve soil structure and fertility, and decrease erosion

crop tolerance: ability of a crop to endure treatment with a chemical or injury from an environmental factor with only minor adverse effects; amount of a pesticide that is legally allowable on or in a crop at harvest time

crown: compacted series of nodes from which shoots and roots arise at the base of the stem of cereals and forage species and from which tillers or branches arise; in woody plants, the root-stem junction; upper portions of a tree bearing leaves, flowers, and fruit

cull: (n.) volume of sawtimber that is not usable because of decay or other factors; discarded plant products; (v.) to collect, gather, or remove

culm : stem of grasses and bamboos (usually hollow except at the swollen nodes)

cultivar (cv.) : cultivated variety; assemblage of closely related plants of common origin within a species that differ from other cultivars in certain minor details (e.g., form, color, flower, or fruit), which, when reproduced sexually or asexually, retain their distinguishing features. *See* variety

culture : (v.) to artificially grow and propagate microorganisms or plant tissue on a prepared nutrient medium; (n.) growth of one organism or a group of organisms artificially maintained on such food material

culture plate : shallow, circular, covered dish of thin glass or plastic (e.g., petri dish) used to grow bacteria and fungi in pure culture in the laboratory

cuticle (adj. cuticular) **:** thin, waxy, protective, continuous, noncellular membrane over most epidermal cells of higher plants consisting mostly of waxes and cutin that prevents excessive water loss (the cuticle is broken only by natural openings, e.g., stomata and lenticels); (of fungi) a layer of specialized hyphae on the pilear or stipe surface of a basidiocarp; (of nematodes) the outermost noncellular covering

cutin : clear or transparent waxy material related to cellulose, very impermeable to water, that comprises the inner layer of the cuticle of plants

cutinolytic : of certain enzymes able to digest cutin

cv. : *see* cultivar

cylindrical : round in cross section and of equal width throughout the length

cyto- (prefix) **:** pertaining to cells

cytology : science or study of cells and their components, structure, organic processes, and functions

cytoplasm : all living substance (protoplasm) of a cell, except the nucleus and cell wall, consisting of a complex protein matrix or gel and including essential membranes and cellular organelles

damage threshold : lowest pest population density at which economic or aesthetic damage occurs

damping-off : disease, generally caused by soil- and seed-borne fungi or by a few bacteria, that results in decay or death of seeds or seedlings at the time of germination in soil (preemergence damping-off) or of young, emerged seedlings that suddenly topple over and die from a rotting at the stem base (postemergence damping-off)

dark-field illumination : method of microscope illumination in which the illuminating beam is a hollow cone of light formed by an opaque stop at the center of the condenser large enough to prevent direct light from entering the objective (the specimen is seen with light scattered or diffracted by it)

dark-field microscopy : microscopic examination in which the microscopic field is dark and any objects, such as organisms, are brightly illuminated

decay : disintegration or decomposition of plant tissue or other substrates by bacteria, fungi, and possibly other microorganisms

decline : reduced vigor of perennial plantings as a result of chronic symptoms of disease; the gradual reduction in health and vigor of a plant or planting that is in the process of slowly dying

decomposition (v. decompose) **:** degradation into simpler compounds; rotting of colonized plant tissue, usually caused by microorganisms

defoliate : to lose or become stripped of leaves

dehydrate (n. dehydration) **:** to reduce water content; to become dry

delignification : chemical, usually enzymatic, removal of lignin from xylem, leaving primarily a cellulosic residue

deltoid : triangular in shape

dematiaceous : pigmented more or less darkly

dendriform, dendritic, or **dendroid :** having repeated, irregular, treelike branching; treelike

denitrification : conversion of nitrogenous compounds (nitrates and nitrites) in soil into gaseous nitrogen or nitrogen oxides by denitrifying bacteria

dentate or **dentoid :** toothed; tooth-shaped projections

desiccant : substance that accelerates drying of plant or other tissues or maintains dryness of materials

determinate : not continuing indefinitely at the top of an axis, e.g., pertaining to the cessation of growth of a structure when a fixed size or shape is attained; clearly marked; definite; (of conidiophores) pertaining to the cessation of growth with the production of terminal conidia; (of flowering) pertaining to the uniform flowering of plant species within certain time limits, allowing most of the fruit to ripen at about the same time

detoxification : inactivation or destruction of a toxin by alteration, binding, or breakdown of the toxic molecule to nontoxic (noninhibitory) products

Deuteromycotina : class of asexual (anamorphic, imperfect) fungi, some of which possess teleomorphs (sexual states) in Ascomycotina and Basidiomycotina. *See* Fungi Imperfecti

dextrinoid : pertaining to spores, hyphae, etc. that turn yellowish brown, red, or bright reddish brown in Melzer's iodine reagent; pseudoamyloid

diagnosis (pl. diagnoses) **:** identification of the nature and cause of a disease problem or a shortened, Latin version of a taxonomic description of a species or other taxon

diagnostic : pertaining to a distinguishing characteristic important for identification of disease or other condition

diaphanous : transparent or nearly so

dichotomous key : natural key with two or more choices to consider at any point leading to the next choice and, eventually, to the name of the taxon (supposedly reflecting a degree of phylogenetic relationship among taxa)

dicotyledon or **dicot** (pl. dicotyledonae, adj. dicotyledonous) **:** flowering plant that has two cotyledons (seed leaves) in an embryo compared with monocotyledons (e.g., grasses and cereals), which have one. *See* monocotyledon

dieback (v. die back): progressive death of shoots, branches, or roots generally starting at the tip, which may be caused by cankers, stem or root rots, borers, nematodes, winter injury, deficiency or excess of moisture or nutrients, stress complex, or some other factor

differential host: special species, cultivar, inbred, or isogenic line of plant varying in susceptibility to a given disease agent, such that its distinctive symptoms in response to the agent facilitate its identification, especially as to physiologic race. *See* indicator plant

dikaryon (adj. dikaryotic): cell or strain of fungal cells with two genetically distinct haploid nuclei ($n + n$) (a specialized condition common in members of Ascomycotina and Basidiomycotina)

dilution, serial: successive dilution of a specimen (a 1:10 dilution equals 1 ml of specimen plus 9 ml of a diluent, such as sterile water; a 1:100 dilution equals 1 ml of a 1:10 dilution plus 9 ml of diluent)

dimorphic (n. dimorphism): having two distinct shapes or forms (e.g., yeast and mycelium); producing two types of zoospores; pertains to some male and female nematodes

dioecious aphid: one whose whole life cycle alternates between a primary and secondary host plant. *See* monoecious aphid

diplo- (prefix): two; twice; double

diploid: pertaining to the $2n$ condition, in which there are twice as many chromosomes in the nucleus of a cell as in the haploid condition ($1n$) (characteristic of the sporophyte generation)

disbudding: pinching off of unwanted buds to control branching or to produce larger flowers from remaining buds

disciform or **discoid**: flat and circular; resembling a disk or plate, usually with a central attachment

Discomycetes: the cup fungi, a subclass of Ascomycotina, that produce asci in apothecia

disease, plant: any disturbance of a plant that interferes with its normal growth and development (e.g., structure and function), economic value, or aesthetic quality and leads to development of symptoms; continuous, often progressive, condition in which any part of a plant is abnormal (e.g., structure, function, or economic value) or in which the normal activity of the plant's cells or organs is interfered with (injury, in contrast, results from a momentary damage)

disease cycle: sequence of events involved in disease development, including the stages of development of the pathogen and the effect of the disease on the host; the sequence of events between the time of infection and the final expression of disease

disease pyramid: the four factors needed simultaneously for a plant disease to occur: a susceptible host, a virulent pathogen, an environment favorable for infection, and time for the disease to develop

disease range: geographic distribution of a disease

disease tolerance: capacity of a plant to maintain fairly normal vigor, without excessive injury or loss in yield, even though a pathogen is established within the plant

disinfectant (syn. germicide): chemical or physical agent that kills pathogenic microorganisms (e.g., bacteria, fungi, or nematodes) once a plant, or any of its parts, has become infected or infested

disinfestant: chemical or physical agent that removes, kills, or inactivates disease-causing organisms before they can cause infection and may be applied to the surface of a seed, other plant part, or tool or incorporated into the soil

disorder: harmful, nonpathogenic deviation from normal growth; abiotic disease. *See* disease, plant

disperse: to break up compound particles, such as aggregates, into the individual component particles; to distribute or suspend fine particles, such as clay, in or throughout a dispersion medium

distal: remote from the point of attachment or origin; away from the center of the body

distribution: spread of a pathogen to areas outside its previous geographical range ("geographical distribution" is synonymous with "range")

diurnal: occurring daily or during daylight hours, as opposed to nocturnal

DNA (deoxyribonucleic acid): any of various nucleic acids that occur in nuclei of plant and animal cells and are made up of molecules composed of repeating subunits of deoxyribose (a five-carbon sugar), phosphoric acid, and four nitrogenous bases (adenine, cytosine, guanine, and thymine); dsDNA = double-stranded DNA; ssDNA = single-stranded DNA. Every inherited characteristic is determined by the individual's complement of DNA. dsDNA is held together by hydrogen bonds between base pairs of nucleotides (adenosine, guanosine, cytidine, and thymidine); base pairs form only between adenosine and thymidine and between guanosine and cytidine. It is possible to determine the sequence of either strand, therefore, from that of its partner. DNA is a linear chain of deoxyribonucleotides, which is double stranded in some DNA viruses and single stranded in others. In dsDNA, the two strands are antiparallel in a double helix. Purine bases of one strand form hydrogen bonds with pyrimidine bases of the other strand.

DNase: deoxyribonuclease, an enzyme that breaks down DNA by hydrolysis

dolioform: barrel- or cask-shaped to broadly subglobose

dolomitic limestone: native limestone rich in magnesium carbonate, $CaMg(CO_3)_2$

dominant gene: one that is fully expressed in the heterozygote to give the characteristic phenotype; the gene (or the expression of the character it influences) that, when present in a hybrid with a contrasting gene or allele, completely dominates in the development of the character. *See* recessive gene

dormant (n. dormancy): being in a state of reduced physiological activity (occurs in seeds, buds, bulbs, corms, and tubers); resting; quiescent

dormant period: time during which no growth occurs

dorsal: pertaining to the back or upper surface of an organism or the surface facing away from the axis. *See* ventral

dose or **dosage** (syn. rate): quantity of a substance applied per unit of plant, soil, or other surface

double resistance or **durability:** long-lasting resistance

drupe: simple, fleshy, indehiscent fruit derived from a single carpel

dry matter percentage: percentage of the total fresh plant material left after water is removed, calculated by weighing a sample of fresh plant material, oven drying the sample, and then reweighing the dried sample (dry matter % = [dry weight/fresh weight] × 100)

dwarfing: underdevelopment of a plant or plant organs, which may be caused by any type of disease agent under certain conditions, faulty nutrition, other unfavorable environmental conditions, or chemical or mechanical factors

ear: fruiting spike or head of a cereal or grass plant, including the kernels and protective structures

eccentric: one-sided; off center; asymmetrical in growth

ecdysis (pl. ecdyses): act of shedding the outer cuticular layer of an insect, together with internal cuticular linings, that occurs between each of its growth stages or instars; a molt

echinulate: pertaining to spores, scales, etc. that have small, sharply pointed spines

ecology: (plant disease) study of the effect of environmental factors on the occurrence, severity, and distribution of plant diseases or pathogens

economic injury level: lowest population density of pests (including pathogens) that will cause economic damage

economic threshold level: density of pests or pathogens at which control measures should be implemented to prevent an increasing population from reaching the economic injury level; pest or pathogen population density at or above which the value of crop losses, in the absence of management efforts, would exceed the cost of management practices (especially the use of pesticides)

ecosystem: community of living things and their environment; the interacting, functional system comprised of all living organisms in an area and their nonliving environment

ecotype: genetic variant within a species that is adapted to a particular environment yet remains interfertile with all other members of the species; part of a population of a species showing morphological, chemical, or physiological characteristics that appear to be genetically determined and correlated with particular ecological conditions but that are thought to have no taxonomic significance

ecto- (prefix): outside; external

ectoparasite: parasite living on the outside of its host. *See* endoparasite

edaphic: of or pertaining to soil as it influences plant growth

edema or **oedema** (adj. edematous): watery swelling of plant organs or parts; an abnormal accumulation of serous fluid in connective tissue or in a serous cavity

effuse: stretched out; flat and spreading; pertaining to a filmlike growth

electron: negatively charged particle that forms a part of all atoms

electron microscopy: use of an electron microscope, in which a focused beam of electrons produces a greatly enlarged image of minute objects, such as virus particles (similar to the process by which light forms an image in a compound microscope)

element: substance that cannot be divided or reduced by any known chemical means to a simpler substance (92 naturally occurring elements are known)

ELISA: enzyme-linked immunosorbent assay, an immunoassay system that employs enzyme bound to antibody as the immunologic probe for determining the extent of antigen-antibody reaction; a method for detecting and quantifying specific antibodies

elliptic, ellipsoid, or **ellipsoidal:** pertaining to spores, etc. that have the shape of an ellipse; elliptical in optical section

eluviation: removal of soil material in suspension (or in solution) from a layer or layers of a soil (loss of materials in solution is usually described as "leaching")

emasculate: to remove anthers from a bud or flower before pollen is shed (a normal preliminary step in hybridization to prevent self-pollination)

embargo: government order prohibiting transport of plants or plant parts into a country or region to prevent spread of disease, pathogens, or vectors

embryo: organism in the early stages of development; a nematode before hatching from the egg; a young sporophytic plant, before the beginning of its rapid growth; a rudimentary plant

emergence: appearance of the shoot above the soil surface

emulsion: dispersion system in which droplets of one liquid are suspended in another liquid, both liquids being immiscible, e.g., oil droplets in water

enation: small, abnormal outgrowth of host tissue or eruption from a plant surface often from veins (mostly from leaves, petioles, and flowers) usually induced by certain virus infections (literally "small leaf")

encyst (n. encystment): to form a cyst or encasement

endemic: pertaining to a persistent low and steady level of normal disease occurrence, to a disease native to or restricted to a certain country or geographic region, and to plant species native to a particular environment or locality. *See* enphytotic and epiphytotic

endo- (prefix): inside; within

endodermis: single layer of cells with thick walls and no intercellular spaces that surrounds the vascular tissues of stems and roots; innermost layer of the cortex, most conspicuous in roots

endomycorrhiza (pl. endomycorrhizae; syn. vesicular-arbuscular mycorrhiza): fungus-root association in which fungal hyphae invade cortical cells of the root

endoparasite: parasite that enters and infects its host. *See* ectoparasite

endophyte (adj. endophytic): plant, such as a fungus, living inside another plant

endosperm: nutritive tissue formed within the seeds of angiosperms and on which the embryo feeds while germinating (develops from the sexual fusion of a sperm cell with two polar nuclei of the embryo sac and is thus triploid [$3n$])

endospore: highly resistant and thick-walled asexual spore, formed within a bacterial or fungal cell, capable of surviving exposure to high temperatures or long periods of dryness; inner wall of a spore

enphytotic (syn. endemic): pertaining to plant disease that regularly causes a constant amount of damage in a restricted area each year. *See* epiphytotic

entomology: science or study of insects

enzyme: complex, high molecular weight protein produced in living cells by protoplasm that catalyzes a specific biochemical reaction but does not enter into the reaction

epi- (prefix): upon

epicormic sprout: sprout arising abnormally along a trunk or limb as the result of release of dormant buds or the differentiation of buds from callus; water sprout

epicotyl: portion of an embryonic axis or seedling above the attachment of the cotyledon(s) (the growing apex of the epicotyl forms the young stem of a plant)

epidemic: widespread and rapidly developing outbreak of an infectious disease of humans in a community (used loosely of plants and animals). *See* epiphytotic

epidemiology: study of the factors influencing the initiation, development, and spread of infectious causes of disease; (in plant pathology) the study of epiphytotics; the study of disease in plant populations

epidermis (adj. epidermal): outermost layer of cells of leaves, young stems, roots, flowers, fruits, and seeds but is absent from the root cap and apical meristem; (of nematodes) the outer cylinder of cells consisting of a single layer of epithelium that secretes the cuticle (hypodermis)

epinasty: abnormal downward twisting, curling, or bending of a leaf blade or stem

epiparasite: organism parasitic on another that parasitizes a third

epiphyte: organism growing (not for parasitic purposes) on the surface of another, e.g., a plant growing on a larger plant from which it gains physical support

epiphytology: science or study of epiphytotics (epidemics)

epiphytotic: sudden and destructive development of a plant disease, usually over large areas (corresponds to an epidemic of a human disease)

eradication (n. eradicate): control of plant disease by destroying or removing the pathogen or pest after it is already established in a given area or by eliminating the plants carrying the pathogen

erosion: the wearing away of the surface soil by wind, moving water, or other means

erumpent: breaking through to the host surface or substratum in the course of development; bursting forth

escape: susceptible plant that avoids infectious disease through some character of the plant or its location

-esce or **-escent** (suffix): denoting beginning; slightly

etiolation (v. etiolate): blanching or yellowing of tissue (absence of chlorophyll); elongation of stems (spindliness) and failure of normal leaf development caused by reduced light or complete darkness

etiologic agent: parasite causing a disease in a plant or animal

etiology or **aetiology**: science of the causes or origins of disease, together with the relations of the causal factor to the host; the study of the causal factor, its characteristics, and relations with the host

eukaryote (adj. eukaryotic): organism having membrane-limited nuclei that divide by mitosis (all organisms except viruses, bacteria, and blue-green algae). *See* prokaryote

evapotranspiration: total loss of moisture through the combined processes of evaporation from the soil surface and transpiration from plants for a given area and during a specified period of time

ex- (prefix): from, out of; without; not having

exclusion: control of disease by excluding from disease-free crop-production areas pathogens or plant material from infected crop-production areas (e.g., by quarantines and embargoes)

excurrent: pertaining to erect shoot growth in which there is a single main stem, usually with many lateral branches, as in a pine; alternate, narrowly elongated

exerted: sticking out; protruding beyond

exo- (prefix): outside

exocarp: outermost layer of the fruit wall (pericarp)

exotic: pertaining to an organism introduced from another country or area; not indigenous

exotoxin: diffusible toxic substance produced by a microorganism and then secreted into the surrounding medium

explant: to transfer living cells or tissue from a plant into an artificial medium for tissue culture

extrude (n. extrusion): to push out, emit to the outside

exudate: substance, often a liquid ooze or slime, discharged or secreted from diseased, injured, or healthy plant tissue; metabolic by-products discharged from roots into surrounding soil

F: *see* Fahrenheit

F$_1$: first generation progeny of a cross between any two parents

F₂: second filial generation, obtained by crossing two members of the F_1 generation or by self-pollinating plants of the F_1 generation

facultative: sometimes; not necessarily; pertaining to an organism able to live under a variety of conditions

facultative parasite: organism that is normally saprophytic but that can live as a parasite under certain conditions and be cultured on laboratory media. *See* obligate parasite and parasite

facultative saprophyte or **saprobe**: organism that is normally parasitic but that can live saprophytically during part of its life cycle

Fahrenheit (F): thermometer scale in which the freezing point of water (ice point) is set at 32° and the boiling point (steam point) at 212°

fallow: pertaining to cropland not cultivated or not planted for one or more seasons; maintained essentially without plant growth

family: taxonomic grouping of plant or animal genera according to resemblance and similarity and denoting natural relationships; category above a genus and below an order

fasciation: distortion of a plant caused by an injury, infection, or genetic factor that results in thin, flattened, and sometimes curved or curled shoots (the plant may look as if several of its stems are fused); the condition of being bound or bundled together

fasciculate or **fascicled**: having growth in fascicles or bundles

fastidious organism: organism that is difficult to isolate or cultivate on ordinary culture media

fastigiate: having leaves or branches more or less parallel and pointed upward; having massed, upright branches; bunched, clustered

fat: a food insoluble in water and composed of carbon, hydrogen, and oxygen, with proportionately less oxygen than that in carbohydrates. *See* lipid

fatty acid: organic compound containing carbon, hydrogen, and oxygen that combines with glycerol to make a fat

fermentation: oxidation (decomposition) of certain organic substances, especially carbohydrates, in the absence of molecular oxygen by the enzyme action of microorganisms; chemical changes in organic substrates caused by enzymes, usually those of living yeasts. *See* aerobic respiration and anaerobic respiration

ferruginous or **ferrugineous**: rusty reddish brown or rust colored

fertilization: fusion of two sex nuclei (gametes) resulting in doubling of the chromosome number to form a zygote; the act (process) of rendering soil fertile, especially by use of needed plant nutrients, such as nitrogen, phosphoric acid, potash, and others

fertilizer: any organic or inorganic material of natural or synthetic origin containing nutrients (elements) essential to the growth of plants (to be labeled and sold as such, it must be state licensed and have the analysis printed on the package label)

fetid: bad-smelling; of a nauseating odor

fiber: elongated, tapering plant cell with thick walls (in xylem and phloem), which is dead at maturity

fibril: minute, thin, threadlike fiber

fibrous: containing, consisting of, or resembling fibers (tough, stringlike tissue)

field capacity or **field moisture capacity**: maximum amount of water the soil can hold against the force of gravity

field immune: pertaining to plants that do not become infected by a pathogen in the field, although they are susceptible under experimental conditions

field resistance: resistance observed under natural field conditions but usually not detected under experimental conditions. *See* tolerance

firing: drying and dying of leaves, especially of maize (corn) and other grasses

fission: transverse splitting in two of bacterial cells; asexual form of cell division; becoming two by division of the complete organism; (of conidial liberation) secession by separation of a double septum. *See* budding

fixation: preservation of biological structures for microscopic examination by killing in a suitable chemical or physical condition so as to avoid changes in structure; (of soil) the process by which certain chemical elements essential for plant growth convert from a soluble or exchangeable form to a much less soluble or nonexchangeable form

flag or **flagging**: branch with drooping, yellowing, or dead leaves on an otherwise healthy-appearing shrub or tree

flag leaf: leaf originating from the first culm node below the rachis; uppermost leaf on cereals and grasses

flagellate: having one or more flagella or whiplike processes

flagellum (pl. flagella, adj. flagellar): flexible hair-, whip-, or tinsel-like filament projecting from a cell of certain bacteria and zoospores of the lower fungi that enables them to swim through a liquid (functionally similar to a cilium but longer)

fleck: small or minute (0.2–2 mm), white to tan lesion that is often translucent and visible through a leaf (characteristic response of tobacco and numerous other plants to toxic amounts of ozone in the atmosphere)

flexure: turn or fold

floccose: cottony soft, downy; woolly or with soft, very small, cottony or woolly tufts; like cotton flannel

flocculose: having small tufts of soft, woolly hair; delicately cottony

floret: small flower, usually part of a dense cluster

flower break: break or stripe in flower color

fluorescence (adj. fluorescent): visible emission of light when a substance or organism is placed in ultraviolet (UV) or other radiation

fluorescence microscope: compound light microscope arranged to emit radiation of specific wavelengths, such as UV, to the spectrum, which then fluoresces

flush: sudden development of growth

focus (pl. foci): site of initial disease infection from which secondary spread may occur

foliar: pertaining to leaves

foliar burn: dehydration of shoot tissue caused by contact with high concentrations of chemicals, e.g., certain fertilizers, pesticides, or salt

foliose: leaflike

foot-candle (ft-c): standard measure of light (English system); the illumination of a standard light source of one candle, 1 ft away, falling at all points on a 1-ft^2 surface; 1 ft-c = 10.76 lux. *See* lux

foot rot: rot involving the lower part of the stem-root axis

-form (suffix): shape

form genus or **species**: genus or species based on morphology and not on evolutionary relationships, e.g., in Deuteromycotina (Fungi Imperfecti) and rusts (Uredinales)

forma specialis (pl. formae speciales; abbr. f. sp. or ff. sp.): special form; a biotype (or group of biotypes) of a species of pathogen that differs from others in its ability to infect selected genera or species of susceptible plants; an infraspecific taxon characterized from a physiological standpoint (especially host adaptation) but scarcely or not at all from a morphological standpoint. *See* physiologic race

foundation planting stock: stock of high quality, grown separately for genetic purity, and very carefully rogued to produce daughter plants for sale

foundation seed: seed stocks increased from breeder seed and handled to closely maintain the genetic identity and purity of a cultivar

frass: wet or dry, sawdust-like material excreted by insects (usually evident at their exit holes)

freeze-drying (syn. lyophilization): preservation of living microorganisms (or concentration of macromolecules with little or no loss of activity) by removing water under a high vacuum while tissue remains in a frozen state

friable: easily crumbled into small aggregates or powder when handled, as soil

frost crack: linear separation of a woody stem along radial lines caused by unequal shrinkage in the tangential and radial planes

frost rib: structure of callus tissue that forms over a frost crack

frost tolerance: ability of plants to withstand a lower temperature than they originally could by being conditioned through repeated exposures to temperatures just above the freezing point

fruit: matured ovary or cluster of matured ovaries

fruit body or **fruiting body**: complex, multicellular fungal structure (fructification) that contains or bears spores, e.g., acervulus, apothecium, ascocarp, ascoma, basidio-carp, coremium, pycnidium, perithecium, sporangium, sporocarp, and sporodochium

f. sp.: *see* forma specialis

fungal: of, or having to do with, fungi

Fungi Imperfecti (Deuteromycotina): form class erected to contain asexually reproducing fungi with affinities to the Ascomycotina and Basidiomycotina. Two classes are recognized: 1) Hymenomycetes—forms that produce sterile mycelia (agonomycetes) or conidia that are borne on separate hyphae or aggregations of hyphae (as synnemata or sporodochia) but not within discrete conidiomata; and 2) Coelomycetes—forms that produce conidia in acervuli, pycnidia, pycnothyria, or in cupulate or stromatic conidiomata.

fungus (pl. fungi): eukaryotic organism that lacks chlorophyll, usually reproduces by sexual or asexual spores, and commonly produces mycelium; a nonchlorophyllous plant whose vegetative body (thallus) consists of threadlike filaments (hyphae) usually aggregated into branched systems (mycelia)

funiculose: having hyphae occurring in ropelike strands or bundles

funiculus (adj. funicular): cordlike stalk of an ovule

furcate: forked; branching like a fork

fusiform: spindle-shaped; slender and tapered toward each end

fusoid: somewhat fusiform; tapering toward each end

gall: more or less spherical, abnormal swelling or localized outgrowth (tumor) composed of unorganized cells on leaves, stems, roots, and other plant parts produced by host plant cells in reaction to attack by a fungus, bacterium, nematode, insect, mite, or other agent

gametophyte: phase in the life history of a plant that arises from a haploid spore resulting from meiosis in a diploid sporophyte (plants have haploid nuclei during the gametophyte phase); a haploid or sexual plant; haplont or haplophase. *See* sporophyte

gelatinous: resembling gelatin or jelly

gene: one or more forms of the DNA molecule in a chromosome that determines or conditions one or more hereditary characters; the smallest functioning unit of genetic material on a chromosome; bearer of an inherited factor; the fundamental unit of inheritance; an ordered sequence of nucleotides that specifies the manufacture of a single type of protein (or for some genes, certain RNAs) (a gene that has small observable effects upon the phenotype is termed a "minor gene," and one that has large observable effects is a "major gene")

gene-for-gene hypothesis: concept that corresponding genes for resistance and virulence exist in the host and pathogen, respectively

gene pool: all the genes of an interbreeding program

generation time: length of time necessary for an organism to complete its life cycle

genetic: pertaining to heredity or the heritable characteristics as influenced by germ plasm

genetic engineering or **genetic manipulation:** alteration of the genetic composition of a living cell by various tissue culture procedures (e.g., transformation and protoplast fusion) so that the cell can produce more or different chemicals or perform completely new functions

genetics: science or study of patterns of inheritance of specific traits

geniculate: bent like a knee

genotype: aggregate of genes in an organism that influences the phenotype; class or group of individuals sharing a specific genetic makeup; the total genetic constitution, expressed and latent, of an organism

genus (pl. genera)**:** group of structurally or phylogenetically related species

germ: historically popular word for a pathogen or disease-causing microorganism (microbe); the living portion of a seed

germ plasm: material capable of transmitting heritable characteristics sexually or asexually; the total genetic variability available to an organism represented by the pool of germ cells or seed

germicide: substance that kills microorganisms. *See* disinfectant

germinate (n. germination)**:** to begin growth, as of a seed, spore, sclerotium, zygote, or other reproductive body, starting with imbibition of water

girdle: to encircle and cut through a herbaceous stem or the bark and outer few annual wood rings of a woody plant, disrupting the phloem and xylem, resulting in death from root starvation; encircle

globoid, globate, globose, globular, or **globulose:** globe-shaped; spherical or nearly so

glucose: simple, six-carbon sugar (monosaccharide, $C_6H_{12}O_6$), the basic building block of carbohydrates and other plant chemicals

glume: one of the two sterile, chaffy bracts at the base of a grass spikelet

glycolysis: anaerobic process of glucose dissimilation to pyruvic or lactic acid (an early stage of respiration)

glycoprotein: conjugated protein in which the nonprotein group has at least one carbohydrate covalently attached to an amino acid

gnotobiotic culture: culture in which all the living components are known. *See* axenic culture

grain: simple, indehiscent fruit of a grass, chiefly a cereal or grain crop, whose ovary walls are fused to the seed

gram-positive or **gram-negative:** pertaining to bacteria that stain violet or red, respectively, after treatment with Gram's stain

gramineous: of or relating to a grass; grassy

granular, granulate, or **granulose:** pertaining to a surface covered with very small particles (granules)

green manure: crop plowed under while still green and growing to improve the soil

gregarious: in groups or companies scattered in a restricted area; closely associated but not joined together or tufted

group: any taxonomic unit; taxon

growing medium: soil, soil mix, or soil substitute prepared by combining materials such as peat moss, vermiculite, perlite, sand, or composted organic material and used for growing potted plants and cuttings and for germinating seeds

growing season: period between commencement of growth in the spring and cessation of growth in the fall

growth inhibitor: natural substance that inhibits the growth of a plant

gyrate or **gyrose:** curved backward and frontward in turn; folded and wavy; convoluted like a brain

gyre: circular or spiral motion or form

habitat: natural environment of an organism

halo: area of diseased tissue, often discolored or water soaked, that surrounds a lesion

haploid: having a single complete set of unpaired chromosomes ($1n$) in each cell nucleus (characteristic of the gametophyte generation); (of mycelium) composed of haploid cells

hardening or **hardening-off:** adapting plants to unfavorable outdoor conditions by withholding water and lowering the temperature or nutrient supply to hasten maturation of tissues for increased tolerance

hardpan: impervious layer of soil or rock in the lower A horizon or upper B horizon that prevents downward drainage of water and root growth (a "chemical hardpan" is caused by the presence of toxic ions or excessively high concentrations of ions and often is related to extremes in soil pH)

head: type of inflorescence, typical of the composite family (Asteraceae), in which numerous small flowers are densely crowded upon a usually disk-shaped, common receptacle, e.g., sunflowers (*Helianthus*)

heartwood: central cylinder of nonconducting xylem tissue in a woody stem or trunk that contains no living cells and is darker than the sapwood

helical or **helicoid:** in a coil; in the form of a spiral; coil-like

helminthology: branch of zoology dealing with helminths, especially parasitic worms

hemi- (prefix)**:** half; in part

herb: botanically, a plant with fleshy, not woody, stems commonly used for flavoring, fragrance, or medicinal purposes

herbicide: any chemical or physical agent used for killing or inhibiting the growth of plants where they are not wanted; a weed or grass killer; a type of pesticide

hetero- (prefix)**:** other; not normal; different

heterogeneous: differing in structure, tissues, qualities, or composition. *See* homogeneous

hexagonal: six-angled or six-sided

hibernate: to lie dormant

hip: rose fruit formed by a group of achenes surrounded by a receptacle and hypanthium

hirsute: covered with coarse, stiff, long fibers or hairs; (of nematodes) rough with hairs or bristles

hispid: covered with coarse, stiff, erect hairs or bristles visible to the naked eye

histochemical: pertaining to chemical reactions in tissues

histology: science or study of the microscopic structures of plant or animal tissues

homo- (prefix): one and the same; common; like

homogeneous: similar in certain characteristics, such as chemical or physical properties

homonym: name that, under the International Code of Botanical Nomenclature, must not be used because of an earlier use in a different sense (i.e., the names are the same but the types are different)

honeydew: sweet, sticky secretion from aphids, mealybugs, scales, whiteflies, and other sucking insects in which sooty mold fungi grow

hopperburn: marginal yellowing, scorching, and curling of leaves (e.g., of potato and dahlia) caused by the feeding of certain leafhopper species

horizon, soil: layer of soil, approximately parallel to the soil surface, with distinct characteristics produced by soil-forming processes. *See* A, B, and C horizons

horizontal resistance: partial resistance that is effective against most or all races (genetic variants) of a pathogen and is usually determined by multiple genes

horny: (of textures) hard and brittle, homogeneous, and difficult to section

horticulture: art and science of fruits, vegetables, flowers, ornamentals, shrubs, and amenity trees

host: living organism (e.g., a plant) harboring or invaded by a parasite and from which the parasite obtains part or all of its nourishment; (in virology) an organism or cell culture in which a given virus can replicate. *See* suscept

host range: complete range of plants that may be attacked by a given pathogen

hotbed: plant bed similar to a cold frame but provided with a source of heat (e.g., electric cables, fermenting organic matter, steam pipes, or hot air flues) to supplement that from the sun. *See* cold frame

humidity, relative: weight of water vapor in a given quantity of air compared with the total weight of water vapor that the air is capable of holding at a given temperature (expressed as a percentage)

humus: more or less stable, decomposing organic matter from any source (e.g., vegetable refuse, leaf mold, manure, peat, or animal matter) that may become fine, rich, black soil

hyaline: transparent or nearly so; translucent; clear or colorless

hybrid (v. hybridize): offspring of two individuals differing in genetic character or in one or more heritable characteristics; offspring resulting from a cross between two species or genotypes

hydrogen ion concentration (pH): measure of the H$^+$ concentration of a solution (expressed in terms of the pH of the solution). *See* pH

hydroponics: growing of plants in an aerated water solution containing all the essential mineral elements rather than in soil

hydrosis: *see* water soaked

hygiene: *see* sanitation

hygroscopic: capable of readily absorbing water and expanding; becoming soft in wet air, hard in dry; (of a fruit body) opening and discharging spores in dry air

hyper- or **hypero-** (prefix): above, beyond, over

hyperplasia (adj. hyperplastic): plant overgrowth (e.g., gall, enation, tumor, or witches'-broom) caused by increased cell division; excessive, abnormal, usually pathological multiplication of the cells of a tissue or organ. *See* hypertrophy

hypersensitivity: increased sensitivity, e.g., the rapid death of host tissue at the point of attack by a pathogen so that infection does not spread

hypertrophy (adj. hypertrophic): plant overgrowth (gall or tumor) caused by abnormal cell enlargement; excessive, abnormal, usually pathological enlargement of individual cells in a tissue or organ. *See* hyperplasia

hypha (pl. hyphae, adj. hyphal): basic vegetative unit of structure and function of most fungi; a usually microscopic tubular filament that increases in length by growth at its tip (new hyphae arise as lateral branches)

Hyphomycetes: division of Deuteromycotina, the members of which bear conidia on conidiophores not organized into a fruit body

hypo- (prefix): under; lower

hypocotyl: that portion of a developing embryo or seedling just below the point at which the cotyledons are attached (the root meristem is situated at the tip of the hypocotyl)

hypoplasia (adj. hypoplastic): underdevelopment (malformation) of plant tissue resulting from decreased cell division. *See* hyperplasia

hypotrophy: underdevelopment of plant tissue resulting from reduced cell enlargement

ID: infective dose, i.e., the number of pathogens or propagules of a pathogen required to infect a host

imbibition: process by which solid (chiefly colloidal) particles absorb liquids or vapors into the ultramicroscopic spaces, in materials such as cellulose, and swell

immune (n. immunity): not affected by or responsive to disease; exempt from infection

imperfect flower: flower lacking either stamens or pistils; unisexual

in vitro: pertaining to biological reactions taking place in an artificial environment or in culture; outside the living host, as opposed to in vivo; Latin for "in glass"

in vivo: pertaining to biological reactions taking place in the host or within a living organism; applied to laboratory testing of agents within living organisms; used in contrast to in vitro; Latin for "in living"

incidence: (of disease) number of plants affected within a population

incipient: pertaining to the early period in the development of a disease or condition

incised: as if cut into; deeply and sharply notched

incompatibility: (floral) failure to obtain fertilization and seed formation after pollination, usually because of slow pollen tube growth in the stylar region; (graft) failure of the two graft components (scion and stock) to unite and develop into a successfully growing plant

incompatible: pertaining to different kinds or varieties of plants that do not successfully cross-pollinate or intergraft; not cross-fertile; not able to be cross-mated; pertaining to the failure of a pathogen to infect a host plant or to an interaction between a plant and a pathogen that does not result in disease. *See* compatible

incomplete flower: flower that lacks one or more of the four floral organs (sepals, petals, stamens, and pistils). *See* complete flower

incrusted: lightly to densely covered with crystalline material

incubate: to hold in a steady environmental state to await development; to allow a microorganism to grow undisturbed under a given set of conditions

incubation period or **stage**: time between penetration of a host by a pathogen and the first appearance of symptoms and/or signs on the host; a period of rest of a pathogen prior to infection; period of maintaining inoculated plants or pathogens in an environment favorable for disease development

indentate: having an irregular margin. *See* serrate

indexing: any procedure for demonstrating the presence of a virus or other pathogen in susceptible plants

indicator: substance that changes color as conditions change (e.g., pH indicators reflect changes in acidity or alkalinity)

indicator plant: plant that reacts to certain pathogens or environmental factors by producing specific symptoms and is used to detect and identify these pathogens or factors

infect (n. infection): to enter (invade or penetrate) and establish a pathogenic (parasitic) relationship with a host plant; to enter and persist in a carrier

infection type: gross appearance or size of lesions, as from infection by a rust fungus

infectious: capable of infecting and spreading disease (a pathogen) from plant to plant

infectivity assay: bioassay in which mechanical sap transmission is used to quantitatively determine the amount of infectious virus

infested (v. infest; n. infestation): containing or covered with large numbers of pests, especially insects, mites, or nematodes, as an area or field; contaminated or inoculated with a microorganism (e.g., bacterium or fungus) without infection or parasitism

inflected or **inflexed** (v. inflect): curved, bent, or directed abruptly inward, downward, or toward the body axis

inflorescence: axis bearing a flower or cluster of flowers, e.g., an umbel, spike, or panicle

ingress: the act, by a plant pathogen, of gaining entrance into the tissues of a plant

injury: momentary (transitory) damage by a causal agent, e.g., insect feeding, action of a chemical, physical, or electrical agent, or an adverse environmental factor

inner bark: interior, living portion of the bark; secondary phloem

inoculation (n. inoculate): artificial introduction of a pathogen at the site of infection of a host (the infection court) to induce disease; introduction of a microorganism into a culture medium; treatment of seeds of leguminous plants with bacteria (*Rhizobium* spp.) to achieve nitrogen-fixation in the roots

inoculation feeding period (syn. test feeding period): length of time a vector feeds on a test plant during a pathogen-transmission experiment

inoculation threshold period: minimum feeding period an inoculative vector needs on a susceptible test plant to transmit a disease agent

inoculative insect: infective insect that will transmit a pathogen during a given test access period

inoculum (pl. inocula): pathogen or its parts (e.g., fungal spores or mycelium, nematodes, or virus particles) used for inoculation to produce disease

insect: member of the class Hexapoda (phylum Arthropoda) (in the adult stage, true insects have six walking legs, wings, and three body segments)

integrated control or **integrated pest management (IPM)**: use of all available methods (biological, chemical, genetic, physical, and cultural) to manage plant diseases and pests for best control results but with the least cost and damage to the environment. *See* plant health care

intercostal: between veins; interveinal

invasion (v. invade): growth or spread of a pathogen into a plant and its establishment within the plant

ion: atom, group of atoms, or molecule bearing an electrical charge and formed by the dissociation of molecules (loss of electrons [cations] or the gain of electrons [anions]) of certain substances called electrolytes

IPM: *see* integrated pest management

irradiation: exposure of pollen, seed, or other plant parts to radiant energy of various types (X-rays or other short-wavelength [gamma] radiation) to increase mutation rates

isogametes: gametes, presumably of opposite sex or mating types, that are morphologically indistinguishable

isolate: (v.) to separate a fungus, bacterium, nematode, or virus from an infected host or substrate and establish it in pure culture; (n.) the pure culture itself and the subcultures derived from it; the process of getting an organism into pure culture; a sample (e.g., a virus) from a defined source

isometric: pertaining to a structural form that has three equal axes at right angles to one another; equally long, as a virus particle with all axes of equal length (essentially spherical)

jointing: growth stage in grasses manifested by rapid culm elongation

karyogamy: fusion of two sex nuclei after cell fusion, i.e., after plasmogamy

kernel: whole seed (caryopsis) of a cereal; a fruit seed; the inner, softer part of a seed, fruit stone, or nut

klendusity: special kind of disease escape, in which a susceptible plant or variety avoids disease because of an intrinsic property of the plant or cultivar itself that greatly reduces the chances of its being inoculated by the pathogen

knot: knoblike overgrowth on roots or stems resulting from an imperfect vascular system; a localized abnormal swelling; a gall

labium (pl. labia, adj. labial)**:** lip

lacerate: as if deeply split or finely torn

lacteous: milky

lamina (pl. laminae)**:** flat, expanded part (blade) of a leaf; a layer

lanceolate: lance- or spear-shaped; oblong and tapering to a point

languid: hanging down; drooping; weak

larva (pl. larvae)**:** juvenile; immature growth stage(s) between the embryo or egg stage and the adult stage in insects and nematodes

latent: present but not manifested or visible. *See* masked

latent infection or **latency:** infection in a plant without visual symptoms

latent period or **latent phase:** period after a vector has acquired a pathogen (e.g., a virus) and before it can transmit it

lateral: at the side

latex: rubberlike, milky secretion produced by various kinds of plants

latticed: like a network; cross-barred

layering: form of vegetative propagation in which an intact branch develops roots as the result of contact with the soil or another rooting medium

LD: lethal dose

leafhopper: active insect (order Homoptera, family Cicadellidae) with sucking mouthparts (often a vector of pathogens, especially viruses, and also a cause of direct plant injury by feeding)

lemma: lower and larger of the two bracts enclosing a grass flower

lenticel: small pore (natural opening) on a stem, tuber, root, or fruit made up of loosely arranged cells in the periderm through which carbon dioxide and other gases pass

lenticular: shaped like a biconvex lens

lesion: well-marked, localized area of diseased or disorganized tissue; a wound

lethal: capable of causing death

leucoplast: colorless plastid involved in the formation or storage of starch in many types of plant cells

life cycle or **life history:** cyclical progression of stages in the growth and development of an organism (plant, animal, or pathogen) that occur between the appearance and reappearance of the same stage of the organism

light soil: coarse-textured, sandy soil (hence, easy to till)

lign-, ligni-, or **ligno-** (prefix)**:** wood

lignification (adj. lignified)**:** hardening of mainly xylem tissue from deposition of lignin in the cell wall

ligule: in grass leaves, an outgrowth from the upper and inner side of the leaf blade where it joins the sheath; a scalelike projection or row of hairs

lime: in agriculture, material containing the carbonates, oxides, and/or hydroxides of calcium and/or magnesium used to increase soil pH and neutralize soil acidity and to make Bordeaux mixture

linear: long and narrow; resembling a line

lipid: any of a group of organic compounds greasy to the touch, insoluble in water, and soluble in fat solvents (e.g., alcohol or ether); any of the fats and other esters with analogous properties and constituents; with proteins and carbohydrates, the chief structural component of living cells. *See* fat

lipo- (prefix)**:** fat

liquefaction: transformation of a gel to a liquid

litmus: blue amphoteric dye from depside-containing lichens used as an indicator for pH (turns red in the presence of acids and back to blue in alkalis) and oxidation or reduction

loam: mellow, textural class for soil composed of about equal parts silt and sand and less than 20% clay

lobed: pertaining to leaves that are shallowly or deeply divided by clefts or sinuses, as maple (*Acer*) or oak (*Quercus*)

locular: containing chambers or hollows

lodging: lying down, as stems, culms, stalks, etc. that have been beaten down by wind, rain, disease, or insect attack

lucid: shining; clear

lumen (pl. lumina)**:** central cavity of a cell or other structure (e.g., canal or duct of the esophagus in nematodes); hollow center of the culm (grass plant)

lunate, luniform, or **lunulate:** crescent-shaped; like a new moon

lurid: pale yellow, wan, sallow, ghostly; the color of glowing fire; brown tinged with red

luteous: yellow; dull egg yellow

lux: standard measure of light (metric system); illumination impinging on a surface of 1 m², each point being 1 m away from a standard light source of one candle; a photometric unit expressing the illumination of 1 lumen/m²; 1 lux = 0.093 ft-c. *See* foot-candle

lyophilization (v. lyophilize; syn. freeze-drying): technique by which water is removed under high vacuum while the preparation or tissue is frozen (used for long-term preservation of bacteria, fungi, viruses, antisera, and other types of pathogens). *See* freeze-drying

lysis: cellular breakdown (dissolution) of tissues or destruction of cell walls by enzymes (e.g., a lysin) or viruses and subsequent rupture of cell membranes and loss of cell contents

macro- (prefix): long, large, or great

macromolecule: molecule with a molecular weight from several thousands to many millions

macronutrients: essential chemical elements required in relatively large quantities (usually >1 ppm) for the growth of plants, e.g., nitrogen (N), phosphorus (P), and potassium (K)

macroscopic: visible to the unaided eye

maggot: growing stage or larva of a fly

male-sterile: having nonfunctional or no male sex organs (a condition in some plants where pollen is not formed or does not function normally, even though the stamens may appear normal)

malignant: pertaining to a cell or tissue that divides and enlarges autonomously; i.e., its growth can no longer be controlled by the organism on which it is growing

mandibles: horny, jawlike mouthparts of insects used to bite and chew food

manual transmission: spread or introduction of inoculum to infection courts by hand manipulation

marbled: containing irregularly colored streaks

masked: pertaining to symptoms that are absent under certain environmental conditions but appear under other conditions (e.g., in an infected host that does not exhibit disease symptoms because the environment is not favorable for disease expression and/or development). *See* latent

mealybug: small, oval insect (family Pseudococcidae, superfamily Coccoidea) with sucking mouthparts and a cottony, scalelike covering (unlike scale insects, mealybugs possess functional legs and reproduce by producing eggs or living young)

mechanical injury: injury of a plant part by abrasion, mutilation, or wounding; physical damage

mechanical transmission or **inoculation**: artificial spread or introduction of inoculum (e.g., a virus) to an infection court (especially a wound) by hand manipulation accompanied by physical disruption of the host tissue; in the field, transmission of a pathogen from one plant to another by contact of foliage or roots with one another

median: in the middle

medium or **culture medium** (pl. media): chemical environment (substrate) providing nutrition for microbial growth in the laboratory (may be solid or liquid, organic or inorganic, defined or undefined)

meg-, mega-, or **megalo-** (prefix): large; of great size. *See* macro-

megapascal: 10^6 pascals; approximately 7.5×10^3 mm of mercury

meiosis: reduction division; process by which chromosome numbers are reduced by half; process by which a zygote ($2n$) divides and produces haploid ($1n$) gametes

melanin or **melanose** (adj. melanoid): brown black pigment

membranaceous or **membranous**: membranelike; like a thin skin or parchment; thin, compact, but pliant

meq: milliequivalent weight

meristem (adj. meristematic): layer or zone of undifferentiated plant tissue that functions principally in cell division and is thus responsible for the first phase of growth; a mass of growing cells capable of frequent cell division

meristem tip: meristem dome of cells and one or two pairs of primordial leaves (0.5–1 mm in diameter), which comprise the explant removed from a bud and grown in tissue culture to produce a disease-free plant

mesocarp: middle layer of a fruit wall (pericarp)

mesophyll: chlorophyllous parenchyma tissue in leaves between the two epidermal layers

met- or **meta-** (prefix): changed in form or position; between, among; with; after

metabolism: sum total of the chemical processes (physiological activities) occurring in the body of a living organism; the process by which organisms utilize nutrients to build structural components and break down cellular material to obtain energy and simple substances for special functions

methylene blue: photoreactive dye used as a vital stain for cells and for the detection of nucleic acids in gel electrophoresis

micro- (prefix): small; minute

microbe: microscopic organism composed of one to many cells and capable of reproduction by growth and division

microbiology: science or study of microorganisms, e.g., bacteria, actinomycetes, fungi, algae, viruses, and protozoans

microculture: culture of an organism under the microscope, as in a hanging drop (in a van Tieghem cell) or similar microscopic techniques

microgram (μg): 10^{-6} g; one-millionth of a gram

micrometer: disk or slide of glass ruled with lines and forming a metric scale for measuring objects under a microscope in microns

micron (1 μ or **μm)**: one-millionth of a meter (m) or one-one-thousandth of a millimeter (0.001 mm)

micronutrients: essential chemical elements required in only minute amounts (<1 ppm) for the growth of plants, e.g., boron, chlorine, copper, iron, manganese, molybdenum, and zinc

microorganism (syn. microbe): microscopic organism such as a bacterium, fungus, nematode, or protozoan

microscopic: too small to be seen except with the aid of a microscope

microtome: instrument for making thin sections of tissue or cells

middle lamella: thin, cementing layer between adjacent primary cell walls generally consisting of pectinaceous materials (especially calcium pectate), except in woody tissues, where pectin is replaced by lignin

midrib: central, thickened vein of a leaf

mildew: plant disease caused by a fungus and characterized by a thin coating of mycelial growth and spores on the surfaces of infected plant parts

milli- (m) (prefix): one-thousandth; 10^{-3}

millimicron (mμm): one-thousandth of a micron or 10 Å; more commonly, 1 nm

mineral soil: soil whose makeup and physical properties are largely those of mineral matter

mite: minute (1/64- to 1/32-in.-long) animal (order Acarina, families Tetranychidae and Eriophyidae) without evident body divisions that is six legged as a larva and eight legged as an adult

mitosis: usual process of nuclear cell division in which each chromosome duplicates, pulls apart, and produces two daughter nuclei of the same genetic constitution as the parent nucleus (usually, but not always, followed by cellular division [cytokinesis])

MLO: *see* mycoplasmalike organism

moisture basis or **dry basis**: moisture content (%) of a product, calculated as (net weight of water lost by drying/dried weight of the material) × 100

mold or **mould**: fungus with conspicuous, profuse, or woolly superficial growth (mycelium and/or spore masses) on various substrates; especially an economically important saprobe

molecular weight (MW): weight of a molecule expressed as the sum of the atomic masses of its constituent atoms

molecule: unit of matter; the smallest portion of an element or a substance composed of one or more atoms that retains the chemical identity with the substance en masse (usually consists of two or more atoms, although some organic molecules contain hundreds to millions of atoms)

mollicute: one of a group of prokaryotic microorganisms lacking a cell wall and bounded by flexuous membranes (plant-pathogenic mollicutes are now called phytoplasmas). *See* phytoplasma

molt or **moult**: to shed or cast off a cuticle or body encasement during a phase of growth

mon- or **mono-** (prefix): one

monocotyledon or **monocot** (pl. monocotyledonae, adj. monocotyledonous): one of a subclass of flowering plants, including the grasses and cereals, with one cotyledon (seed leaf) in the embryo and characterized by parallel-veined leaves and fibrous roots. *See* dicotyledon

monoculture: continuous planting of a single crop, cultivar, or noncrop species on the same site

monoecious aphid: one that spends its entire life cycle on a single plant species. *See* dioecious aphid

monogenic: containing or controlled by one gene; (of nematodes) producing offspring of only one sex

monokaryon: individual with one haploid nucleus per cell; haplont

monoxenic culture: growth of an organism in the presence of another organism

morphometrics: body measurements

mosaic: symptom of many viral diseases characterized by a patchy mottling of the foliage or by variegated patterns of dark and light green to yellow caused by disarrangement or unequal development of the chlorophyll

motile: exhibiting or capable of independent movement from place to place

mucilaginous: viscous; slimy; sticky when wet

muck: highly decomposed black soil in which the original plant parts are not recognizable (similar to peat soil but often having a lower percentage of organic materials)

mucus or **mucous** (adj. mucoid or mucilaginous): slime (e.g., gelatinous matrix of nematodes or colonies produced by highly encapsulated bacteria)

mulch: protective layer of some nonliving substance such as straw, dry leaves, wood-bark chips, compost, peat moss, plastic film, gravel, small stones, ground corncobs, cocoa or peanut shells, buckwheat hulls, or sawdust spread over the soil surface to catch rainfall, prevent splashing and soil crusting, retard water loss, control weeds, keep the soil temperature down during the summer, protect roots against cold, keep produce clean, and improve soil structure and fertility

mult- or **multi-** (prefix): many; much; a great number

multiallelic: having a series of alleles at the locus or loci

multiline: cultivar with a mixed population of two or more genotypes of plants

multinucleate: having more than one nucleus per cell; polynucleate

multiple-cycle disease: disease whose pathogen possesses a secondary infection cycle

mummy: dried, shriveled fruit; plant part or organ partly or completely replaced with fungal structures; desiccated plant part or organ

mutagen: substance that causes an increased mutation rate

mutation (adj. mutant): abrupt heritable or genetic change in a gene or an individual as a result of an alteration in a gene or chromosome or an increase in chromosome number; an individual species, or the like, resulting from such a departure and the process by which such changes occur

mutualism: relationship between two species of organisms in which both benefit from the association (e.g., alga and lichen)

mutually exclusive: pertaining to a condition in which two or more parasites or pathogens cannot exist together in the same host cell or tissue

MW: molecular weight

myc-, mycet-, myceto-, or **myco-** (prefix): pertaining to fungi

Mycelia sterilia: members of Deuteromycotina (Fungi Imperfecti) that do not produce spores, except for chlamydospores in some species; Agonomycetales

mycelial: referring to mycelium

mycelial fan: mass of mycelium, usually between bark and wood, with feathery branching strands

mycelial felt: mass of mycelium that fills shrinkage cracks in decayed wood, usually becoming compactly interwoven to form tough, leathery, feltlike sheets

mycelioid: resembling mycelium

mycelium (pl. mycelia): strands, group, or mass of hyphae making up the vegetative body (thallus) of a true fungus; "spawn" (the mycelia of fungi show great variation in appearance and structure); (in bacteriology) long filaments of segmented cells

mycology: science or study of fungi

Mycoplasma: genus of prokaryotic, parasitic bacteria of the class Mollicutes whose members are smaller (0.1–1.0 μm in diameter) than "walled" bacteria and much larger than viruses, have a confining unit membrane but lack a rigid cell wall, are highly variable in shape (pleomorphic), contain both DNA and RNA, and reproduce by budding or fission

mycoplasmalike organism (MLO): nonhelical bacterium with apparent features of a mycoplasma but not proven to be a mycoplasma found in the phloem and phloem parenchyma of diseased plants and assumed to be the cause of the disease. *See* phytoplasma

mycorrhiza (pl. mycorrhizae or mycorrhizas, adj. mycorrhizal): usually symbiotic association of the mycelium of a typically nonpathogenic or weakly pathogenic fungus with the roots of a higher plant that may aid in the uptake of certain nutrients, primarily phosphorus, by the plant host (most flowering plants cannot grow normally without the presence of mycorrhizal fungi)

mycotrophic: pertaining to green plants with mycorrhizae

myxamoeba (pl. myxamoebae): zoospore (swarm cell) after becoming amoebalike, without a definite shape; an amoeboid cell, particularly one of the myxomycetes

myxomycete: true slime mold; member of a class (Myxomycetes) of funguslike organisms characterized by amoeboid, vegetative protoplasts, multinucleate plasmodia, and brightly colored, sterile filaments (capillitia) among the spores

n.: noun

nanism: dwarfism

nanometer (nm): unit of length equal to 10^{-6} mm, 1 mμm, 10 Å, or 0.000000001 m (10^{-9} m)

natural openings in plants: stomata, lenticels, hydathodes, and nectarthodes

natural preservative: preservative substance, such as tannin or resin, that occurs naturally in wood and bark and tends to reduce attacks by fungi and insects

natural selection: process in nature by which the best-fitted types of organisms survive and produce offspring and the unfit are eliminated

navicular or **naviculate:** boat-shaped; cymbiform; with one sharply tapered end and one broad end

neck: peduncle; culm area just below the head and above the uppermost leaf of a grass plant

necrosis (pl. necroses, adj. necrotic): death of plant cells or plant parts, usually resulting in the tissue turning brown or black because of oxidation of phenolics; a symptom of disease (commonly of fungus, nematode, virus, or bacterial infection)

nectar: sticky, often sweet secretion, especially of the spermagonia (pycnia) of rust fungi, in which spores or spermatia may be freed and that attracts insects; a sweet fluid secreted by the nectaries of plants that is the chief raw material of honey

nectarthode: opening at the base of a flower from which nectar exudes

nectary (pl. nectaries): floral, nectar-secreting gland

needle blight: disease caused by fungal attack of evergreen foliage of conifers of any age and therefore possible anytime there is coincidence of high relative humidity or free moisture and spores (blighted foliage is sometimes cast after spore release). *See* needle cast

needle cast: a large drop of needles by conifers, generally caused by fungal species of Rhytismatales (syn. Phacidiales) (only young needles of the current year are generally attacked)

nematode or **nema:** eelworm; an unsegmented, wormlike animal (phylum Nematoda) parasitic in or on plants or animals or free-living in soil, decaying matter, or water; a generally microscopic roundworm with a cuticle and a hydrostatic skeleton abundant in many soils

nematology: science or study of nematodes

neutral soil: soil in which the surface layer is neither acid nor alkaline in reaction. *See* pH

nig-, nigri-, or **nigro-** (prefix): black

nitrification: conversion of ammonia and ammonium compounds into nitrites and nitrates through bacterial action in soils

nm: nanometer; 10^{-9} m; 1 mμm

nocturnal: active or occurring during the night

node: enlarged, usually solid joint on a stem; site from which a leafy bud and branch arise

nodule: lump, knot, or tubercle often on the roots of certain plants; an enlargement within which nitrogen-fixing bacteria live

nodulose or **nodulous:** having very small rounded knobs, knots, or nodules; (of spores) having broad-based, blunt, wartlike excrescences

nonaggressive strain: strain of fungus limited in pathogenicity

noninfectious disease: disease (or disorder) caused by an environmental factor (abiotic agent), a macroscopic animal, or a human and not by a pathogen; a disease that cannot be transmitted from a diseased plant to a healthy plant

nonpathogenic: incapable of causing disease

normal saline solution: sodium chloride (8.5 g) and water (1,000 ml)

no-till (syn. stubble culture): cultural system most often used with annual crops in which the new crop is seeded or planted directly in a field on which the preceding crop plants were cut down and the tops harvested or destroyed by a nonselective herbicide rather than having been removed or incorporated into the soil, as is common in preparing a plant bed

nucellus: tissue (megasporangium) found in the integuments of the young seed (ovule) enclosing the megagametophyte and in which the embryo sac develops (in some species it is absorbed as the seed matures)

nucleic acid: compound of high molecular weight that consists of a pentose sugar (ribose or deoxyribose), phosphoric acid, and nitrogen bases (purines and pyrimidines) in a long chain of repeating units; genetic material consisting of DNA or RNA present in all living cells and nuclear and viral materials (the infectious parts of plant viruses consist of nucleic acids)

nucleoside: a sugar (ribose or deoxyribose) joined to a nitrogen base molecule in a nucleic acid. *See* nucleic acid

nucleotide: phosphorylated sugar of a nucleoside (nucleotides are the building blocks of DNA and RNA)

nucleus (pl. nuclei): dense, usually spherical or ovoid protoplasmic body that contains the chromosomes that transmit hereditary characteristics, is present in most living cells of plants and animals, and is essential in all synthetic, developmental, and reproductive activities of a cell

nut: dry, indehiscent, one-seeded fruit with a hard, woody pericarp (shell) generally produced from a compound ovary, such as that of walnut (*Juglans*) or pecan (*Carya*)

nutlet: seedlike fruit segment of the mint and related families (when the fruit is split longitudinally into four sections, each section, shaped like a quarter section of an apple, is a nutlet)

nutrients, plant: essential elements available to plants through soil, air, and water that are utilized in metabolism and growth. *See* macronutrients and micronutrients

nymph: juvenile insect (e.g., aphid or grasshopper) that superficially resembles the adult

ob- (prefix): inversely, oppositely, or reversely

obclavate: shaped like an upside down club; inversely clavate; thickened towards the base

obconic or **obconical:** like an inverted cone, having a very narrow base and a broad apex

objective: set of lenses in a compound microscope nearest the object being viewed

obligate: necessary; essential; obliged

obligate parasite: parasite that can grow and multiply in nature only on or in living tissue and that cannot be cultured on an artificial medium. *See* facultative parasite and parasite

obovate or **obovoid:** reversely ovate (ovoid), narrowest at the base; egg-shaped in two directions (e.g., spores)

obpyriform: reversely pear-shaped

obtuse: rounded or blunt; greater than a right angle

occluded: closed

ochreous or **ocherous:** yellowish buff; cinnamon buff to brownish yellow; ochre yellowish

oct- or **octo-** (prefix): eight

odontoid: dentate, toothlike, as a hymenial surface with short, often compound teeth

oedema: *see* edema

oidium (pl. oidia): asexual spore formed by the fragmentation of vegetative hyphae into short, cylindrical segments

oleoresin: resin, pitch; a viscous, aromatic mixture of terpenes, resin acids, and fatty acids produced by various conifers

olivaceous: color of a green olive; yellowish green

oo- (prefix): egg

ooze: droplets or strands composed of bacteria or fungal spores mixed with host fluids found on the surfaces of lesions or on the cut ends of diseased stems, fruits, or leaves

opalescent: reflecting an iridescent light like an opal

opaque: dull; not shining

opportunistic fungi: naturally saprobic and often common but on occasion able to cause disease in a host plant, human, or higher animal rendered susceptible by one or more predisposing factors

opportunistic parasite: normal saprophyte that takes advantage of lowered host resistance to cause disease

opposite: bearing two leaves or two buds at a node on opposite sides of a stem

oral capsule or **cavity:** stoma or buccal capsule

orbicular or **orbiculate:** spherical or ringlike; round

organ: one of the major parts of a plant body (leaf, stem, and root) composed of various cells and tissues and adapted to perform certain specific functions

organelle: usually a membrane-delimited structure or body within a cell (e.g., Golgi apparatus or mitochondria) having a specialized function

organic matter: any material of, pertaining to, or derived from plants or animals

organism: any form of life that maintains various vital processes

ornamental plant: one grown for accent, attraction, beautification, color, pleasure, screening, specimen, or other aesthetic reasons

osmosis (adj. osmotic): diffusion of fluids (usually water) through a differentially permeable membrane from the side of higher concentration to the side of lower concentration of the fluid

oval: broadly elliptical to nearly circular; egg-shaped, with the larger end at the base

ovary: female reproductive structure of higher plants that produces or contains the egg or ovule; the basal, generally enlarged part of a pistil within which seeds develop (the mature ovary is a fruit)

ovate, oviform, or **ovoid**: shaped like an egg with one end narrower than the other (compare with obovate)

overseason: to survive from one planting season to the next

ovule: small egg, especially one in an early stage of growth; enclosed structure consisting of a female gametophyte, nucellus, and one or two integuments that, after fertilization, becomes a seed

oxidation: chemical reaction in which oxygen combines with another substance or in which hydrogen atoms or electrons are removed from a substance

ozone (O_3): highly reactive form of oxygen (a photochemical oxidant) that in relatively high concentrations may injure plants, humans, and animals

palatable (n. palatability): agreeable or attractive to the palate or taste

palea: upper and smaller of the two bracts that enclose a grass flower

pallid: light-colored; pale

PAN: peroxyacetylnitrates, toxic air pollutants produced by photochemical reactions in daylight air originating from the exhausts of internal combustion engines and injurious to plants

pandemic: widespread and destructive epidemic (epiphytotic) that occurs over an extended geographical area or areas

panicle: loose, branched, open-flower cluster common in the grass family (Poaceae); a compound raceme

papilla or **papillum** (pl. papillae, n. papilloid): minute, rounded, blunt, or nipple-shaped projection

parasexualism (adj. parasexual): recombination of genetic characters based on mitosis without sexual processes

parasite: organism (bacterium, fungus, nematode, insect, mite, etc.) or virus living with, in, or on another living organism (host) and obtaining food from it without giving benefit in return and frequently causing disease; a biotroph. *See* facultative parasite, obligate parasite, and pathogen

parasitism: association whereby one organism (the parasite) grows at the expense of another (the host)

parenchyma: physiologically active, soft tissue of higher plants composed of thin-walled, often isodiametric cells that commonly store food or perform other functions, usually retain meristematic potential, and have intercellular spaces between them

parthenocarpy: natural or artificially induced development of fruit without sexual fertilization (pollination)

(such fruits are seedless)

parthenogenesis (adj. parthenogenetic or parthenogenic): type of asexual reproduction in which an egg (female gamete) develops into a new individual without fertilization by a sperm (male gamete)

particle size, soil: effective diameter of a particle measured by sedimentation, sieving, or micrometric methods

parts per billion (ppb): method of expressing the concentration of chemicals or other proportions; 1 ppb = 1 in. in nearly 16,000 miles, 1 drop in 20,000 gal, 1 oz in 753 million gallons, and two crystals of sugar in 1,000 lb

parts per million (ppm): method of expressing the concentration of chemicals or other proportions; 1 ppm = 1 in. in nearly 16 miles, 1 lb in 500 tons, two crystals of sugar in 1 lb, 1 oz of salt in 62,500 lb of sugar, 1 min of time in about 2 years, a 1-g needle in a 1-ton haystack, 1 oz of sand in 34.25 tons of cement, 1 oz of dye in 7,530 gallons of water, 1 mg/liter or kg, 0.0001%, 0.013 oz by weight in 100 gallons, and one penny of $10,000

pascal (Pa): unit of pressure equivalent to 0.00014504 lb/in^2; 1 kPa = 0.14504 lb/in^2, and 10^6 Pa is approximately 7.5×10^3 mm of mercury

pasteurization: method of freeing a medium of selected pathogenic microorganisms by using heat

path- or **patho-** (prefix): suffering (hence, disease)

pathogen: organism or agent (e.g., bacterium, fungus, nematode, virus, or viroid) capable of causing disease in a particular range of hosts (suscepts)

pathogenesis: sequence of processes in disease development from the time of infection to the final reaction in the host; production and development of disease

pathogenic: causing or capable of producing disease

pathogenicity: ability of a pathogen to cause (incite) disease; the state or condition of being pathogenic

pathology, plant: science or study of diseases, their nature and effects on plants, causes, and control

pathotoxin: toxin or poison produced by a pathogen in the infected host that functions in the production of disease but is not itself the initial inciting agent. *See* vivotoxin

pathovar (pv.) (syn. subspecies): strain or group of strains of a bacterial species differentiated by pathogenicity in one or most hosts (species or cultivars); pathotype

PDA: potato-dextrose agar, a common culture medium for growing fungi

peat: unconsolidated soil mass high in semicarbonized organic materials consisting of partially decomposed plant tissue and formed in the water of marshes, bogs, or swamps, usually under conditions of high acidity (e.g., sphagnum peat moss)

pectin: methylated polymer of galacturonic acid found in the middle lamella and the primary cell walls of plants; jelly-forming substance found in fruit

pectinase (adj. pectolytic): enzyme that breaks down pectin

ped: unit of soil structure such as a soil aggregate, crumb, prism, block, or granule formed by natural processes

pedicel or **pedicle**: small, slender stalk bearing an individual flower, inflorescence, or spore

pedicellate or **pediculate**: having a stalk or pedicel; borne on a pedicel

peduncle: stalk or main stem of an inflorescence; part of an inflorescence or a fructification; stalk or pedicel

pellucid: translucent or transparent

peltate: shield-shaped, like a round plate with a stalk from the center of the underside, e.g., a leaf of garden nasturtium (*Tropaeolum majus*)

pendent or **pendulous**: hanging down or suspended by a slender dorsal attachment

penetration: initial invasion (entrance) of a host by a pathogen

pentose: sugar with five carbon atoms, e.g., ribose

penultimate: next to the last

per- (prefix): through or sometimes completely

perennial: plant that continues growth more or less indefinitely from year to year and usually produces seed each year; (of fungi) a fruit body, e.g., a basidiocarp, that persists in an active, living state for more than one growing season

perfect flower: one having both stamens and pistils (carpels); a hermaphroditic flower

perforate: to pierce through

peri- (prefix): near, around, about

perianth: external envelope of a flower consisting of sepals and petals (calyx and corolla)

pericarp: outer tissue layer (modified walls) of a ripened ovary (fruit); a covering; the entire fruit body

pericycle: layer of cells in a stem or root located between the endodermis and the vascular cylinder from which branch roots arise

peridium (pl. peridia): outer layer of protective tissue in a stem or root; a collective name for cork, cork cambium, and phelloderm; a tissue found on the inner surface of the cork cambium

permanent wilting percentage: amount of water in the soil at the time plants become permanently wilted. *See* wilting point

permeability (adj. permeable): quality or condition allowing a fluid or substance in a fluid to pass or diffuse through a membrane

permutate: to change completely

pest: any organism (e.g., some bacteria, fungi, insects, mites, nematodes, rodents, and weeds) injurious or detrimental to a beneficial plant, plant product, animal, human, or the environment

petal: one of the structural units of the corolla of a flower, frequently conspicuously colored

petiole: stem of a leaf; the stalk that attaches a leaf blade to a stem

petri plate or **dish**: *see* culture plate

pH: measure of acidity and alkalinity within the range of 0 to 14, where pH 7 is neutral, values less than 7 indicate acidity, and values greater than 7 indicate alkalinity (the scale is logarithmic, pH = $-\log$ [H$^+$], so that a change of one pH unit is equal to a 10-fold change in the hydrogen ion concentration)

phage: general term for viruses isolated from prokaryotes, including bacteria, blue-green algae (cyanobacteria), and mollicutes (mycoplasmas and spiroplasmas)

-phage (suffix): one that eats

-phagous (suffix): eating, feeding on

phase-contrast microscope: compound microscope with an annular diaphragm in the front focal plane of the substage condenser and a phase plate at the rear focal plane of the objective to make visible differences in the phase or optical path in transparent or reflecting media

phelloderm: secondary tissue produced by cork cambium and found on the inner surface of the cork cambium (resembles the cortical parenchyma in morphology); the innermost layer of the periderm

phellogen: cork cambium; lateral meristem forming the periderm, a protective tissue in stems and roots (phellem [cork] is produced toward the surface of some plants and phelloderm toward the inside)

phenotype: external, visible physical characteristics of an organism determined by its genotype and modified by the environment

pheromone: any of a class of hormonal, odor-generating substances secreted to the outside of an individual as a sexual or aggregation stimulus and received by a second individual of the same species (first discovered in insects)

-phile, -philic, or **-philous** (suffix): inhabiting, living upon; a liking for a substance or state

phloem: inner bark; a complex, vascular, food-conducting tissue consisting of sieve tubes (through which synthesized nutritive and other materials move from the leaves), companion cells, fibers, and phloem parenchyma; **secondary phloem**: phloem formed by the vascular cambium

photochemical oxidants: highly reactive compounds formed by the action of sunlight on less toxic precursors

photodegradation: degradation caused by light, usually sunlight

photoperiod: relative duration of night and day to which plants are exposed; the length of day or period of daily illumination required for the normal growth and sexual reproduction of some plants

photosynthesis: complex fundamental process by which green plants make carbohydrate (sugar) from water and carbon dioxide in the presence of chlorophyll(s) using light energy and releasing oxygen

phototropism: growth movement (curvature) induced by the stimulus of light (the response, which is auxin regulated, is a bending toward the strongest light)

phycomycete (adj. phycomycetous): fungus whose mycelium has no cross walls; an old term used for "lower fungi" (those other than members of Ascomycotina, Basidiomycotina, and Deuteromycotina)

phyllody: transformation of floral parts (petals) to leaflike structures caused by certain pathogens, insects, or mites

physiogenic (physiological) disease: disease (or disorder) produced by some unfavorable physical or environmental factor (e.g., excess or deficiency of light, water, temperature, soil nutrients, a chemical, etc.). *See* noninfectious disease

physiologic race: subdivision within a species, the members of which are alike in morphology but differ from other races in virulence, symptom expression, biochemical and physiological properties, or host range. *See* biotype

physiology, plant: science of the functions and activities of living plants

phyto- and **-phyte:** combining forms meaning plant

phytoalexin: antibiotic or fungistatic metabolite arising from host-parasite interaction and inhibitory to microorganisms attacking plants; a substance that inhibits the development of a microorganism, formed mainly when host plant cells come in contact with the parasite (response to certain stimuli); low molecular weight, antimicrobial compound that is both synthesized by and accumulated in plants after exposure to microorganisms

phytopathology: plant pathology; science or study of plant diseases

phytoplasma: nonhelical, plant-pathogenic member of the class Mollicutes; mycoplasmalike organism (MLO)

phytosanitation: any measures involving the removal or destruction of infected plant material likely to be a source of reinfection by a pathogen

phytotoxic (n. phytotoxicity): poisonous, injurious, or lethal to plants (usually used to describe a chemical)

phytotoxin: toxin injurious to plants; a toxin produced by a plant

pigment: colored compound, such as chlorophyll (some plant pigments are water soluble and found mainly in the cell vacuole)

pinching back: removing the tip of a stem (apical meristem), an extra flower bud, or terminal bud (by using fingernails, knife, or shears) to stimulate branching

pinna (pl. pinnae): one of the primary divisions of a pinnate leaf or frond

pinnate: having branches, lobes, leaflets, or veins arranged in a featherlike manner; attached or arranged on two sides of a stem

piriform or **pyriform:** pear-shaped

pistil: female or seed-bearing organ of the flower in a seed plant, typically consisting of an ovary and one or more styles and stigmas (a simple pistil consists of a carpel, while a compound pistil has two or more partly or wholly fused carpels)

pistillate flower: one that contains pistils (female parts) but no stamens (male parts)

pith (adj. pithy): soft, spongy (loosely packed), thin-walled, parenchyma cells in certain plant stems and roots

pitted: having depressions or cavities often caused by death of small, localized areas in fleshy or woody tissue beneath healthy-appearing external tissue

pl.: plural

placenta (pl. placentae): small mass of ovary tissue in a fruit to which the seeds (ovules) are attached

plane: a flat or even surface

plano- (prefix): motile or flat

plant disease: *see* disease, plant

plant-disease interaction: the concurrent parasitism of a host by more than one pathogen in which the symptoms or other effects produced are of greater magnitude than the sum of the effects of either pathogen acting alone (an example of synergism)

plant growth regulator: substance that affects plants or plant parts physiologically rather than physically by accelerating or retarding growth, prolonging or breaking a dormant condition, promoting decay, or inducing other physiological changes

plant health care (PHC): holistic approach to plant care that focuses on the health and growth of plants; giving a plant proper care to prevent it from undergoing stresses

plant pathology: science or study of plant disease; phytopathology

planthopper: small, leaping, homopterous insect (family Delphacidae) with piercing-sucking mouthparts known to transmit a number of plant-infecting viruses and mycoplasmalike organisms

plasmalemma: outer membrane composed of protein and lipid, bounding the protoplast next to the cell wall; cell, cytoplasmic, or plasma membrane

plasmid: generally a small, covalently closed, circular piece of nonchromosomal, double-stranded DNA, found in certain bacteria and fungi, that carries generally nonessential genetic information and is self-replicating

plasmodesma (pl. plasmodesmata): fine cytoplasmic strand passing through the cell wall of adjoining cells, interconnecting the two living protoplasts

plasmodium (pl. plasmodia, adj. plasmodiophoraceous): naked, amoeboid-like mass of protoplasm, generally reticulate, without cell walls, and containing multiple nuclei, that moves and feeds in amoeboid fashion and is the result of fusion of uninucleate amoeboid cells; the vegetative growth phase of the myxomycetes and some lower fungi

plasmolysis: contraction and shrinkage of the cytoplasm from the cell wall in a living cell caused by loss of water

plastid: any of the various cytoplasmic organelles (chloroplasts, leucoplasts, etc.) of photosynthetic cells that serve in many cases as centers of special metabolic activities

pleomorphic (n. pleomorphism): able to assume various shapes (and perhaps sizes); having more than one independent form or spore stage in the life cycle; polymorphic

plerome: procambium strands; inner tissue of the root meristem; plant tissue inside the cortex

pleur- or **pleuro-** (prefix): pertaining to a side; at the side

pleurogenous: borne or formed along the sides

plow layer: surface soil layer ordinarily turned over during tillage

plow pan: hardpan layer of soil formed by continual plowing at the same depth

plumose or **plumate:** finely feathery

plumule: bud- or shoot-forming meristem of an embryo or that portion of the young shoot above the cotyledons. *See* epicotyl

plur- or **pluri-** (prefix): several, many

plurivorous: attacking a number of hosts or substrates; not specialized

pocket rot: localized decay in trunks or roots of trees caused by wood-decay fungi

poison (syn. biocide): substance that, when absorbed or taken orally, can cause illness, death, retardation of growth, or shortening of life

polar transport: directed movement of compounds (usually hormones) within plants, mostly in one direction (overcomes the tendency for diffusion in all directions)

pollen or **pollen grains:** minute, yellow bodies (male sex cells) produced within the anthers of flowering or seed plants

pollination: transfer of pollen from a stamen (or staminate cone) to a stigma (or ovulate cone)

poly- (prefix): many; a large number

polyclonal: derived from many clones

polymorphic: having different forms; pleomorphic

polynucleate: having more than one nucleus per cell

polypeptide: chain of amino acids linked together by peptide bonds from the synthesis or partial hydrolysis of a protein

polyploid (n. polyploidy): having more than two sets of $2n$ chromosomes per nucleus

polysaccharide: large, long-chain, organic molecule consisting of two to many monosaccharides (simple sugars)

pome: simple, fleshy, indehiscent fruit derived from several carpels with a fleshy receptacle and outer pericarp and a papery inner pericarp, e.g., apple, pear, and quince fruits

pore: small opening

postemergence: period after the appearance of a specified weed or crop plant

postemergence spray: pesticide applied after crop plants have emerged from the soil

potting mixture (syn. soil mix): combination of several ingredients, e.g., soil, peat moss, sand, perlite, or vermiculite, designed for starting seeds or cuttings or growing plants in containers

ppb: *see* parts per billion

ppm: *see* parts per million

predator: animal that feeds by preying on other animals

predispose: to make prone to infection

predisposition: increase in the susceptibility of a plant to disease caused by nongenetic factors such as pathogenic or environmental stresses

preemergence: period before the emergence of a specified weed or crop plant

preemergence spray: pesticide applied after planting but before crop plants have emerged from the soil

preservative: substance used for killing or stopping the growth of microorganisms in or on any kind of substrate

primary: first; first-formed

primary cycle: of plant disease, the first cycle to begin in a given year

primary host: plant on which the sexual forms of an aphid mate and lay eggs to overwinter; the main host of a pathogen. *See* secondary host

primary infection: first infection by a pathogen after going through a resting or dormant period

primary inoculum: survival structures of a pathogen, usually from an overwintering source, that cause initial rather than secondary outbreaks of disease

primary root: root that develops directly from the mesocotyl (radicle) of an embryo rather than from a crown or node

primary symptom (syn. local symptom): symptom produced soon after infection at the site of entry, in contrast to a secondary symptom, which follows more complete invasion (colonization) of the host. *See* secondary symptom

primary tissue: tissue developed by an apical meristem during growth in length

primordium (pl. primordia, adj. primordial): rudimentary or initiating portion from which any organ, structure, or individual is formed; first in order of appearance; the earliest stage of development

pristine: early; original; primitive; free from soil or decay; fresh and clean

pro- (prefix): before

procambium: primary meristem, which gives rise to vascular bundles

process: outgrowth or projection from a surface

procumbent: nearly prostrate; spreading

profile, soil: vertical section of the soil through all the horizons and extending into its parent material

prokaryote (adj. prokaryotic): an organism lacking membrane-limited nuclei and not exhibiting mitosis (e.g., bacteria); organism with a genome of a simple, circular, double-stranded DNA free in the cytoplasm and not in organelles. *See* eukaryote

pronate: inclined to grow prostrate

propagule: any part of an organism capable of initiating independent growth when separated from the parent body

prophylaxis (adj. prophylactic): prevention

protein: complex, high molecular weight, organic, nitrogenous substance (polymer compound) composed of amino acids joined by peptide bonds and constituting a major essential part of the organic materials in living protoplasm

proto- (prefix): first; primitive; primordial

protophloem: conductive tissue of actively growing parts of the plant (its sieve tubes function for a brief period and are replaced by other phloem elements)

protoplasm: living material within a cell on which all vital functions of nutrition, secretion, growth, and reproduction depend; essential semifluid, viscous, translucent colloid of all plant and animal cells

protoxylem: undifferentiated wood element in a fibrovascular bundle; the first-formed xylem, with annular, spiral, or scalariform wall thickenings

protuberant: bulging out; prominent

proximal or **proximate**: nearest; next to or nearest the point of attachment or origin; opposite of distal

pruning: judicious removal of leaves, shoots, twigs, branches, or roots of a plant to control its size and shape, increase its usefulness, vigor, and productivity, or remove dead, diseased, or insect-infested material

pseudo- (prefix): false; spurious

pseudoparenchyma (adj. pseudoparenchymatous): isodiametric or oval fungus cells organized into a tissue in which the individual hyphae have lost their identity; aggregate of closely interwoven hyphae forming a definite body

psi: pounds per square inch

psyllid: jumping plant lice (Homoptera) of the family Psyllidae (some species are vectors of mycoplasmalike organisms)

pubescence (adj. pubescent): outer covering of short, soft, silky hairs or down, as on the surfaces of leaves, stems, and fruiting bodies, etc.; the state of being so covered

pulverulent: powdery, as though dusted

pulvinate: cushion-shaped; circular and usually strongly convex

punctate: marked with dots, very small spots, scales, or hollows

punctiform: dotlike but visible with the naked eye

pungent: having an acrid or biting flavor

punky: soft and rather tough, as decaying wood

pupa (pl. pupae): quiescent, nonfeeding stage between the larva and the adult of certain insects

pupate: to be inactive, as with certain insects in cocoons or pupal cases, before ultimate maturation into an adult

pure culture: culture in which only one species or biotype of an organism is growing

pustule: small, blisterlike or pimplelike elevation from which a fruiting structure of a fungus erupts

putrefaction: anaerobic decomposition of organic substances, especially proteins, by microorganisms that produces disagreeable odors

pv.: *see* pathovar

pyriform or **piriform**: pear-shaped

quadi- or **quadri-** (prefix): four, fourfold

quarantine: legislative control of the transport, import, export, and/or sale of plants or plant parts, usually to prevent spread of pathogens, insects, mites, weeds, or other pests; a law placing material in quarantine; holding of imported plants or plant parts in isolation for a period to ensure their freedom from diseases and pests

quiescence (adj. quiescent): dormant, quiet, rest period caused by external conditions unfavorable for germination or growth

quinque- (prefix): five

quintal: unit of weight in the metric system equal to 100 kg and equivalent to 220.462 avoirdupois pounds

race: subgroup or biotype of pathogens within a species, variety, or pathovar distinguished by behavioral differences, such as virulence, symptom expression, or to some extent host range, but not by morphology; a genetically and often geographically distinct mating group within a species. *See* physiologic race

raceme: type of inflorescence in which the main axis is elongate and unbranched and the flowers are borne on pedicels that are of about equal length

rachilla: internal axis of a spikelet of grasses and sedges

rachis: main axis of a grass or cereal head; the elongated axis of an inflorescence; an extension of the petiole of a compound leaf that bears the leaflets; (of fungi) a geniculate or zigzag, spore-bearing extension of a conidiogenous cell resulting from sympodial development; (of nematodes) a central strand of nonnucleated tissue to which oogonia are attached in the ovary and whose function is uncertain

radial section: section of a stem or root cut longitudinally on a radius

radicle: rootlet; the first or embryonic stem of a seedling plant or both; the first stem and the primary root; the root primordium of an embryo

ramify: to branch; to separate into divisions; to split into branches or constituent parts; to send forth branches or extensions

range: (of a plant pathogen) geographical region or regions in which it is known to occur; (of land and native vegetation) vegetation that is predominantly grasses, grasslike plants, or shrubs suitable for grazing by animals

rate: amount of active ingredient of a pesticide applied to a given surface; dose or dosage

ray: corolla of a marginal flower of a complete inflorescence; a vascular medullary ray, the tissue or narrow group of cells (usually parenchyma, but sometimes also tracheary cells) that extends radially in the secondary xylem and phloem of a woody plant

re- (prefix): back, backward; again

receptacle: more or less expanded terminal portion of the stem on which flower parts are borne; enlarged upper end of a pedicel or peduncle to which flower parts are attached; (of fungi) an axis having one or more organs; any hymenium-supporting structure; the inflated tips of certain brown algae within which gametangia are borne

recessive: pertaining to one of a pair of contrasting characters that is masked, when both are present, by the other, or dominant, character; also refers to the genes determining such characters

recessive gene: one that is not expressed in the phenotype of the heterozygote. *See* dominant gene

recurrence: return of symptoms after a period of absence

reduction: chemical process involving the removal of oxygen, the addition of hydrogen, or a gain of electrons

reduction division: stage of meiosis in which two daughter nuclei each receive half the chromosomes of the parent nucleus

registered seed: progeny of foundation or other registered seed produced and handled so as to maintain satisfactory genetic identity and purity and approved and certified by an official certifying agency (normally grown for the production of certified seed)

regular flower: one in which the corolla is made up of similarly sized and shaped petals equally spaced and radiating from the center of the flower; star-shaped flower; actinomorphic flower

relative humidity (RH): ratio of the quantity of water vapor pressure in the atmosphere to the quantity that would saturate it at the same temperature

renewal or **replacement spurs:** grape canes near the trunk cut back to two buds to provide new fruiting wood in a desired location

reniform or **renarious:** kidney-shaped; fabiform

renovate (n. renovation): to invigorate or rejuvenate; to thin plants, remove weeds, and form new plants

replication (v. replicate): process by which a virus particle induces the host cell to reproduce the virus; the process by which a DNA or RNA molecule makes an exact copy of itself; a folding or bending back, as a replicated leaf; repetition of an experiment or procedure at the same time and place; one of several identical experiments, procedures, or samples

reproduction, sexual: development of new plants by seeds (except in apomixis)

residue, pesticide: amount of chemical that remains on or in a harvested crop

resin: sticky to brittle plant product derived from essential oils and often possessing marked odors (used in varnishes, incense, medicines, etc.)

resinosis: exudation of oleoresin (pitch) from a wound or the result of infection of a conifer

resistance: ability of an organism to exclude or overcome to some degree or completely the effect of a pathogen or other damaging factor. *See* immune

resistant: possessing qualities that hinder the development of a given pathogen (infected little or not at all)

respiration: the oxidation of food by plants and animals to yield energy for cellular activities

retro- (prefix): backward

retrorse: backward

rev/min or **rpm:** revolutions per minute

RH: *see* relative humidity

rhizoplane: surface of a root

rickettsia (pl. rickettsiae): pleomorphic or rod-shaped, small, fastidious bacterium with scalloped cell walls belonging to the Schizomycetes group (an intracellular parasite in the xylem and phloem of a diseased plant)

ring spot: circular area of chlorosis with a green center (characteristic symptom of many virus diseases)

RNA: ribonucleic acid, a nucleic acid made up of nucleotides, the sugar ribose, and the bases adenine, guanine, cytosine, and uracil, found in the nucleus and cytoplasm, and involved in protein synthesis (the only nucleic acid [genetic material] of many plant viruses)

rogue: (v.) to remove from a population and destroy undesired individual plants on the basis of disease infection, abnormal growth habit or rate, insect infestation, or other reason; (n.) a variation from the standard varietal type

root devitalization: nematode-induced cessation of root elongation resulting from feeding at or near the root tip and causing stubby-root, coarse-root, and/or curly-tip symptoms

root hair: threadlike, single-celled outgrowth from a root epidermal cell through which water, nutrients, and other substances are absorbed into a plant

root knot: nematode-caused disease characterized by galls on the roots

root rot: decay in roots of living plants sometimes caused by pathogenic fungi

rootstock (syn. understock): trunk or root material into which buds or scions are inserted during grafting

rosaceous, rosellate, roseate, or **roseous:** pinkish or rose-colored

rot: softening, discoloration, and often disintegration of plant tissue by enzymes produced by fungal or bacterial pathogens (may be hard, soft, dry, wet, black, brown, white, etc.) *See* decay

rugae: wrinkles

rugose mosaic: severe mosaic accompanied by deformation such as leaf crinkling, curling, or roughening of the leaf surface

rugulose: finely wrinkled

rupture, irregular: of conidiomata, dehiscence by irregular breakdown of the upper wall or disorganized fracture of overlying tissues

s: second of time

saccate, sacciform, or **saccular:** like a bag, sac, or pouch

saccule: small sac or pouch

sanitation: removal and burning (usually) or composting of infected plants or plant parts; decontamination of tools, equipment, hands, etc.

sap rot: fungal decay of the dead sapwood of dead standing or fallen trees, stumps, and slash

sapr- or **sapro-** (prefix): rotten

saprobe: saprophyte

saprogenesis: part of the life cycle of a disease-producing organism in which it is not directly associated with a living host. *See* pathogenesis

saprophyte or **saprophile** (adj. saprophytic, saprogenic, saprogenous, saprophilous, or saprobic): organism that feeds on dead organic matter commonly causing its decay; a necrophyte on dead material that is not part of a living host. *See* parasite

scab: roughened, hyperplastic, crustlike, diseased area on the surface of a plant organ; a disease in which such areas form

scale: small, inconspicuous insect of the family Coccidae that is immobile throughout most of its life

schizocarp: simple, dry, indehiscent fruit composed of two or more united carpels that split apart at maturity, each part usually with one seed

schizogenetic or **schizogenous**: formed by cracking, cleavage, or splitting

scientific method: approach to a problem that consists of stating the problem, establishing one or more hypotheses as solutions to the problem, testing these hypotheses by experimentation or observation, and accepting or rejecting the hypotheses

scion: detached bud, piece of a twig, or shoot inserted by grafting into another shoot (the shoot that receives a scion is called the stock or rootstock)

scion-stock interaction: effect of a rootstock on a scion and vice versa, in which the scion on one kind of rootstock performs differently than it would on its own roots or on a different rootstock

scler- or **sclero-** (prefix): hard

sclerenchyma (adj. sclerenchymatous): supportive or protective plant tissue composed of hard, short cells or elongated cells with thickened, often lignified, cell walls (may or may not be devoid of protoplasm at maturity)

scleroid, sclerosed, or **sclerotic**: having a firm, hard texture; more or less lignified

sclerotium (pl. sclerotia): firm or hard, frequently rounded, usually darkly pigmented, sterile, vegetative resting body of a fungus composed of a compact mass of usually thick-walled, special-sized hyphal cells with an outer, usually darkened and sclerotized rind and an inner parenchymatous medulla with or without the addition of host tissue or soil (may remain dormant for long periods and is capable of surviving under unfavorable environmental conditions)

scorch: sudden browning and death (necrosis) of large, indefinite areas on a leaf, fruit, or stem from infection, lack of some element, chemical injury, or unfavorable weather conditions

scurf (adj. scurfy): a deposit of thin, dry scales or branlike flakes on the surface of a plant part

scutellum (pl. scutella, adj. scutellar or scutellate): single cotyledon of a grass embryo; the rudimentary, leaflike structure at the base of the first node of the embryonic culm of a grass plant

secondary cycle: of plant disease, any cycle initiated by inoculum generated during the same season

secondary host: host plant on which the asexual aphid form occurs; alternative host for a pathogen (do not confuse with alternate host, which is the second of two hosts necessary for completion of the life cycle of a fungus). *See* alternate host, alternative host, and primary host

secondary infection: infection resulting from the spread of infectious material (inoculum) produced after a primary infection (the first infection by a pathogen after a resting period) or from subsequent infections without an intervening inactive period

secondary inoculum: inoculum resulting from primary infections

secondary organism or **pathogen**: organism that multiplies in already diseased tissue; not the primary pathogen

secondary symptom: a symptom that appears after the primary symptom and follows more complete colonization of the host. *See* primary symptom

secondary xylem: xylem cells formed by activity of the vascular cambium (development of secondary xylem results in the so-called annual rings seen in most trees)

sectoring: in plate culture, "mutation" (genetic instability) results in one or more sectors having a changed form (e.g., color) or rate of growth

sedentary (syn. sessile): remaining in a fixed location; stationary

sedimentary rock: rock formed from material originally deposited as a sediment and then physically or chemically altered by compression and hardening while buried in the earth's crust

seed: mature ovule of a flowering plant containing an embryo, sometimes an endosperm, and a seed coat; **breeder seed**: seed, or sometimes vegetative propagating material, directly controlled by the originator, the sponsoring plant breeder, or institution that provides the source for the initial increase of foundation seed; **certified seed**: progeny of foundation, registered, or other certified seed that is so handled as to maintain satisfactory genetic identity and/or purity and that has been approved and certified by a certifying agency; **foundation seed**: seed stocks so handled as to maintain specific genetic identity and purity, such as may be designated by an agricultural experiment station (the source of certified seed, either directly or through registered seed); **registered seed**: progeny of foundation or other registered seed so handled as to maintain satisfactory genetic identity and purity and approved and certified by a certifying agency; **true seed**: fertilized, mature ovule consisting of an embryonic plant, stored food material, and a protective seed coat

seed infection: colonization of seed tissues by pathogens with or without conspicuous symptoms

seed infestation: contamination by pathogen propagules carried on the outside of seeds or associated with seed lots

seed pathology: study of seed diseases and microorganisms and agents carried on or with seeds

seed piece: vegetative unit of a plant used for propagation

seed primordia: meristematic tissues differentiable into seed

seed transmission: passage of inoculum from an infected or infested seed to a plant

seed treatment: chemical or nonchemical treatment used to control pathogen propagules associated with true seeds and seed pieces

seedbed: soil that has been prepared for planting seeds or transplants

seedborne disease: one showing symptoms on infected seeds or on seedlings produced from infected seeds

seedborne inoculum: propagules of a pathogen capable of causing disease of the seed itself or the seedling or plant derived from the seed

seedborne microflora: minute particle indicating the association of bacteria, fungi, nematodes, mycoplasmas, or, incorrectly, viruses with seeds

seedborne pathogen: infectious agent associated with seeds and having the potential to cause disease of a seedling or older plant

seedling disease: disease of young plants that may occur before or after emergence and may result in a reduced yield. *See* damping-off

selection: process of isolating and preserving certain individuals or characters from a group of individuals

seleniform: shaped like a full moon; round

self-fertile (n. self-fertility): capable of fertilization and production of viable seed after self-pollination (sexual reproduction occurs as a result of the fusion of eggs and sperms produced by the same individual)

self-incompatible: self-sterile; incapable of producing viable seed by self-pollination

semi- (prefix): half, partial

semipermeable: pertaining to a membrane, as of a cell, that allows some fluids or solutes but not others to pass through; differentially permeable

senescence (v. senesce, adj. senescent): decline or degeneration of tissues caused by maturation or a physiological aging process (often hastened by environmental stress, disease, or insect attack); the process of growing old

sensitive (n. sensitivity): pertaining to a diseased organism that reacts with severe symptoms to the attack of a given pathogen

sepal: one of the divisions of the outer floral envelope or calyx; one of the modified leaflike structures, usually green, collectively forming the calyx

separation: type of propagation based on the natural breaking apart of plant segments, e.g., bulbs or corms

sept- or **septi-** (prefix): partition

serology: method of detecting and identifying antigenic (protein) substances and the organisms that carry them by using the specificity of the antigen-antibody reaction; the study, detection, and identification of antigens, antibodies, and their reactions

serrate: having edges with sharp notches or teeth; like a saw blade

serrulate or **serratulate:** minutely toothed

sessile: pertaining to a leaf, leaflet, flower, floret, fruit, ascocarp, basidiocarp, etc. without a stalk, petiole, pedicel, stipe, or stem; (of nematodes) permanently attached; not capable of moving about

setose: covered with bristles or setae

severity: of disease, measure of damage done by a plant disease (as distinguished from disease incidence, a measure of the number of individuals affected)

sexual: participating in or produced as a result of a union of nuclei in which both fertilization and meiosis are involved; (of fungi) pertaining to the teleomorph or perfect state

sexual propagation: an increase in numbers by seed

sexual reproduction: reproduction requiring the union of two compatible nuclei or gametes and involving nuclear fusion and meiosis

shade tolerant: pertaining to plants that can grow in reduced sunlight

sheath: lower tubular part of a grass leaf that clasps the culm; a membranous cover

sidedress: fertilizer or other chemical applied in a band alongside an established crop

sieve tube: long tube formed by a series of thin-walled phloem cells through which food materials are transported

sign: any visible indication of the pathogen or its parts (e.g., spores, mycelium, exudate, or fruit bodies) on a host plant. *See* symptom

silt: class of textural soil having particles between 0.002 and 0.05 mm in diameter

silviculture: science of developing, caring for, or cultivating forests

single-cycle disease: disease whose pathogen lacks a secondary infection cycle; monocyclic

single-stranded: pertaining to nucleic acid molecules (DNA and RNA) with only one chain of nucleotides

sinuate or **sinuous:** serpentine, crooked, wavy, or flexuous

slash: residue of unmerchantable cull logs, tops, branches, and other material left after timber harvesting, thinning, pruning, or natural phenomena such as windstorms or snow and ice damage

slime: wet, generally sticky substance; mucus

slime flux: fermented sap exuded through a wound in a limb or trunk; a thick liquid from the stem or branches of trees made up of, or having a connection with, yeasts, fungi, or bacteria. *See* wetwood

slime mold: one of the fungi of the classes Acrasiomycetes and Myxomycetes; the superficial "disease" caused by one of these organisms on low-growing plants and mulches

slip: herbaceous or softwood cutting

sod: generally 2.5–7.5 cm (1–3 in.) of soil permeated and held together by grass roots or grass-legume roots

softwood: wood of gymnosperms; a wood lacking wood fibers

soil: solid portion of the earth's crust in which plants grow (composed of mineral material, air, water, and both living and dead organic matter)

soil inhabitant: microorganism that is usually strongly competitive with other normal microflora of the soil and that often survives many years as a saprophyte in the complete absence of host plants

soil invader: microorganism that is poorly competitive with normal soil microflora and seldom survives more than 1 or 2 years in the soil in the complete absence of hosts

soil pasteurization: treatment of soil with heat, usually steam at 60–70°C (140–160°F), to destroy most harmful pathogens, nematodes, insects, and weed seeds without generally affecting the saprophytic soil flora (a less severe temperature treatment than soil sterilization)

soil salinity: amount of soluble salts in a soil, expressed as parts per million (ppm), millimhos/cm, or other convenient ratios

soil sterilant: chemical that, when present in the soil, prevents the growth of plants, microorganisms, etc.

soil sterilization: treatment of soil with heat (usually steam at 100°C [212°F]), with chemicals, or by gaseous fumigation to kill all organisms living in it

soil texture: relative percentages of sand, silt, and clay in a soil

soil tilth: physical condition of soil as related to its ease of tillage, fitness as a seedbed, and suitability for plant growth

soilborne: pertaining to a soil source or origin of a pathogen; pertaining to a microorganism living and surviving in the soil

solubility: maximum amount of a liquid or solid that will dissolve in a liquid

solute: dissolved substance

solution: homogeneous mixture in which the molecules of the dissolved substance (the solute) are dispersed among the molecules of the solvent

solvent: substance, usually a liquid, that can dissolve other substances (solutes)

somatic: pertaining to the vegetative phase in plants as distinguished from the reproductive phase

sooty mold: one of the fungi (members of Capnodiales, Atichiaceae, or several families of Dothideales) that grow on the honeydew secreted by aphids, mealybugs, scales, and whiteflies and form a dense, superficial, sooty coating of dark hyphae on foliage, stems, and fruit

source of inoculum: place or object on or in which infective propagules are produced

sp. (pl. spp.)**:** species (a genus name followed by "sp." indicates that the particular species is undetermined; "spp." after the genus name indicates that several species are grouped together without being named individually). *See* species

sp. nov., sp. n., or **n. sp.:** new species

spathe: large bract or pair of bracts sheathing a flower cluster (e.g., in Jack-in-the-pulpit [*Arisaema triphyllum*] or palms)

spathulate or **spatulate:** shaped like a spoon, spatula, or paddle; oblong with a narrowing base

spear: stylet of a nematode

species (sp.) (pl. species)**:** any one kind of life subordinate to a genus but above a race, strain, or variety; usually the smallest unit in the classification of organisms; a group of closely related individuals of the same ancestry, resembling one another in certain inherited characteristics of structure and behavior and relative stability in nature (the individuals of a species ordinarily interbreed freely and maintain themselves and their characteristics in nature). *See* genus and race

specimen: individual taken as a representative of a larger group, e.g., a single plant; a portion or sample of material (from such a specimen, a culture can often be made and one or more strains [isolates] isolated)

spermoplane: surface of a seed

spermosphere: microhabitat around a seed in soil

sphaeroid or **spheroidal:** nearly spherical, as spores

spicate or **spiciform:** having the form of a spike (tail)

spiculate: having a small, erect point

spike: elongated and unbranched inflorescence that bears possible sessile flowers

spikelet: spike appendage comprised of one or more reduced flowers and associated bracts arranged alternately along an unbranched axis (the rachilla); unit of inflorescence in grasses; a small spike

spine: narrow, sharply pointed process

spinose, spiny, or **spinous:** having spines; aculeate

spinule: small spine

spiroplasma: member of a genus (*Spiroplasma*) of small, pleomorphic, prokaryotic bacteria that lack rigid cell walls; flexuous, often helical mycoplasma present in the phloem of diseased plants (there are also nonpathogenic spiroplasmas that live on plant surfaces)

spongy mesophyll: tissue composed of loosely arranged, chlorophyll-bearing parenchyma cells of diverse form found just inside the lower epidermis of the leaf

spongy parenchyma: layer in a leaf between the palisade parenchyma and the lower epidermis made up of loosely packed, thin-walled cells

spora or **air spora:** population ("spore flora") of airborne particles of plant or animal origin

spore: (of fungi) one- to many-celled, microscopic, reproductive body that can develop into a new plant (may be sexually or asexually produced and of a wide variety of sizes, shapes, colors, and origins); (of bacteria) more correctly endospore, a resting phase, a body within the main part of the vegetative microbial cell

spore horn: *see* cirrhus

sporodochium (pl. sporodochia): fruiting structure, typical of members of Tuberculariaceae, consisting of a cluster of short conidiophores woven together on a mass of hyphae

sporogenesis: spore development; the production of spores

sporophyll: spore-bearing leaf

sporophyte: diploid or asexual phase in the life cycle of a plant; diplont; diplophase. *See* gametophyte

sporulate (n. sporulation): to form or produce spores

spot treatment: application of a chemical (e.g., a fertilizer or pesticide) to a restricted or small area

spp.: species (plural)

sprig or **sprigging**: single turfgrass stem (stolon, rhizome, or tiller), usually with some attached roots, that is used in vegetative propagation

springwood: early part of the yearly xylem growth ring in woody plants consisting typically of cells that are larger than those formed later in the season (summerwood)

spur: short, woody stem (branch), the principal fruiting area of many fruit trees

spurious: false

squamose or **squamous**: covered with or consisting of scales

squamule: small, scalelike structure

ss: single-stranded (nucleic acid)

staghead: antlerlike formation of a tree dying from the top downward; defoliated, dead, or dying major branches in the crown of a tree often caused by a wilt or root problem

stamen (adj. staminate): male reproductive structure of a flower consisting of an anther (pollen-bearing portion) borne on a stalk or filament; a modified microsporophyll

staminate (male) flower: one that bears stamens (male parts) but no pistils (female parts)

stand: number of established individual plants or shoots per unit area

starch: white, complex, water-insoluble polysaccharide consisting of glucose units; the principal food storage substance of plants

stele (adj. stelar): central cylinder of vascular and closely associated tissues in the stems, roots, and other parts of higher plants; xylem, phloem, and (when present) pith, pericycle, and interfascicular parenchyma

stellate or **stelliform**: star-shaped; (of crystals) arranged in radiating or star-shaped patterns

stem: main body of a plant, usually the ascending axis, whether above or below the ground, as opposed to the descending axis or root (stems, but not roots, produce nodes and buds); (in nomenclature) the part of a name to which the suffix appropriate to a category is added

sterile: infertile; free from living microorganisms; uncontaminated; not spore producing

sterilization: elimination of living cells or organisms (including pathogens) by heat, chemicals, light, or other means from soil, containers, etc.; reduction of sexual structures or processes in an individual or race

sterilized: free from living cells (usually microorganisms)

stigma: part of a pistil, usually the apex, that receives pollen and upon which pollen grains germinate; a darkened area in the upper margin near the ends of the wings of certain insects

stipule: small, leaflike structure or appendage, usually paired, found at the base of leaf petioles in many species of plants

stock: portion of the stem and associated root system into which a scion is grafted; sometimes applied to an artificial breeding group or a production planting; strain, race, or group of genetically similar organisms

stolon: elongated, slender stem ("runner") that grows horizontally above or under the soil surface and is capable of developing roots and stems at the nodes and ultimately of forming an independent plant

stoloniferous: bearing or developing stolons

stoma or **stomate** (pl. stomata or stomates, adj. stomatal): regulated microscopic opening (pore) in the epidermis of leaves and stems, controlled by two guard cells, that functions in the passage of gases and water vapor and sometimes serves as a point of entry for pathogens

stool: (horticultural) collectively, sprouts that arise from the base of the plant below ground and become roots (used for vegetative propagation)

stool or **stooling**: (agronomic) collectively, shoots that arise from below ground at the base of a plant

strain: biotype; race; form; isolate; an organism or group of similar organisms that differ in minor aspects from other organisms of the same species or variety; descendants of a single isolate in pure culture

stratified: arranged in layers

streak: elongated lesion with irregular sides

stress, water: condition that results when plant(s) are unable to absorb enough water to replace that lost by transpiration (may cause wilting, cessation of growth, or death of the plant or plant parts)

stringy rot: type of white rot in which the decayed wood in advanced stages has a fibrous appearance

strip cropping: practice of growing crops that require different types of tillage, such as row and sod, in alternate strips along contours or across the prevailing direction of the wind

stripe: elongated necrosis of tissue between vascular bundles in leaves or stems of cereals and other grasses

stubble: stem bases and crowns of harvested plants still rooted in soil

stunted: reduced in size and vigor as a result of unfavorable environmental conditions or a wide range of pathogens or abiotic agents

style: stalklike structure that connects the stigma and ovary in the pistil of most flowers and through which pollen tubes grow toward the ovule

stylet: relatively long, pointed, stiff, slender, hollow feeding organ in the mouth portion of plant-parasitic nematodes and some insects (e.g., aphids, leafhoppers) for piercing and withdrawing nutrients from plant cells

styletborne : pertaining to a noncirculative virus borne on the stylet of its insect vector

sub- (prefix) **:** under; below; almost qualified; slightly; somewhat; more or less

subculture : culture (e.g., of bacteria or fungi) derived from another culture

subcuticular : beneath a cuticle; between the cuticle and upper epidermal wall

subepidermal : beneath the epidermis

suberin : waxy, water-insoluble, hydrophobic deposit found in the walls of cork and certain other cells

suberize (adj. suberized) **:** to convert exposed plant surfaces into tough cork tissue; to harden cell walls by conversion to cork (suberin)

subhyaline : nearly hyaline but somewhat colored

subirrigation : application of water from below the soil surface, usually from a ditch or perforated hose or pipe or by placing a potted plant on a constantly moist surface

submicroscopic : too small to be seen with a light (compound) microscope

subsoil : soil below the plow layer

subsoiling : breaking up of compact subsoils, without inverting them, by use of a special knifelike chisel pulled through the soil, usually at depths of 30–60 cm (12–24 in.) and at spacings of 60–150 cm (2–5 ft)

subspecies (subsp.) : infraspecific population defined on the basis of more than one sharply defined character (morphology for most organisms) that distinguish its members from typical representatives of the species and often possessing a distinct habitat

substrate or **substratum** (pl. substrata) **:** surface or medium on or in which a microorganism is growing (attached) or living and from which it may get its nourishment; chemical substance acted upon, often by an enzyme

subtend : to bear above, as a bract or leaf that bears a flower in its axil

subulate or **subuliform :** rather slender and tapered to a point; awl-shaped

succulent (n. or adj.) **:** plant or plant part with tender, juicy or watery tissues

sucker : shoot arising from a root

summer fallow : tillage of uncropped land during the summer to control weeds and other pests and store moisture in the soil for the growth of a later crop

summerwood : part of the yearly xylem growth ring in woody plants formed late in the growing season and consisting of cells smaller than those of springwood

sunscald : damage to the surface plant tissues, including those of unshaded fruits and house plants, resulting from the action of intense sunlight; high-temperature injury caused by intense sun's rays warming the trunks and major limbs of woody plants, usually during the winter, that often results in bark cracking and splitting

super- or **supra-** (prefix) **:** above, in either position or extent

superimposed : overlapping

suppressive soils : soils in which certain diseases are reduced (decreased in severity or incidence) as a result of the presence of microorganisms antagonistic to the pathogen or pathogens

suscept : any living organism attacked or susceptible to a given disease, pathogen, or toxin (an abbreviated form of "susceptible plant" or "susceptible species"). *See* host

susceptibility : inability of a plant to resist the effect of a pathogen or other damaging factor

susceptible : not immune; lacking resistance; prone to infection

sustainable agriculture : appropriate use of crop and livestock systems and the agricultural inputs supporting their activities to maintain economic and social viability while preserving the high productivity and quality of the land

suture : (of higher plants) groove or seam along which the carpels of a fruit separate at maturity

sward : carpet of grasses or other ground covers such as clovers

sym- (prefix) **:** *see* syn-

symbiont or **symbiote :** one member of a symbiotic relationship

symbiosis : the living together of unlike organisms; a mutually beneficial association of two or more different kinds of organisms

symmetrical : of equal morphology about a designated axis

symptom : indication of disease by reaction of the host; the visible effect produced in or on a plant by the presence of a pathogen or abiotic factor. *See* sign

syn- or **sym-** (prefix) **:** with; (in compounds, etc.) growing together; aggregation; adhesion

syn. : synonym(s)

syndrome : totality of effects produced in a plant by one disease, whether all at once or successively; pattern or sequence of disease development; a combination of symptoms and signs of one or more causal agents; (of viruses) symptoms caused by the joint infection of two or more viruses, which are greater than the sum of those caused by the individual viruses

synergism (adj. synergistic) **:** mutual association (living together) of two or more organisms or environmental factors acting at one time and eliciting a host response that one alone could not (the magnitude of host response to concurrent pathogens exceeds the sum of the separate responses to each pathogen); (of pesticides) the action of a chemical that, when mixed with a pesticide, increases its toxicity (the synergist may or may not have pesticidal properties of its own)

synnema (pl. synnemata) **:** conidia-bearing structure composed of a more or less compacted group of erect and sometimes fused conidiophores bearing spores at the apex only or on both the tip and sides

synonym : (in biology) another name for an organism (species or group), especially an earlier or later illegitimate name; a rejected scientific name, other than a homonym

synoptic key: artificial key in which each taxon is often given a number (all characters of the taxa are listed, with genus numbers entered for those that are positive, and by eliminating successive characters, at random if desired, the single number remaining gives the name of the taxon)

systemic: pertaining to a chemical, pathogen, or infection that generally spreads internally throughout the plant body as opposed to remaining localized

tangential (face) section: section of a cylindrical organ, such as a stem, cut lengthwise and at right angles to a radius of the organ

tannin: one of a heterogeneous group of bitter (astringent) polyphenol derivatives widely distributed in certain plant tissues, such as bark, heartwood, and others, which condense with proteins to form a leatherlike substance insoluble in water

taproot: primary, elongated, deeply descending root of a plant from which the secondary or lateral roots arise

tawny: dull yellowish brown, about the color of a lion

taxonomy: science dealing with the systematic describing, naming, and classifying of organisms on the basis of their natural relationships

teleomorph: sexual or perfect state of a fungus, i.e., the form involved in producing meiotic spores

tenuous: delicate

terminal: the end

terrestrial, terricole, or **terricolous:** growing on the ground; living on land as opposed to water

test plant: plant species, variety, or cultivar used for diagnosing, or evaluating the infectivity of, a pathogen

testa: outer coat of a true seed developed from the integument(s) of an ovule

tetra- (prefix): four

tetraploid: having four sets of chromosomes ($4n$) per nucleus (twice the usual diploid number)

texture: (of soil) relative proportions of the variously sized groups of individual soil grains in a mass of soil; proportions of sand, silt, and clay in a given amount of soil; (of plants, fungi, etc.) arrangement of the components of different tissues as compact, loose, etc.

thallophyte: lower plant whose vegetative (somatic) phase is devoid of true stems, roots, or leaves and that usually propagates by means of spores (e.g., a bacterium, alga, fungus, slime mold, lichen, or moss)

thallus (pl. thalli): relatively simple, vegetative body of a thallophyte (nonvascular plant) that has comparatively little cellular differentiation and is without true stems, roots, and leaves

thatch: tightly intertwined layer of undecomposed or partially decomposed organic residues situated above the soil surface and generally below the green portions of the turf

thermal death (syn. inactivation point): lowest temperature at which death of an organism occurs (usually by heating) after a selected interval of time, usually 10 min; the temperature (varying from 45 to 95°C) at which virus particles lose their infectivity or an enzyme its activity

thermostable: relatively resistant to heat (to temperatures of 100°C)

thorax: insect body part between the head and abdomen that bears the legs and wings

thrips: small, slender insect (order Thysanoptera) with rasping-sucking mouthparts, a 10-segmented abdomen, and usually four long, narrow wings (a vector of fungal and bacterial pathogens and the tomato spotted wilt virus)

tiller: lateral shoot, culm, or stalk arising from a crown bud

tilth: state of soil aggregation or consistency ("good tilth" implies a porous, friable texture). *See* soil tilth

tipburn or **tip necrosis:** necrosis of the leaf tip that affects only a small percentage of the entire leaf (may be caused by internal water stress usually resulting from wind desiccation, salt accumulation, or a canker, wilt, or root pathogen)

tissue: group of cells, usually of similar structure, that perform the same or related functions

tolerance (adj. tolerant): ability of a plant to endure an infectious or noninfectious disease, adverse conditions, or chemical injury without serious damage or yield loss. *See* resistance, resistant, and susceptible

tomentose or **tomentous:** downy or woolly; having a covering of short, soft, densely matted hairs or hyphae that produce a texture like that of a woolly blanket

topdressing: prepared soil mix added to the soil surface and usually incorporated into the soil by raking or irrigating (materials such as fertilizer or compost are applied to the soil surface while plants are growing)

topworking or **top-grafting:** partially or completely changing the cultivar (or cultivars) of a tree by inserting buds or grafts on its main scaffold branches

torsive: twisted spirally

toruloid, torulose, or **torulous:** cylindrical and contorted but having swellings at intervals like a chain of beads; moniliform

toxic: of, caused by, or acting as a poison; capable of causing injury to a living organism

toxicant: poisonous or toxic substance

toxicity: relative capacity of a substance to produce injury to a living organism

toxin: poisonous secretion produced by a living organism; a nonenzyme metabolite produced by one organism that is injurious to another

transgressive segregation: genetic mechanism transferring characters (to progeny) not expressed by parents

translocation: movement of water, minerals, food, pathogens, wastes, etc. within or throughout a plant; segmental exchange between nonpaired (nonhomologous) chromosomes to form two new chromosomes

translucent: so clear that light rays may pass through; transmitting light without being transparent

transmission: the transfer or spread of a virus or other pathogen from plant to plant or from one plant generation to another (transmission by physical contact, vector, or pollen is described as horizontal; transmission from mother to progeny, e.g., seed transmission of many plant viruses, is vertical)

transovarial: through ovaries and eggs

transpiration: diffusion and loss of water vapor from aerial plant parts, chiefly through stomata in the leaves

transverse: crosswise; at right angles to the longitudinal axis

transverse section: section cut at right angles to its longitudinal axis

treehopper: small, leaping, homopterous insect of the family Membracidae with piercing-sucking mouthparts (some species are known to transmit a number of plant-infecting viruses

trenching: physical separation of soil in a vertical plane to sever grafted roots between trees

tri- (prefix): three, triple

trichome: single- or many-celled outgrowth from a plant epidermal cell, usually hairlike, glandular (secretory), or spiny

trifoliate leaf: leaf with three leaflets on a petiole

trifurcate: bearing three branches or forks

trilaminar: consisting of three thin, parallel layers

tripartite: three-parted; divided into threes

triploid: having three sets of chromosomes ($3n$)

triradiate: having three radiating arms or branches

triturate: to grind or pulverize with a mortar and pestle or blender

tropism: movement or curvature caused by an external stimulus that determines the direction of movement

truncate: ending abruptly as though the end was cut off squarely

trunk rot: decay in the main trunk of a living tree above the stump level

tuber: short, fleshy, much enlarged, mostly underground stem having numerous buds or "eyes"

tubercle: small, wartlike process

tuberous: round and swollen like a tuber

tubule: small tube

tumor: overgrowth of tissue or tissues that grows independently of surrounding tissues and may invade those tissues

turbid: cloudy; not clear

turgid (n. turgidity): pertaining to cells or tissues that are swollen, distended, plump, or rigid because of internal water pressure

turgor or **turgor pressure:** inflation of a plant cell by fluid contents; pressure within a cell resulting from the absorption of water into the vacuole and the imbibition of water by the protoplasm (lack of turgor pressure causes plants to wilt)

turgor movement: plant movement resulting from changes in water pressure within certain tissues

twig: 1-year-old stem or branch of a woody plant

tylosis (pl. tyloses, adj. tylose): balloonlike extrusion of a parenchyma cell through a pit into the lumen of a contiguous xylem vessel or tracheid that partially or completely blocks it; the process of tylose formation

type: (of nomenclature) the specimen or specimens on which the name of a taxonomic group (taxon) is based; an object that serves as the base for the name of a taxonomic category (e.g., a specimen is the type for a species, a species the type for a genus, a genus the type for a family, etc.)

-ule or **-ula:** suffix forming diminutives

ultra- (prefix): excessive or extreme

ultramicroscopic: too small to be seen with a compound (light) microscope; submicroscopic

umbel: type of inflorescence in which flowers are borne at the end of a common stalk forming a more or less flattened or rounded cluster

umber: dull olive-brown or dark smoky brown

umbilicate: having a small, rounded central depression or hollow

un- (prefix): with a verb, expresses reversal of action or removal; with an adjective, adverb, or noun means "not," giving a negative or opposite force

unavailable water: water held by the soil so strongly that plant roots cannot absorb it

uncinate or **uncate:** hooked

understock: portion of the stem and associated root system onto which a scion is grafted. *See* rootstock

undulate: wavy; not flat or uniformly curved

uni- (prefix): one, single

unicellular: having one cell, as an organism the entire body of which consists of a single cell

unisexual: pertaining to an organism that produces eggs or sperms but not both; pertaining to a flower that bears stamens or pistils but not both

urniform: urn-shaped; having a swollen base and narrow median portion expanded at the apex

v.: verb

vacuole: relatively clear, bubblelike, membrane-lined space or cavity in the cytoplasm of a plant cell filled with an aqueous solution of sugars and other plant products and by-products

variety (var.): (botanical) group of closely related plants of common origin within a species that differ from each other in distinct morphological characters such as form, color, flower, and fruit; a subgroup within a species given a Latin name according to the rules of the International Code of Botanical Nomenclature (a taxonomic variety is known by the first validly published name applied to it so that nomenclature tends to be stable); (cultivated) *See* cultivar

vascular: pertaining to conductive (xylem and phloem)

tissue or a parasite that grows and/or moves in these tissues

vascular bundle or **strand**: distinct group of elongated conducting cells in a root, stem, or leaf, consisting of primary xylem for water conduction and primary phloem for food conduction and frequently enclosed by a bundle sheath of parenchymatous tissue or fibers for mechanical support

vascular ray: ribbonlike aggregate of cells that extends radially in stems through the xylem and phloem

vascular tissue: conducting tissue; xylem and phloem

vector: living organism (insect, mite, nematode, parasitic plant, human, etc.) able to carry or transmit a pathogen (virus, bacterium, fungus, nematode, etc.) and disseminate disease

vector resistance: resistance of a host plant to the vector of a virus, of which there are three basic types: 1) antibiosis—the growth and multiplication of the vector on the host is inhibited; 2) nonpreference—the host is one that a vector prefers not to feed on (this may cause the vector to probe-feed several times shortly after landing before moving to another plant, which may reduce the transmission of a persistent virus but often increases the spread of a nonpersistent virus); and 3) tolerance—the ability of a host plant to withstand a vector's attack without suffering severe damage (this type does not control virus spread). *See* resistance and tolerance

vegetation: the plants that grow in a region; formed of species that make up the flora of the area

vegetative reproduction: asexual reproduction by a root, stem, leaf, or other primarily vegetative part of a plant body

vein: vascular strand of xylem and phloem in a leaf, petal, or other plant part

velvety: like velvet; thickly covered with close, short, soft, delicate hairs (hyphae)

venation: arrangement of veins in a leaf blade

ventral: front or lower surface, as opposed to dorsal

verdant: grass green color

vernal: of or belonging to spring

vernalization: the process by which floral induction in some plants is promoted by exposing the plants to chilling for a certain length of time

verrucose or **verrucous**: adorned with small, rounded, wartlike structures; appearing as a minutely roughened wall (commonly applied to a variety of roughened surfaces between echinulate and reticulate)

verruculose: delicately or minutely verrucose

versicolor: changeable in color

verticillate: having a whorl of three or more branches or sporogenous cells arising at the same level; arranged in verticils or whorls

vesicular-arbuscular: pertaining to mycorrhizae that possess both vesicles and arbuscules (the penetrating hyphae in the cortical cells may be coiled [pelotons] or finely branched [arbuscules] haustorial branches)

vesiculiform or **vesiculate**: having the form or structure of a vesicle; saclike or bladderlike; having a swollen appearance

vessel: series of water-, mineral nutrient-, and sap-conducting dead xylem tubes (cells) in the stem and root that are attached end to end

vestige (adj. vestigial): trace or visible sign left by something vanished or lost; a minute remaining amount; a small and degenerate or imperfectly developed body part or organ that remains from one or more fully developed organs in an earlier stage of the individual, in a past generation, or in closely related forms

viable (n. viability): state of being alive; capable of growth and development (applied usually to seeds and spores)

vicid: slimy, sticky, mucilaginous, glutinous, lubricious, viscous. *See* gelatinous and viscid

villose or **villous**: covered with long, soft, silky hairs that are not matted. *See* tomentose

vinaceous: having pale lavender, purplish, or wine-red tints

violaceous: of a violet hue

virescence (adj. virescent): state or condition of turning green, as a normally white or colored tissue that develops chloroplasts and turns green

viridant: becoming grass green or verdant

virion: complete virus particle consisting of nucleic acid (RNA or DNA) and a protein coat; the infectious unit of a virus

viroid: infectious pathogen much smaller than a virus; unencapsidated, circular, single-stranded ribonucleic acid (ssRNA); small, low molecular weight (247–374 nucleotides) RNA that can infect certain plant cells, direct its own replication from host components, and cause disease

virulence: quantitative term describing the severity of symptoms or the extent to which a pathogen damages a host

virulent: pathogenic; having the capacity to cause slight or severe disease

virus particle: morphological form of a virus (in some viruses, the particle consists of a genome surrounded by a protein capsid; other virus particles have additional structures such as envelopes, appendages, tails, etc.)

viscera: internal organs

viscid, viscose, or **viscose**: sticky, gluey, slimy, clammy, half-liquid. *See* vicid

vivotoxin: substance reproduced internally in an infected host or by the pathogen within it and responsible for some or all of the harmful changes produced in the course of the disease. *See* toxin

volatile (n. volatility): able to evaporate or vaporize readily at ordinary temperatures or upon exposure to air

walling-off: separation of diseased from healthy tissues by barrier tissues produced by the diseased plant

water soaked: describing a diseased area of a plant that appears wet, dark, and usually sunken and translucent

water table: upper surface of ground water or that level below which the soil is saturated with water

waterlogged: without soil aeration, resulting from a lack of or poor soil drainage

watersprout: small, rapidly growing shoot or branch on a large stem, developed from adventitious tissues as a response to a shift in auxin concentration after the death of apical meristems

weed: any plant growing where it is not wanted

wetwood: discolored, water-soaked, diseased tissue of the heartwood and older sapwood of certain trees that bleeds freely when wounded and is sometimes forced out through weak areas of previously unwounded bark by built-up pressure. *See* slime flux

white rot: type of wood decay resulting from enzymatic action of fungi that degrade all components of wood and also bleach the decayed wood in the later stages

whitefly: small (about 1/12 in. long), four-winged, homopterous insect (family Aleyrodidae) that has sucking mouthparts and is covered with what appears to be a fine, white powder (known to transmit a number of plant-infecting viruses)

wilt or **wilting:** lack of rigidity and drooping of plant parts generally caused by insufficient water in the plant

wilting point or **permanent wilting point (PWP):** moisture content of soil at which plants wilt and fail to recover when placed in a humid atmosphere

wind burn: death and browning of tissues, most commonly on the leaves farthest from the roots, caused by atmospheric desiccation. *See* winter desiccation

windbreak: a planting of trees or shrubs, usually perpendicular or nearly so to the principal wind direction, to protect soil, crops, homes, roads, etc. against the effects of winds

windthrow (syn. windfall): the overthrowing and uprooting of a tree by the wind

winter desiccation: death of leaves, other plant parts, or entire plants caused by drying during winter dormancy

winter hardiness: ability of a plant to tolerate severe winter conditions

winterkill: any injury to plants that occurs during the winter period

witches'-broom: disease symptom characterized by the abnormal, massed, brushlike development of many weak shoots or roots of mainly woody plants arising at or close to the same point or resulting from the proliferation of buds (caused by mites, MLOs, viruses, fungi, false mistletoes, nematodes, etc.)

wood: technically, nonfunctioning xylem; popularly, the xylem of trees and shrubs

woody: hardy, tough, and fibrous; nonherbaceous

wound: injury to a plant in which the surface is cut, scraped, torn, or otherwise broken

wound parasite: pathogen that can enter a host only through wounds

xanthophyll: pigment usually associated with chlorophyll in chloroplasts; yellowish orange (oxygenated) carotenoid pigment of plants that is insoluble in water

xer- or **xero-** (prefix): dry; drought

xerophile or **xerophyte:** plant growing in soil with a scanty water supply or in soil where water is absorbed only with difficulty

xylem: complex, supporting, water- and mineral-conducting tissue of vascular plants consisting of tracheids, vessel members, parenchyma cells, ray cells, and wood fibers; wood

xylem parenchyma: living parenchyma cells associated with the xylem

yellowing: (distinct from chlorosis) making or rendering tissue yellow that once was green (may be caused by a canker, wilt, root pathogen, etc.)

yellows: plant disease characterized by yellowing and stunting of the host plant or affected parts, usually caused by xylem-limited, fastidious bacteria or MLOs

zonate (n. zonation): marked with concentric zones (bands or lines) of different colors and/or textures; targetlike; pertaining to any leaf or stem development appearing in concentric rings

zoo- (prefix): animal

zyg- or **zygo-** (prefix): yoke, pair

zygote: diploid cell ($2n$) or fertilized egg resulting from the union of two haploid gametes (isogametes or heterogametes); a new organism beginning via sexual reproduction

Sterilizing Agents

Sterilization is fundamental to all microbiological procedures. Pure culture techniques are required for isolation, identification, and multiplication and otherwise working with fungi and bacteria. All require sterile equipment and working areas. The methods commonly used are dry or moist heat, gas, and ultrafiltration of liquids (Dhingra and Sinclair, 1995; Sykes, 1965; Tuite, 1969).

I. Heat Sterilization

A. Dry Heat

1. Flame is used to sterilize things such as inoculating or transfer needles; the tips of chromium-plated or stainless steel forceps, scalpels, and scissors; the lips of flasks and culture tubes; microscope slides; and coverslips. Heat metal objects to redness after first dipping them into 75–95% ethanol. Pass glass objects through a Bunsen flame several times. (To avoid cracking, do not immediately place heated glass on a cool surface.)
2. Sterilize dry glassware, metal instruments, certain plastics, and other heat-stable compounds in a hot-air oven heated with electricity or burning gas. Wrap the apparatus in heavy paper (for lower temperatures), or place it in a sealable metal or other heat-resistant container to retain sterility after treatment. Air should circulate around each item so the heat will penetrate to all contaminated parts. Sterilization should be complete after the following times and temperatures: 12–16 h at 120°C; 3–4 h at 140°C; 2–3 h at 160°C; 2 h at 170°C; and 1 h at 180°C. Start counting the exposure time when the objects to be sterilized have reached the desired temperature inside the oven.

B. Moist Heat Under Pressure

Most culture media and screw-cap, cotton-plugged, or rubber-stoppered bottles and flasks are sterilized by heating them in an autoclave or domestic pressure cooker (equipped with pressure gauges, thermometer, automatic pressure-control valves, and exhaust valves) at 10–15 lb/in². Autoclave pressures provide the following approximate temperatures: 5 lb, 107°C; 7 lb, 110°C; 10 lb, 115°C; 15 lb, 121°C; and 20 lb, 126°C. For most purposes, a sterilization time of 15 min at 121°C or 30 min at 115°C is suggested. Culture media are altered by heat treatment; use no more heat than necessary. Excessive autoclaving partially hydrolyzes agar, which may inhibit growth of microorganisms. Additives such as antibiotics, hormones, growth substances, and other compounds, which may be destroyed by heating, are usually sterilized by filtration and added after autoclaving when the medium has cooled to 45–50°C. Agar is also acidified with sterile lactic acid, acetic acid, or other chemicals *after* autoclaving.

TIPS FOR USING AN AUTOCLAVE

Always follow the instructions and precautions given in the manual provided with the unit. The following suggestions are for general use:

1. All vents, the exhaust and safety valves, and the chamber should be kept clean.
2. Use loosely packed, nonabsorbent cotton plugs or loosely screwed caps, which should be firm enough to allow steam to enter and air to exhaust during decompression.
3. Containers should be no more than one-half to two-thirds full to prevent boiling over and plugging the trap.
4. Check that sufficient water is present before loading.
5. Screw down the autoclave lid, and tighten the diametrically opposed wing nuts in pairs to clamp the lid evenly on the gasket.
6. Open the steam valve or cock, and where needed, light the gas or switch on the electricity.
7. Let the steam issue for about 5 min to replace *all* air in the chamber before closing the exhaust valve or cock. This is important to obtain the desired temperature.
8. Adjust the pressure, and turn down the steam or gas. Rely on a thermometer, not the pressure gauge.
9. Keep the thermostatic trap clear of plugging materials; check the trap if the temperature is not being reached.
10. Operate for the required time, and turn off the steam, gas, or electricity. (Do not fill the autoclave with materials requiring widely different sterilization times.)

11. Do not leave paper-wrapped items at the bottom of the autoclave.
12. Do not autoclave cellulose nitrate tubes; they may explode.
13. Let the pressure *slowly* drop to zero by cooling, and then open the steam valve or cock. Exhausting the steam slowly will prevent blowing of stoppers or plugs and the boiling over of liquids.
14. Wait 5 min and then open.

C. Steaming Without Pressure

When routine autoclave heating cannot be done, "discontinuous steaming" at atmospheric pressure for about 20 min each day for three consecutive days is often satisfactory.

II. Surface Sterilization

A. Household Bleach

Immerse plant material in a fresh solution of 5.25% liquid household bleach diluted 1:4 with cold water. Leave the material in the solution for a few seconds (leaves) to 5 min (wood). Liquid bleach or a quaternary ammonium product (e.g., Hyamine 1622) can be used to disinfect bench tops and other work surfaces.

B. Propylene Oxide (PO)

Filter paper, cotton plugs, and similar material can be sterilized by exposure to fumes of PO in an airtight (chilled) vessel chamber or polyethylene bag (about 1–2 ml per liter of chamber space) overnight. The relative humidity should be 30–50%. Ventilate well before use.

Agar-filled culture (petri) plates can also be sterilized. Add 1 ml of cooled PO to each plate of hardened agar using a chilled pipette with a bulb. Leave the plates at room temperature for 24 h in an exhaust hood. If the plates are to be stored or small numbers of several media are being sterilized, stack 15–20 plates in a polyethylene bag together with a cotton wad soaked with 10 ml of chilled PO. Knot or seal the bag securely, and leave it for it at least 24 h.

PO is very volatile, flammable, and somewhat poisonous. Use PO chilled in small quantities, and remove it with care using a chilled pipette with a bulb. Keep it refrigerated and tightly sealed, and handle with great care.

C. Ethylene Oxide (EO)

Ethylene oxide is even more poisonous, flammable, and explosive than PO but is more effective and has greater penetrating power. It is also a direct-contact skin irritant.

EO is available commercially in metal containers, cans, or glass bottles as mixtures of 10% EO and 90% carbon dioxide. The mixture in glass bottles must be kept refrigerated.

Use 0.5–1 g of EO per liter of chamber space. Sterilization requires a minimum of 3 h (6 h at 25°C for heavily contaminated material). The relative humidity should be between 30 and 50%. For storing culture plates, stack 15–20 in a polyethylene bag, add 3 ml of EO, seal tightly, and leave for at least 24 h.

III. Liquid Sterilization by Filtration

Filtration is superior to other sterilization methods since no changes occur in the properties of the filtrate. It is used for organic solutions, heat-sensitive oils, and aqueous solutions to physically separate fungi, bacteria, and other microorganisms, cells, and debris from liquids (but is *not* used for viruses, viroids, or metabolic by-products). Commonly used filters include sintered glass, unglazed porcelain, asbestos pads, diatomaceous earth disks or candles, and cellulose ester membranes. Microorganisms and other large particles are retained on the filter. For the advantages, disadvantages, and precautions of using filters, see Dhingra and Sinclair, 1995, and Tuite, 1969. Since filters are easily clogged, all liquids should first be clarified by passing them through a coarse filter (e.g., filter paper) before filtering for sterilization.

Culture Media, Formulas, and Procedures

When fungi and bacteria are grown in the laboratory, diagnosticians normally try to obtain them in pure culture. The most common method of cultivation is to grow these organisms on sterile nutrient gel (usually agar plus added nutrients) in a culture dish (petri dish), test tube, or bottle. Cotton plugs or metal closures are normally used to seal test tubes and flasks.

Melt the culture medium in a flask, and then sterilize it in an autoclave for 15–20 min (up to 500 ml) after the pressure has reached 15 lb/in^2. Allow the pressure to slowly return to ambient level to avoid wetting the plugs. See Appendix A for details concerning sterilizing methods and procedures.

Most fungi and bacteria may also be grown in liquid media, either on the surface of the liquid or, when on a shaker, throughout the medium. Fungi and bacteria can also be grown on various solid materials, e.g., twigs or other stems, grains or other seeds, leaves, roots, fruits, straw, and soil.

A number of factors affect the growth of fungi and bacteria and the sporulation of fungi in culture. The most important are the culture medium, temperature, light, pH, and aeration. There is no universal set of conditions for culturing fungi and bacteria.

Culture Medium. Hundreds of culture media have been described. The medium is the single most important factor influencing the cultivation of microorganisms. The concentration of ingredients making up the medium determines the amount and quality of growth and, for fungi, whether vegetative growth or sporulation will dominate. A good culture medium for fungi is one in which sporulation is high and mycelial growth is low. A medium low in carbon or nitrogen (or both) generally stimulates sporulation and restricts vegetative growth. Fructification and growth are usually greater on natural media containing plant parts, such as decoctions, extracts, juices, powdered plant material, or gas-sterilized green leaves, stems, roots, flowers, straw, or seeds. Plant material placed on the agar surface of the culture plate stimulates the growth of many pathogenic bacteria and fungi and the production of fruiting structures of many fungi.

Temperature. All microorganisms have minimum, optimum, and maximum temperatures for growth. The optimum temperature for mycelial growth of fungi may be different than that for fructification and sporulation. Some fungi grow and sporulate best at a constant temperature; others require diurnal fluctuations.

Light. For most fungi the presence of light (sunlight, incandescent white lamps, or near-ultraviolet fluorescent lamps) stimulates both asexual and sexual reproduction when nutrition and temperature are favorable. Some fungi sporulate best under continuous light, while others require alternating dark and light periods (Dhingra and Sinclair, 1995; Tuite, 1969). The interaction of light and temperature is important for fungi that require alternating light and dark periods for sporulation.

pH. The optimum pH range for mycelial growth of fungi is broader than the optimum for sporulation. The pH of culture media is usually adjusted by the addition of hydrochloric acid (HCl), potassium hydroxide (KOH), or sodium hydroxide (NaOH) before autoclaving and lactic acid or acetic acid after autoclaving. The pH often changes after autoclaving. In preparing highly acidic agar media (pH < 4), adjust the pH after autoclaving, since the agar does not gel upon cooling. Buffering agents are often used in bacterial and fungal cultures to maintain a constant pH during growth. For a list of buffers, many of which are nontoxic and nonnutritive, see Appendix J.

Aeration. The presence of excess ammonia or carbon dioxide often inhibits the growth and sporulation of some microorganisms. Sufficient gaseous exchange, however, usually occurs in culture plates, cotton-plugged test tubes, and bottles.

Helpful Suggestions

1. Fungi usually prefer a slightly acidic medium (pH 6–6.5), whereas bacteria generally grow best at pH 7.0 or slightly above.
2. Many bacteria do not grow on laboratory media commonly used for fungi. King's medium B agar, tetrazolium chloride medium, nutrient agar, nutrient broth–yeast extract agar, yeast extract–dextrose–calcium carbonate agar, and 523 medium are often preferred for isolating and growing most plant-pathogenic bacteria. A few bacteria (e.g., *Xanthomonas* spp., *Erwinia carotovora, Clavibacter (Corynebacterium) tritici, Agro-*

bacterium tumefaciens, and *Streptomyces* spp.) require special media. Sugarcane pathogens prefer glucose-peptone agar.

3. Some carbohydrates and proteins in acid or alkaline solutions are decomposed by heat. Do not oversterilize. Add the heat-sensitive ingredients separately after sterilization.

4. Agar is the solidifying agent of choice. It slowly (during 1–2 h of boiling) dissolves or becomes hydrated, so it can be advantageous to completely dissolve the agar in half the final volume of the medium, dissolve the nutrients in the other half, and then mix the two. In some laboratories, agar is made up in a concentrated standard solution for adding later to media.

5. Agar does not solidify well in solutions that are very acidic or very alkaline. The lower the pH of the medium, the more agar is required.

6. Peptone may generally be omitted from mycological media.

7. Tap water contains useful trace elements and therefore is sometimes preferred to glass-distilled water. In some areas, however, the tap water may be somewhat toxic with heavy metal elements, especially to species of *Phytophthora.*

8. Antibacterial supplements are often added to mycological media. Three of the most widely used are the following:

   a. Chloramphenicol

      The medium is prepared in the usual way. Just before autoclaving, add 0.05 g of chloramphenicol (suspended in 10 ml of 95% alcohol) per liter of boiling medium. Mix well, and autoclave at 118°C for 10 min.

   b. Penicillin

      Add one vial (60 mg, or 100,000 units) to 200 ml of sterile distilled water, and refrigerate until it is to be used. Add 1 ml of this solution to 10 ml of molten agar cooled to 45–50°C before pouring culture plates. The concentration is 50 units/ml.

   c. Streptomycin

      Add one vial (1 g, at 745 units/mg) to 750 ml of sterile distilled water, and refrigerate until it is to be used. Add 1 ml of this solution to 10 ml of molten medium cooled to 45–50°C before pouring culture plates. The concentration is 100 units/ml.

   Antibiotics, as well as certain sugars and vitamins, are commonly sterilized separately by filtration through sintered glass or a membrane (millipore) filter and then added aseptically prior to using the medium.

9. Unless otherwise stated, autoclaving of all media is performed at 121°C for 15 min (at sea level). The composition of all media is in grams per liter of glass-distilled water, unless otherwise noted.

10. In media containing mineral salts, dissolve the salts one at a time. Some diagnosticians prefer to dissolve the chemicals separately and then add them as solutions. If $MgSO_4 \cdot 7H_2O$ is included, it should be dissolved first or added separately as a solution.

11. Chelates such as EDTA are being used increasingly in place of inorganic salts, which precipitate more readily with a change in the pH of the medium or by autoclaving.

12. Buffers are often suggested, especially in bacteriological media; they inhibit some fungi, especially water molds and other lower fungi.

13. Calcium carbonate or magnesium carbonate in a medium may be beneficial for acid-producing microorganisms. Some diagnosticians prefer to add these "solid buffers" after autoclaving. Some agitation of the medium may be required, to prevent precipitation before culture plates are poured.

14. Media containing decoctions of plant products (e.g., potatoes, seeds, foliage, or stems) are usually prepared by boiling the materials for 15–40 min and then filtering them through cheesecloth, muslin, or cotton wool to obtain a clearer medium. Other ingredients and agar are added prior to sterilization. Some natural media may be inhibitory when exposed to strong light.

15. Time can be saved by using prepared media. Several firms (e.g., BBL, Difco, and Oxoid) produce media in dehydrated form, ready to dissolve in water and sterilize in a laboratory autoclave or domestic pressure cooker. Most media commonly required in plant diagnostic clinics may be obtained in this form.

    Commercial agars are comparatively clean. This permits their use at a concentration of 1.2–2.5% rather than the 2% often stipulated in older recipes. Noble agar is washed with water and is available from Difco. Oxoid No. 2 agar is purified by an ion-exchange process and is useful in microbiological assays.

I. NCSU Media

The following media for fungi and bacteria are used in the Plant Clinic at North Carolina State University (NCSU). Culture plates (petri plates) are color coded with felt-tipped marking pens while the plates are in a stack.

1 and 2. Selective medium for *Phytophthora* (color code: green) and selective medium for *Pythium* (color code: black) (adapted from Phytopathology 52:771)

Prepare Difco cornmeal agar (17 g/liter), and place 200 ml in each of five 250-ml bottles. To each bottle, add

penicillin G sulfate, 50 ppm (50 mg/liter) (penicillin G
   Na salt, 10,000 ppm)
polymyxin B sulfate, 50 ppm (polymyxin B sulfate,
   10,000 ppm, stored in a desiccation jar)
pimaricin Na salt (Delvocid), 30–40 ppm

To select for *Phytophthora* spp., add Tachigaren (hymexazol) at 30 ppm.

PROCEDURE

1. Prepare stock solutions of penicillin, polymyxin, and Tachigaren, with one tube of stock solution to 200 ml of agar. Estimate how many tubes are likely to be needed over a 6-month period. Combine penicillin and polymyxin in the same solution; Tachigaren must be prepared separately (see step 5).
2. Weigh out 10 mg of penicillin and 10 mg of polymyxin for each 1 ml of solution to be prepared (e.g., 300 mg of each for 30 ml of solution).
3. Add the antibiotics to sterile, fresh, distilled water in a sterilized flask. (Pipette sterile distilled water into a sterilized empty flask, and plug the flask.)
4. Allow the antibiotics to dissolve (this may take up to 2 h). Polymyxin requires intermittent shaking and possibly gentle heating.
5. Weigh out 6 mg of Tachigaren per 1 ml of solution to be prepared (e.g., 180 mg for 30 ml of solution).
6. Dispense 1-ml aliquots of the penicillin and polymyxin mixture into a set of sterile test tubes. Dispense the Tachigaren into a separate set of tubes.
7. Place racks of prepared tubes in a plastic bag, seal the bag, and freeze it until the medium is used.

PIMARICIN SUSPENSION

Make a suspension of 3 g of pimaricin in 60 ml of sterile distilled water in a sterilized 250-ml flask. Divide the suspension between two sterilized brown dropper bottles, and keep them refrigerated when not in use. To obtain approximately 50 ppm, dispense four drops of well-shaken suspension into 200 ml of cooled melted agar just before plates are poured.

PLATE POURING

1. Melt the cornmeal agar (CMA).
2. Cool for about 45 min at room temperature, until the temperature of the agar reaches 45–50°C.
3. Add one tube of the penicillin plus polymyxin mixture and four drops of pimaricin to each bottle (200 ml) designated for *Pythium* (TP). To each bottle designated for *Phytophthora* (TPT) also add one tube of Tachigaren. (Remove the tubes from the freezer to thaw before adding to agar.)
4. Using a circular motion, shake each bottle thoroughly to mix the ingredients and keep the pimaricin in suspension. Shake each bottle just before pouring.
5. Pour culture dishes (petri dishes) thinly and evenly to yield 10–12 plates from a 200-ml bottle.

3. Carrot agar (color code: orange over black)

A selective medium for culturing *Thielaviopsis basicola*.

1. Prepare 2.0% water agar (20 g agar in 900 ml of deionized water).
2. Before autoclaving, add 50 ml of canned carrot juice (thoroughly stirred or shaken) to 900 ml of water agar.

3. Sterilize in five 190-ml units (a total volume of 950 ml).
4. Prepare an antibiotic solution as follows:
    a. To 50 ml of sterile water, add the following, in order, dissolving each ingredient before adding the next:
        42 mg nystatin (250,000 units)
        30 mg chlortetracycline
        500 mg streptomycin sulfate
        1.033 g Truban 30 wettable powder (etridiazole [5-ethoxy-3-trichloromethyl-1,2,4-thiadiazole], 30%)
        1 g calcium carbonate ($CaCO_3$)
    b. Dispense 10 ml into each of five test tubes.
    c. Freeze the capped tubes, sealed in a plastic bag, until they are needed.
5. Add one tube of thawed antibiotic solution to each bottle of melted agar. Cool the medium to 45–50°C before pouring it into petri plates.

4. Water agar or broth (color code: blue)

For suspending bacteria.

Dissolve 15.0 g of agar in 1 liter of distilled water. Autoclave for 20 min.

Note. A solution of 15 g of agar per liter is useful for separating bacteria from *Fusarium, Pythium,* and other fast-growing fungi.

5. Potato-dextrose agar (PDA) (color code: yellow over black)

A useful all-purpose medium.

500 g fresh Irish (white) potatoes, thinly sliced or cut into cubes	10 g dextrose 15 g agar

Note. If desirable, 0.2 g of calcium carbonate ($CaCO_3$) and 0.2 g of magnesium sulfate ($MgSO_4 \cdot 7H_2O$) may be added.

Scrub the potatoes clean (do not peel them). Cut them into 12-mm cubes or 1/8-inch-thick slices, and boil them in 1,200 ml of distilled water until cooked (by taste). Strain through cheesecloth or muslin, and adjust the volume of the filtrate to 1,000 ml. Add the dextrose and agar. Stir until dissolved. Autoclave at 15 lb/in² for 20 min.

Adding 1 g of NaCl per liter of distilled water induces sporulation in many fungi, including *Alternaria solani, Botrytis* spp., *Glomerella cingulata, Monilinia fructicola,* and *Rhizopus stolonifer.*

5A. Modified PDA (Phytopathology 52:173)

22 g dehydrated potatoes	15 g agar
10 g dextrose	1 liter water

Add the dehydrated potatoes to 200 ml of hot water. Add this mixture and the dextrose to melted agar, and autoclave.

6. Acid PDA (color code: red)

For isolation of fungi from plant parts and for inhibiting the growth of pathogenic and saprophytic bacteria.

To 200 ml of melted PDA (about 50–60°C) add eight drops of 50% lactic acid solution (see below) and agitate gently to avoid foaming.

Note. Do not remelt the agar after acidifying, or the agar will be too soft.

50% LACTIC ACID STOCK SOLUTION

Pour 50 ml of 85% lactic acid into a graduated cylinder, and add 35 ml of sterile distilled water.

7. Acid water agar (color code: blue)

Prepare water agar (medium 4), and add lactic acid as for acid PDA (medium 6).

8. Nutrient agars

Nutrient agar and nutrient broth base, to which glucose is added, are commercially available, in dehydrated form, from most biological media suppliers. These products are usually preferable to nutrient agars prepared in the laboratory from raw ingredients.

Nutrient agar media are listed as media 8A–G.

8A. Nutrient-dextrose agar (NDA) (color code: blue over green)

Widely used for cultivation of bacteria.

23 g nutrient agar (Difco)	1 liter distilled water
10 g dextrose	

Dissolve the ingredients in water, and autoclave.

8B. Modified nutrient-glucose agar (NGA) and nutrient-glucose broth (NGB)

Most *Erwinia*, *Xanthomonas*, and *Agrobacterium* spp. grow well on NGA. Exceptions include *E. stewartii* and *X. campestris* pv. *oryzae*.

3.0 g beef extract	2.5 g glucose
5.0 g peptone, bacterio- logical	15.0 g agar 1 liter water

To prepare NGB, omit the agar, or add 1% (w/v) D-glucose to Oxoid nutrient agar (CM3) or Oxoid nutrient broth (CM1). Autoclave.

8C. Nutrient agar with yeast extract

This modification of nutrient agar is suitable for the isolation of most plant-pathogenic bacteria, but certain species of *Xanthomonas* (e.g., *X. albilineans* and *X. populi*) require glucose and peptone in the agar. *Clavibacter* (*Coryne-*

bacterium) spp. and *Erwinia* spp., including many in the *E. amylovora* group, grow much better on sucrose-nutrient agar, prepared like nutrient-glucose agar (medium 8B) but with the addition of 50% (w/v) sucrose instead of glucose.

1 g beef extract	15 g agar
2 g yeast extract	1 liter water
5 g peptone, bacterio- logical	

Dissolve the nutrients in water, and adjust the pH to 7.2–7.4 with NaOH. Add agar. Steam to dissolve. Dispense into suitable vessels and loosely cap or plug with cotton. Autoclave. Screw down caps when cool for storage.

This isolation medium requires a dry surface, or else the bacteria move about, and a "carpet" of mixed growth results instead of single colonies. Plates can usually be dried by pouring them at about 50°C a day or two before they are needed and placing them in a stack at an even room temperature.

8D. Beef-peptone agar

3 g beef extract	15 g agar
10 g dextrose	1 liter water
5 g peptone, bacterio- logical	

8E. Nutrient broth–starch agar (Phytopathology 65:1034)

For detection of *Xanthomonas campestris* pv. *campestris* in seeds.

To nutrient agar, add 10 g of soluble starch and 250 mg of cycloheximide.

8F. Nutrient broth–yeast extract agar (NBY) (Appl. Microbiol. 15:1523)

Erwinia stewartii, *Xanthomonas campestris* pv. *oryzae*, and *Clavibacter* spp. are quite characteristic on this medium.

8 g nutrient broth (Difco)	2.5 g glucose
2 g yeast extract (Difco)	15.0 g agar
2 g K$_2$HPO$_4$	1 liter water
0.5 g KH$_2$PO$_4$	

After autoclaving, separately add sterile solutions of 50 ml of 10% glucose and 1.0 ml of 1 M MgSO$_4$·7H$_2$O.

8G. Nutrient-yeast-glucose agar

For growth of moderately fastidious bacteria, such as *Clavibacter michiganense* subsp. *sepedonicum* and *Erwinia tracheiphila*.

10 g peptone	10 g glucose
5 g Lab-lemco (Oxoid)	1 liter water

9. Yeast extract–dextrose–calcium carbonate agar (YDC) (color code: yellow)

For cultivation of bacteria, especially *Erwinia, Xanthomonas,* and *Agrobacterium* spp.

10 g yeast extract	15 g agar
10 g dextrose	1 liter water, preferably
5 g calcium carbonate,	distilled
USP light powder	

Dissolve the ingredients in water, and autoclave. The calcium carbonate must be thoroughly suspended while the medium is being poured. It can be resuspended by gently swirling plates or rolling tubes or bottles, just before the medium solidifies. Most plant-pathogenic bacteria remain viable on this medium for a number of months.

9A. Modified YDC (Phytopathology 57:618)

In this modification of YDC, the calcium carbonate ($CaCO_3$) and dextrose are increased to 20 g each, and the water is reduced to 800 ml. To obtain a milky white medium, the $CaCO_3$ must be finely ground, or it will precipitate. Autoclave at 10 lb/in^2 for 1 h, and mix with 200 ml of an autoclaved 10% dextrose solution. Cool the autoclaved medium to 50°C in a water bath. Before pouring culture plates (petri plates), suspend the $CaCO_3$ throughout the medium by swirling. For test tubes, use a vortex mixer to thoroughly mix the $CaCO_3$ after removing the tubes from the water bath.

9B. Another modification of YDC

In another modification of YDC, the $CaCO_3$ is omitted, the yeast extract is reduced to 7.5 g, and the dextrose is increased to 20 g.

10. King's medium B agar (KB) (color code: yellow over orange) (J. Lab. Clin. Med. 44:301)

Widely used for cultivation of bacteria and for fluorescein pigment production. Not all fluorescent pseudomonads readily produce pigment on this medium, but they may on casamino acid–sucrose–gelatin agar or Oxoid Pseudomonas F medium.

20.0 g proteose peptone	1.5 g K_2HPO_4, anhydrous
(Difco No. 3	1.5 g $MgSO_4 \cdot 7H_2O$
or Oxoid L46)	15.0 g agar
10.0 g glycerol	1 liter water

Adjust the pH to 7.2–7.4, and autoclave. Do *not* filter the medium. Green fluorescent pseudomonad bacteria produce a diffusible greenish yellow, yellow, green, or "brownish" pigment, which fluoresces green or blue in ultraviolet light.

The addition of 10% sterile egg white at 45°C enhances the production of pigment (J. Bacteriol. 94:1296).

11. *Streptomyces ipomoea* growth medium (SPA) (color code: purple)

20.0 g mannitol	0.2 mg CoCl
0.2 g K_2HPO_4	1.0 g yeast extract
0.2 g $MgSO_4 \cdot 7H_2O$	18.0 g agar
5.0 g NaCl	1 liter distilled water
0.2 g $CaCO_3$*	

*This exceeds solubility and can cause the medium to boil over if exhausted too rapidly in the autoclave.

Spread a suspension of *S. ipomoea* spores and mycelia over the surface of the medium in petri dishes. Incubate the plates in the dark at 32–36°C. Smooth-surfaced colonies appear in 1–2 days. At 2–4 days, white aerial mycelium gives the colony surface a cottony appearance. After 5–7 days, the colonies turn bluish green as sporulation occurs. Inoculum should not be collected before this.

12. Tetrazolium chloride medium (TTC or TZC) (color code: purple over yellow)

For isolation of bacteria. Many bacteria form characteristic red tints in colonies. This medium was developed to determine virulence in *Pseudomonas solanacearum.*

10 g peptone	20 g agar
10 g dextrose	1 liter water
1 g casamino acids (Difco)	

Melt the agar in 600 ml of water. Add the peptone, dextrose, and casamino acids to 400 ml of water. Mix the two solutions, and put 200 ml in each of five 250-ml pharmaceutical bottles. Autoclave with the caps loose. After cooling to 50°C, but with the mixture still liquid, aseptically add, to each 200-ml portion, 1 ml of a 0.005% solution of triphenyltetrazolium chloride (TTC) (1% [w/v] aqueous solution), which has been sterilized by autoclaving for 7 min at 121°C. (Autoclaving the medium after combining with TTC will cause the medium to turn red.) Pour 20 ml per plate. Store stock solutions of TTC in dark bottles under refrigeration.

13. Oatmeal agar (Difco) (color code: orange over green)

For general cultivation of fungi.

50 g ground oatmeal	15 g agar
1 ml antifoam*	1 liter water

*Antifom (dimethylpolysiloxane), Kalo Agricultural Chemicals, Overland Park, Kansas.

Add the ground oatmeal and antifoam to 600 ml of water, and heat to 45–55°C. Add the agar, dissolved in 400 ml of water, and autoclave in bottles for 90 min.

There are a number of modifications of this popular medium (see media 13A–F).

13A. Oatmeal agar medium for cultivation of *Phytophthora* spp. (Phytopathology 49:277)

In a commercial blender, blend 75 g of rolled oats in 600 ml of water, and heat to 45–55°C. Then add 20 g of agar dissolved in 400 ml of water. Autoclave for 90 min.

13B. Oatmeal agar medium for general cultivation of fungi (Phytopathology 49:277)

Prepared like medium 13, but with 60 g of oatmeal (rolled oats) and 12 g of agar.

13C. Oatmeal agar for cultivation of *Phytophthora phaseoli* (Phytopathology 43:419)

Soak 50 g of rolled oats for 10 h in 500 ml of water, and then blend in a commercial blender. Mix with 500 ml of water containing 15 g of melted agar. Autoclave.

13D. Oatmeal agar for cultivation of *Streptomyces* (Int. J. Syst. Bacteriol. 16:313)

20.0 g oatmeal	0.1 g $MnCl_2 \cdot 4H_2O$
18.0 g agar	0.1 g $ZnSO_4 \cdot 7H_2O$
0.1 g $FeSO_4 \cdot 7H_2O$	1 liter distilled water

Steam the oatmeal in 1 liter of water for 20 min, filter through cheesecloth, add the trace salts, and restore the volume to 1 liter. Adjust the pH to 7.2 with NaOH, and then steam for 15–20 min.

13E. Oatmeal agar—general medium for culturing fungi

30 g oatmeal	1 liter distilled water
20 g agar	

Boil powdered oatmeal over a water bath, stirring occasionally, for 1 h. Squeeze through cheesecloth or muslin, restore the volume to 1 liter, add the agar, and boil until dissolved. Autoclave at 15 lb/in² for 20 min. If desired, 0.5 ml of wheat germ oil may be added before autoclaving. Use Japanese Kobe agar for sensitive strains.

13F. Oatmeal agar—Clinton's variation

100 g oatmeal	600–700 ml distilled water
15 g agar	

Soak the oatmeal in water for 30 min, and filter through a medium sieve. Then add the agar in distilled water, and boil until dissolved. Add water to restore the volume to 1 liter.

14. Plain cornmeal agar (Difco) (color code: black over purple)

For culturing many fungi; also used as a base for other media, e.g., for cultivation of oomycetes.

17 g dehydrated cornmeal	1 liter distilled water
15 g agar	

Place the cornmeal in 1 liter of water and keep at 58°C (never over 60°C) for 1 h. Filter through filter paper, and then add the agar to the filtrate. Autoclave for 15 min at 115°C.

Modifications include media 14A–E.

14A. Plain cornmeal agar for *Phytophthora* spp., *Pythium* spp., and similarly sensitive fungi

30 g cornmeal (maize meal)	20 g agar
	1 liter distilled water

Place the meal and water in a saucepan. Heat over a water bath until boiling, stirring periodically, for 1 h. Filter through cheesecloth or muslin, add the agar, and boil until dissolved. Autoclave at 15 lb/in² for 10 min.

Wheat germ oil (0.5 ml) may be added. Japanese Kobe agar is preferred.

14B. Cornmeal (maize meal)–pimaricin–polymyxin agar (Phytopathology 54:894)

For isolation and enumeration of *Phytophthora* spp. from soil.

Amend each liter of autoclaved, cooled molten cornmeal (maize meal) agar with the following:

2 mg pimaricin
100 mg quintozene (PCNB, 75% wettable powder)
80,000 units penicillin G
370,000 units polymyxin B sulfate

The concentration of pimaricin is critical. Adjust the pH to 4.6.

14C. Maize meal–pimaricin–vancomycin agar (Phytopathology 65:570)

For isolation of *Pythium* spp. from soil.

Amend each liter of autoclaved, cooled molten maize meal agar with 5 mg of pimaricin and 300 mg of vancomycin.

14D. Maize meal (cornmeal)–pimaricin–vancomycin–quintozene agar (Phytopathology 56:893)

For isolation and enumeration of *Phytophthora* spp. from soil.

Amend each liter of maize meal agar with the following:

10 mg pimaricin
200 mg vancomycin
100 mg quintozene (PCNB, 75% wettable powder)

Autoclave, cool to 45–50°C, and adjust the pH to 6.0.

14E. Filtered cornmeal (maize meal) agar

For single-spore isolations.

To make 2 liters, soak 200 g of shelled corn overnight in 2,250 ml of distilled water. Steam for 1 h, strain through cheesecloth or muslin, add 45 g of agar, and heat to dissolve. Cool to 55°C, and add the stirred (whisked) whites of four or five eggs. Autoclave at 5 lb/in^2 for 1 h. Filter through damp filter paper over a water bath. Place 10 ml in test tubes, and autoclave at 15 lb/in^2 for 20 min.

15. Cornmeal PPP medium (color code: green over red)

For isolation of *Phytophthora* and *Pythium* spp. from plant roots and other tissues.

17 g cornmeal agar (Difco)
5 ml pimaricin Na salt, 10,000 ppm, from a stock
  solution
1 liter distilled water

Prepare cornmeal agar as for medium 14. Autoclave, cool to 55–60°C, and add the stock solution of pimaricin.

II. Other Common Media for Fungi and Bacteria

16. Alfalfa extract agar (Plant Dis. Rep. 47:255)

For inducing sporulation of *Cercosporidium graminis*.
Chop 150 g of alfalfa (*Medicago sativa*) leaves in 500 ml of water in a blender, and strain them through cheesecloth. Add 500 ml of 1–2% melted agar containing 1% dextrose to the filtrate. Autoclave.

17. Alphacel agar (Mycologia 52:47), with modifications

A general-purpose medium for cultivating fungi and for inducing sporulation.

20.0 g alphacel (a non- nutritive cellulose)	1.0 g NaNO$_3$ 50.0 ml coconut milk
1.0 g MgSO$_4$·7H$_2$O	12.0 g agar
1.5 g KH$_2$PO$_4$	1 liter water

Filter the coconut milk through several layers of cheesecloth, and autoclave. Refrigerate at 6°C until needed.

The medium may be modified by adding either 1) 10 g of tomato paste and 10 g of oatmeal or 2) 0.5 g of glucose, 1 g of sucrose, and 1 g of yeast extract to replace the coconut milk. Adjust the pH to 5.6, and autoclave at 20 lb/in^2 for 20 minutes.

18. *Alternaria* sporulation medium (Phytopathology 69:618)

40 g sucrose	20 g agar
30 g CaCO$_3$	1 liter water

Adjust the pH to 7.4 after autoclaving.

19. Apple juice–malt extract–aminoacetic acid solution

For inducing sporulation of *Venturia inaequalis*.

30 g fresh apple juice	1.5 g KH$_2$PO$_4$
4 g aminoacetic acid	0.13 g KNO$_3$
20 g malt extract	1 liter water
0.15 g MgSO$_4$·7H$_2$O	

Autoclave.

20. Arginine–glycerol–mineral salt agar (Appl. Microbiol. 11:75)

For isolation and enumeration of actinomycetes from soil.

1.0 g arginine HCl	10.0 mg Fe$_2$(SO$_4$)$_3$·6H$_2$O
12.5 g glycerin	0.1 mg CuSO$_4$·5H$_2$O
(sp gr 1.249)	0.1 mg ZnSO$_4$·7H$_2$O
1.0 g K$_2$HPO$_4$	0.1 mg MnSO$_4$·H$_2$O
1.0 g NaCl	15.0 g agar
0.5 g MgSO$_4$·7H$_2$O	1 liter water

Adjust the pH to 6.9–7.1 after autoclaving.

21. Asparagine or synthetic *Mucor* agar (Mycologia 46:358)

40.0 g dextrose	0.5 mg thiamine chloride
2.0 g asparagine	15–20 g agar
0.5 g KH$_2$PO$_4$	1 liter distilled water
0.25 g MgSO$_4$·7H$_2$O	

22. Asparagine–casamino acid–pectin agar (Eur. Potato J. 8:181)

For cultivation of *Phytophthora infestans*.

1.00 g asparagine	25.0 g glucose
2.00 g casamino acid	0.5 g KH$_2$PO$_4$
10.00 g apple pectin	0.001 g thiamine HCl
0.25 g MgSO$_4$·7H$_2$O	20.0 g agar
5.00 g yeast extract	1 liter water

23. Asthana and Hawker agar (Ann. Bot. 50:325)

For inducing sporulation of *Sordaria destruens* and for cultivating many fungi.

5.00 g sucrose	0.75 g MgSO$_4$·7H$_2$O
1.75 g KH$_2$PO$_4$	15.00 g agar
3.50 g KNO$_3$	1 liter water

The addition of thiamine and biotin may be needed, depending on the strain of *S. destruens*.

24. B-R medium (Phytopathology 59:288)

For inducing sclerotial production in *Sclerotium rolfsii*.

40 g glucose	200.0 mg MgSO$_4$·7H$_2$O
696 mg K$_2$HPO$_4$	10.0 mg FeSO$_4$·7H$_2$O
149 mg KCl	6.2 mg MnSO$_4$·H$_2$O
8.3 mg ZnSO$_4$·7H$_2$O	20.0 g agar
0.1 mg thiamine HCl	1 liter water
1.0 g NH$_4$NO$_3$	

25. Barley-honey-peptone medium (Phytopathology 58:1040)

For inducing sporulation of *Cytospora cincta.*

Put 70 g of pearl barley kernels with a few glass beads (4 mm) in a 300-ml flask. Autoclave for 1 h. Prepare a solution of 6% honey and 1% peptone. Autoclave the solution separately, and then mix it with the contents of the flask.

26. Bean meal agar (Nature [London] 172:211; Can. J. Bot. 42:311; Trans. Br. Mycol. Soc. 35:248)

For inducing sporangial formation of *Phytophthora* spp.

30 g ground seeds of garden or wax bean (*Phaseolus vulgaris*)	20 g agar 1 liter water

27. *Botrytis* separation agar (Phytopathology 57:795)

A selective medium for distinguishing between *Botrytis aclada* (*B. allii*), which is inhibited by sorbose, and *B. cinerea.*

1.0 g KCl	5.0 g glycerol
1.5 g KH_2PO_4	2.5 g L-sorbose
3.0 g $NaNO_3$	3.0 g yeast extract
0.5 g $MgSO_4 \cdot 7H_2O$	20.0 g agar
5.0 g casein hydrolysate	1 liter tap water

28. *Botrytis* sporulation agar (Phytopathology 67:212)

2.50 g yeast extract	15.0 g agar
7.50 g malt extract	35 g Czapek-Dox broth
0.25 g casein hydrolysate	1 liter distilled water
0.01 g Na nucleate	
1.0 ml microelement stock solution (see below)	

MICROELEMENT STOCK SOLUTION

723.5 mg $Fe(NO_3) \cdot 9H_2O$	2.0 mg H_3MoO_3
203.0 mg $ZnSO_4 \cdot 4H_2O$	2.0 mg $CuSO_4$
2.0 mg H_3BO_3	1 liter distilled water

29. Brown's medium (Ann. Bot. 40:224)

For inducing sporulation of *Athelia rolfsii* (anamorph *Sclerotium rolfsii*).

2.00 g glucose	0.75 g $MgSO_4 \cdot 7H_2O$
1.25 g K_2HPO_4	20.00 g agar
2.00 g asparagine	1 liter water

30. Carnation-dextrose agar (Phytopathology 68:1495)

For cultivation of *Fusarium roseum* 'avenaceum.'

Prepare a hot-water extract of 200 g of chopped carnation tissue, bring the volume to 1 liter, add 20 g of dextrose or glucose and 20 g of Difco agar, and then autoclave.

31. Carnation leaf agar (Phytopathology 72:151)

Very useful for sporulation and identification of many *Fusarium* spp.

Cut off young carnation leaves from actively growing, debudded plants free of pesticide residue. Cut the leaves into 5-mm squares, and dry them in an oven for about 2 h at 45–55°C. The leaves should be green and crisp. Sterilize the leaves in aluminum containers by 2.5 megarads of gamma radiation from a cobalt-60 source. Repeated propylene oxide fumigation may also be used. Place several 5-mm-square pieces in a culture plate or tube, and float them on autoclaved 1.5–2% water agar at 45°C. Store for several days at room temperature, to be sure they are free of contamination, before use.

32. Carrot juice agar (Plant Dis. Rep. 58:300)

For inducing asexual reproduction in *Glomerella cingulata* (*Colletotrichum gloeosporioides*).

125–750 ml carrot juice	15 g agar
100 mg streptomycin or penicillin	1 liter water, make up to

Adjust the pH to 5.1–5.5.

33. Carrot leaf decoction agar (Phytopathology 46:180)

For inducing sporulation of many *Cercospora* spp.

Finely grind 300 g of carrot leaves in a food chopper and mix with 500 ml of distilled water. Steam for 1 h, and strain through a double layer of cheesecloth. Add the filtrate to 500 ml of water in which 12 g of agar has been dissolved. Adjust the volume to 1 liter, and sterilize by steaming for 1 h. Sporulation does not occur if the medium is sterilized under pressure.

34. Carrot leaf–oatmeal agar (Plant Dis. Rep. 59:397)

For inducing sporulation of *Cercospora canescens.*

50 ml carrot leaf juice	20 g agar
50 g oatmeal	950 ml water

Boil the oatmeal in water for 5 min, and filter through cheesecloth. Add the carrot leaf juice to the filtrate, and make up the volume to 1 liter.

35. Casamino acid medium (Conn. Agric. Stn. Bull. 531)

For cultivation of *Fusarium oxysporum* f. sp. *lycopersici.*

15.0 g glucose	10 ml trace elements A–Z from Hoagland's solution
1.0 g yeast extract	
1.5 g casamino acids	tion
1.5 g KH_2PO_4	1 liter water
1.0 g $MgSO_4 \cdot 7H_2O$	

36. Cassava-dextrose agar
(Neth. J. Plant Pathol. 77:134)

Commonly used in the tropics in place of potato-dextrose agar.

Peel cassava (*Manihot esculenta*) tubers, and cut them into small chips. Dry overnight at 55°C, and grind the pieces to a powder that passes through a 30-mesh screen. The powder can be stored for later use. Soak 135 g of the cassava powder in 500 ml of water for 15 min at 60°C. (At temperatures above 62°C the powder swells, making filtration impossible.) Filter through cheesecloth or muslin. Add 20 g of glucose and 12 g of agar to the filtrate. Adjust the volume to 1 liter, and autoclave.

37. Celery leaf–muck soil medium
(Phytopathology 30:623)

For inducing sporulation of *Cercospora apii.*

Place a layer of moist muck soil and sand, 2 cm deep, in a flask. Autoclave. Place celery leaflets on the soil surface, and then reautoclave for 20 min.

38. Cellulose-asparagine medium
(Nature [London] 193:94)

For isolation of cellulolytic fungi from soil (as detected by a clear area).

0.5 g $(NH_4)_2SO_4$	0.1 g $CaCl_2$
0.5 g L-asparagine	0.5 g yeast extract (Difco)
1.0 g KH_2PO_4	10.0 g cellulose
0.5 g KCl	20.0 g agar
0.2 g crystalline $MgSO_4$	1 liter water

Add 10 g of Whatman standard-grade cellulose powder for chromatography to 250 ml of water, and ball-mill for 72 h. Add this to a solution of the other ingredients dissolved in 750 ml of water. Autoclave for 20 min at 110°C. The final pH should be 6.2.

39. Cerelose-nitrate medium, modified
(Phytopathology 39:771)

For cultivation of *Fusarium oxysporum* f. sp. *pisi.*

30.0 g Cerelose	0.5 g KCl
1.0 g KNO_3	trace $FeSO_4$
0.5 g $MgSO_4 \cdot 7H_2O$	1 liter water

40. Chickpea agar (Plant Pathol. 2:103)

For cultivation of *Phytophthora infestans.*

Wash 250 g of chickpea (*Cicer arietinum*) seeds in tap water for 1 h, and then soak them in distilled water overnight. Pour off the water, and mash the seed. Add 1 liter of water, and steam the mixture for 1 h. Filter through cheesecloth. Separately dissolve 15 g of agar and 20 g of sucrose, and add them to the filtrate. Adjust the volume to 1 liter. Autoclave for 15 min at 110°C.

41. Chitin agar (Phytopathology 52:317)

For isolation and enumeration of actinomycetes from soil. To 1 liter of distilled water or mineral salt solution, add 1–2.5 g of colloidal chitin and 20 g of agar.

The mineral salt solution contains (per liter of water)

0.7 g K_2HPO_4	0.01 g $FeSO_4$
0.5 g KH_2PO_4	0.001 g $ZnSO_4$
0.5 g crystalline $MgSO_4 \cdot 7H_2O$	

To prepare colloidal chitin, treat crude chitin several times, alternately, with 1 N NaOH and 1 N HCl, and wash with ethanol until all foreign material is removed. Dissolve the residue in cold, concentrated HCl, and filter through glass wool. Pour the filtrate into an excess of distilled water to precipitate the chitin. Finally, wash the precipitate with distilled water until the wash water has a nearly neutral pH.

42. CMC-maltose medium (Phytopathology 61:54)

For inducing sporulation of *Aureobasidium* (*Kabatiella*) *zeae.*

10.0 g carboxymethyl cellulose	5.0 g maltose
	1 g KH_2PO_4
1.5 g peptone	1 liter water

43. CMC–yeast extract medium (Mycologia 57:962)

For cultivating and inducing sporulation of *Fusarium graminearum* and *Aureobasidium* (*Kabatiella*) *zeae.*

15 g carboxymethyl cellulose*	1.0 g KH_2PO_4
	0.5 g $MgSO_4 \cdot 7H_2O$
1 g yeast extract	1 liter distilled water
1 g NH_4NO_3	

*Dissolve the carboxymethyl cellulose (CMC 7 MP, Hercules Powder Co.) by mixing in warm water in a blender.

44. Corn (maize) leaf agar (Phytopathology 68:131)

For inducing sporulation of *Bipolaris maydis* (teleomorph *Cochliobolus heterostrophus*).

Autoclave 125 g of chopped maize seedling leaves in 875 ml of water. Strain through cheesecloth, and adjust the volume to 1 liter. Amend with 30 g of sucrose and 20 g of agar. Autoclave.

45. Corn (maize) leaf decoction agar
(Phytopathology 73:286)

For growing and inducing sporulation of *Cercospora zeae-maydis.*

Harvest 120 g of green maize leaves or 35 g of dried, senescent leaves, and simmer them for 20 min in 1.5 liter of tap water. Filter through two layers of cheesecloth. To each liter of filtrate, add 4 g of $CaCO_3$ and 19 g of agar. Autoclave.

46. *Cryphonectria* (*Endothia*) *parasitica* sporulation agar (Mycologia 71:213)

1 mg biotin	40 g agar
100 mg DL-methionine	1 liter glass-distilled water

Autoclave and cool before using.

47. Czapek (Dox) agar

For cultivation of many fungi.

2.0 g $NaNO_3$	0.01 g $FeSO_4 \cdot 7H_2O$
1.0 g KH_2PO_4	30.00 g sucrose
0.5 g $MgSO_4 \cdot 7H_2O$	15–20.00 g agar
0.5 g KCl	1 liter water

If plating soil suspensions, use 20 g of agar, and adjust the pH to 4 with H_3PO_4 (1:20). For members of the Mucoraceae, use glucose instead of sucrose.

If using glass-distilled water, add 0.01 g of $ZnSO_4 \cdot 7H_2O$ and 5 mg of $CuSO_4 \cdot 5H_2O$ to 1 liter of water. Heat the full chemical solution, without sucrose, in a water bath for 15 min. After cooling add the sucrose and agar. Melt the agar, and autoclave. Do not filter.

If storing the solution overnight, dissolve the KH_2PO_4 separately, and mix before preparing. Autoclave at 15 lb/in² for 20 min.

48. Czapek-Dox V8 juice agar (Trans. Br. Mycol. Soc. 64:153)

For inducing sporulation of *Pseudoseptoria* (*Selenophoma*) *donacis* and *Stagonospora* (*Septoria*) *nodorum*.

45.4 g modified Czapek-Dox agar (Oxoid)	10 g Oxoid No. 3 agar
3.0 g $CaCO_3$	200 ml V8 juice
10.0 g Oxoid No. 3 agar	800 ml water

49. Dextrose-nitrate agar (Phytopathology 29:1000)

For isolation and enumeration of actinomycetes from soil.

1.0 g dextrose	0.1 g $MgSO_4 \cdot 7H_2O$
0.1 g KH_2PO_4	15.0 g agar
0.1 g $NaNO_3$	1 liter water
0.1 g KCl	

Adjust the pH to 7.0, and autoclave.

50. Dextrose–peptone–yeast extract agar (DPYA) (Soil Sci. 88:112)

For isolation and enumeration of fungi from soil.

5 g dextrose	1.0 g K_2HPO_4
1 g peptone	0.5 g $MgSO_4 \cdot 7H_2O$
2 g yeast extract	trace $FeCl_3 \cdot 6H_2O$
1 g sodium propionate	20.0 g agar
5 g oxgall	1 liter water
1 g NH_4NO_3	

Autoclave and cool to 45–50°C before adding a fresh solution of 20–30 mg of chlortetracycline (Aureomycin) and 30 mg of streptomycin.

51. East and Hamley's X and Y media (Ann. Appl. Biol. 44:410)

For inducing conidial production in *Botrytis elliptica* and *B. fabae*.

MEDIUM X

1.52 g KH_2PO_4	2.00 g peptone
0.52 g $MgSO_4 \cdot 7H_2O$	3.00 g casein hydrolysate
6.00 g $NaNO_3$	0.50 g yeast nucleic acid
0.52 g KCl	20.00 g agar
10.00 g dextrose	1 liter water

MEDIUM Y

5.00 g KH_2PO_4	20.00 g dextrose
1.00 g $MgSO_4 \cdot 7H_2O$	5.00 g peptone
0.20 g $FeCl_3 \cdot 6H_2O$	30.00 g agar
2.00 g KNO_3	1 liter water
0.01 g $CaCl_2$	

52. Egg albumin agar (Soil Sci. 14:279)

For isolation and enumeration of actinomycetes from soil.

1.0 g dextrose	0.25 g egg albumin
0.5 g K_2HPO_4	15.00 g agar
0.2 g $MgSO_4 \cdot 7H_2O$	1 liter water
trace $Fe_2(SO_4)_3$	

Dissolve the egg albumin in water, making it slightly alkaline with 0.1 N NaOH. Add the other ingredients, and adjust the pH to 6.8, or nearly neutral. For isolation of actinomycetes, add 40 mg of cycloheximide just before pouring into culture plates (petri plates) (Can. J. Microbiol. 2:12).

53. Elliott's agar (Am. J. Bot. 4:439)

For cultivation of various fungi.

1.36 g KH_2PO_4	5.00 g dextrose
0.50 g $MgSO_4 \cdot 7H_2O$	15.00 g agar
1.00 g asparagine	1 liter distilled water
1.06 g Na_2CO_3	

54. Elm leaf meal extract agar (Mycologia 74:376)

For cultivation of *Ophiostoma* (*Ceratocystis*) *ulmi* to induce spermatization and production of perithecia.

Grind partially dried American elm (*Ulmus americana*) leaves into particles that pass through a 1-mm mesh, and dry them at 70°C overnight. Boil 10 g of the leaf meal in 800 ml of water, and strain through cheesecloth. Adjust the volume to 800 ml, add 6 ml of linoleic acid, and autoclave. Separately autoclave 15 g of agar in 200 ml of water containing 100 μg of pyridoxine, mix with the leaf extract, and pour into plates.

55. Elm wood extract agar (Mycologia 74:376)

For cultivation of *Ophiostoma* (*Ceratocystis*) *ulmi* to induce spermatization and production of perithecia.

Grind small, air-dried branches (with bark) of American elm into particles that pass through a 4-mm mesh. Boil 60 g of the wood powder for 30 min in 1 liter of water. Strain through cheesecloth, and adjust the volume to 1 liter. Add 15 g of agar, and autoclave. When pouring the culture plates, add 6 ml of linoleic acid, and shake the medium frequently to keep the acid uniformly distributed in the liquid.

56. Emerson's yeast–phosphate–soluble starch agar (YPSs) (Lloydia 4:77; Mycologia 50:589)

For culturing members of the Blastocladiales (e.g., *Allomyces*).

15.0 g soluble starch	0.5 g MgSO$_4$·7H$_2$O
4.0 g yeast extract (Difco)	20.0 g agar
1.0 g K$_2$HPO$_4$	1 liter water

57. Erwin and McCormick's media (Mycologia 63:972)

For inducing oospore formation in *Phytophthora sojae* (*P. megasperma* f. sp. *glycinea*).

200 ml clarified V8 juice	15 g agar
3 g CaCO$_3$	800 ml water

The medium may be amended with 30 mg of β-sitosterol.

58. Filter paper–yeast extract agar (Ann. Bot. 21:465)

For inducing sporulation of *Sordaria* spp.

Blend 12 g of filter paper in 1 liter of tap water, and then add 4 g of yeast extract and 24 g of agar.

59. Flentze's soil extract agar (Trans. Br. Mycol. Soc. 40:322)

For inducing sporulation of *Thanatephorus cucumeris* (anamorph *Rhizoctonia solani*).

1.0 g sucrose	1 liter soil extract
0.2 g KH$_2$PO$_4$	25.0 g agar
0.1 g yeast, dried	

Mix 1,000 g of soil with 1 liter of water, and keep it for a day or two, agitating frequently. Filter through glass wool, and adjust the volume to 1 liter.

60. Fries medium, modified for *Septoria glycines* (Pesq. Agropec. Bras. 21:615)

5.0 g NH$_4$ tartrate	0.13 g CaCl$_2$
1.0 g KH$_2$PO$_4$	0.10 g NaCl
1.0 g yeast extract	30.00 g sucrose
1.0 g NH$_4$NO$_3$	1 liter water
0.5 g MgSO$_4$·7H$_2$O	

61. Frozen lima bean agar (Mycologia 57:85)

For cultivating and inducing oospore formation in *Phytophthora* spp.

Steam 280 g of frozen lima beans in 1 liter of water for 30 min. Filter through cheesecloth. Adjust the filtrate volume to 2 liters, add 20 g of agar, and autoclave.

62. *Fusarium* medium (U.S. Dep. Agric. Tech. Bull. 1219)

For cultivation of *Fusarium oxysporum* f. sp. *lycopersici* and other *Fusarium* spp.

20.000 g sucrose or glucose	0.003 M MgSO$_4$·7H$_2$O
0.022 M KCl	0.008 M KH$_2$PO$_4$
FeCl$_3$, MnSO$_4$, and ZnSO$_4$, each at 0.2 μg/ml	0.0356 M Ca(NO$_3$)$_2$

63. Glucose–asparagine–mineral salt medium (Mycologia 66:1030)

For inducing oospore formation in *Phytophthora* spp.

20.0 g glucose	1.0 mg thiamine HCl
3.0 g KH$_2$PO$_4$	1.0 mg FeSO$_4$·7H$_2$O
0.5 g MgSO$_4$	0.3 mg MnSO$_4$·H$_2$O
2.0 g L-asparagine	0.3 mg (NH$_4$)$_6$Mo$_7$O$_{24}$·4H$_2$O
3.4 mg CaCl$_2$	
1.8 mg ZnSO$_4$·7H$_2$O	30.0 mg ergosterol
0.4 mg CuSO$_4$·5H$_2$O	1 liter water

Prepare all minor elements and thiamine HCl as a 1,000× stock solution. Prepare FeSO$_4$·7H$_2$O at 1.4 mg/ml in a citric acid solution. Autoclave the glucose separately. Dissolve the ergosterol in 30 ml of dichloromethane, add to the KH$_2$PO$_4$ solution containing 0.5 ml of Tween 80, and autoclave separately. Adjust the pH of the combined medium to 6.0 with 1 N KOH.

64. Glucose–glutamic acid medium (Mycologia 56:816)

For cultivating and inducing oospore formation in *Aphanomyces euteiches.*

12.5 g glucose	0.70 mg Zn
3.3 g D-glutamic acid	0.10 mg Mn
150 mg DL-methionine	0.01 mg B
400 mg MgCl$_2$·6H$_2$O	0.02 mg Mo
110 mg CaCl$_2$·2H$_2$O	1 liter water
11.00 mg Fe	

Before autoclaving, adjust the pH to 6.0. Buffer the medium with 16 ml of 0.33 M KH$_2$PO$_4$ and 4 ml of 0.33 M Na$_2$HPO$_4$.

65. Glucose-isoleucine agar (Phytopathology 52:1141)

For inducing the formation of the teleomorph state (*Nectria haematococca*) of *Fusarium solani* f. sp. *cucurbitae*.

10.00 g glucose	0.75 g MgSO$_4$·7H$_2$O
5.00 g DL-isoleucine	20.00 g agar
1.75 g KH$_2$PO$_4$	1 liter water

Adjust the pH to 5.5 before autoclaving.

66. Glucose–mineral salt medium (Can. J. Bot. 50:2097)

For cultivation of *Verticillium* spp.

60.0 g glucose	0.90 g K$_2$HPO$_4$
1.0 g KH$_2$PO$_4$	0.5 mg ZnSO$_4$·7H$_2$O
0.5 g MgSO$_4$·7H$_2$O	0.16 mg CuSO$_4$·5H$_2$O
0.5 mg MnSO$_4$·H$_2$O	20.0 g agar (Difco)
10.0 µg Na$_2$MoO$_4$·2H$_2$O	1 liter water
2.00 g KNO$_3$	

67. Glucose-peptone agar (Mycologia 54:353)

For cultivation of *Colletotrichum musae* (syn. *Gloeosporium musarum*).

10.0 g glucose	0.5 g MgSO$_4$·7H$_2$O
2.0 g Bacto-peptone	15.0 g agar
0.5 g KH$_2$PO$_4$	1 liter water

Grow on test tube slants at 27–30°C.

68. Glucose-phenylalanine agar (Mycologia 45:450)

For cultivating and inducing sporulation of *Ceratocystis fagacearum* (anamorph *Chalara quercina*).

Prepare like maltose–casamino acid agar (medium 99), but replace the maltose and casamino acids with 3 g of glucose and 0.5 g of phenylalanine. Adjust the pH to 6–7.

69. Glucose–sweet potato agar (Phytopathology 56:1322)

For cultivation of *Fusarium oxysporum* f. sp. *batatas*.

Add 10 g of glucose, 18 g of agar, and some fragments of enlarged sweet potato (*Ipomoea batatas*) root or vine or potato tuber to 1 liter of water.

70. Grain or seed medium

For cultivating and inducing sporulation of numerous fungi.

Cook whole grain or seed of barley, corn (maize), lima bean, oats, pea, rice, sorghum, wheat, etc., just until it is soft, pack it in test tubes, bottles, or flasks, and autoclave. Pack the bases of tubes or other containers with absorbent cotton wool to a depth of 2–3 cm, and place a layer of grain or seed above it. Add water to soak the cotton wool and enough to moisten the base of the grain or seed. Autoclave at 15 lb/in^2 for 20 min (longer for bottles and flasks) twice, with an interval of a day between. (Unpolished rice is better than polished.)

71. Green bean agar (Phytopathology 48:79)

For inducing sporulation of *Colletotrichum orbiculare* (syn. *C. lagenarium*).

Finely grind 400 g of cooked green beans, add water to make up a volume of 1 liter, add 20 g of agar, and autoclave.

72. Green bean pod–malt extract agar (BPSMA) (Can. J. Bot. 45:1525)

For inducing sporulation of *Diplocarpon earlianum* (anamorph *Marssonina fragariae*).

Boil 200 g of green bean pods in 700 ml of distilled water. Filter through cheesecloth, and then add 5 g of malt extract and 17 g of agar. Dissolve 5 g of sucrose in 200 ml of water. Autoclave separately and then mix.

73. Guaiacol agar (Phytopathology 65:1325)

For detection of *Bipolaris* (*Helminthosporium*) *oryzae* and *Alternaria* (*Trichoconis*) *padwickii* in rice seeds.

0.125 g guaiacol	5 g agar
(*o*-methoxyphenol)	1 liter water
0.500 g streptomycin	

74. Haglund and King's medium (Phytopathology 52:315)

For inducing oospore formation in *Aphanomyces euteiches*.

5.00 g dextrose	5 mg (each) Fe, Mn, and
2.00 g K$_2$HPO$_4$	Zn chlorides
0.75 g L-asparagine	1 liter water
50 mg MgCl$_2$	
5–20 mg L-methionine	
or L-cystine	

Adjust the pH to 5.5.

75. Hay infusion agar (Thom and Raper, 1945, p. 35)

For isolation, cultivation, and inducing sporulation of aspergilli and many other dark-spored members of the Hyphomycetes.

Autoclave 50 g of decomposing hay in 1 liter of distilled water for 30 min. Filter, and adjust the volume to 1 liter. Add 2 g of K$_2$HPO$_4$ and 15 g of agar. Adjust the pH to about 6.2, and reautoclave for 10 min.

76. Hemp seed agar (Trans. Br. Mycol. Soc. 50:329)

For cultivation of *Phytophthora* and *Pythium* spp. and for inducing sexual reproduction of *Phytophthora* spp. (e.g., *P. capsici* and *P. nicotianae* var. *parasitica*).

Boil 100 g of hemp seed in 1 liter of distilled water, filter, adjust the volume to 1 liter, and add 20 g of agar. Inoculate the agar with small disks, using pairs of isolates about 15 mm apart. Grow for 5–7 days at 21°C (Phytopathology 57:570).

For cultivation of *Pythium middletonii,* grind 20 g of seed in 1 liter of water, filter through cheesecloth or muslin, and remove the seed coats. Add 20 g of agar, and autoclave.

77. Hemp seed–carrot extract agar (Mycologia 64:447)

For cultivating and inducing oospore production in *Pythium* spp.

Boil 20 g of cracked hemp seed in 1 liter of tap water. Add 20 g of agar and extract from grinding 100 g of carrots. Autoclave.

78. Hendrix's agar (Phytopathology 55:790)

For inducing oospore formation in *Phytophthora* spp.

5.4 g glucose	0.5 g $MgSO_4 \cdot 7H_2O$
1.0 g KH_2PO_4	2.0 mg thiamine HCl
1.5 g $NaNO_3$	1 liter water
17.0 g agar	

Adjust the pH to 6.0, autoclave for 10 min, and pour 25 ml into each culture plate. After solidification, spread 20 mg of a sterol, dissolved in 2 ml of ether, over the agar surface in each plate. Use after several hours.

79. Hendrix and Kuhlman's agar (Phytopathology 55:1183)

For isolation of *Phytophthora* spp. from soil.

2.00 g $NaNO_3$	0.05 g $MgSO_4 \cdot 7H_2O$
0.01 g $FeSO_4$	1.00 g KH_2PO_4
0.50 g KCl	30.00 g sucrose
0.50 g yeast extract	15.00 g agar
60.00 mg rose bengal	1 liter water

After autoclaving the ingredients, amend the medium (cooled to 45–50°C) with 100 mg of quintozene (PCNB), 100,000 units of Mycostatin, and 4.8 mg of streptomycin sulfate. Adjust the pH to 4.8 with 50% lactic acid. The basal medium can be prepared in advance and stored in the dark at room temperature. To use, melt the agar, cool to about 45°C, and amend the basal medium.

80. Honey-peptone agar (Mycologia 47:429; Mycologia 49:305)

For cultivation of many fungi, but inhibitory to most bacteria.

Disperse 20 g of agar in 1 liter of water. Add 60 g of honey and 10 g of Bacto-peptone (Difco). Autoclave.

81. Joffe's agar (Mycologia 55:271)

For inducing sporulation of many *Fusarium* spp.

1.0 g KH_2PO_4	0.5 g KCl
0.5 g $MgSO_4 \cdot 7H_2O$	0.2 g glucose
0.2 g starch powder	15.0 g agar
0.2 g sucrose	1 liter water
1.0 g KNO_3	

Autoclave, and then pour the medium into culture plates. Place sterile strips of cellulose lens paper over the agar surface before it solidifies.

82. Kidney bean agar (KBA) (Mycologia 68:511; Phytopathology 52:807)

For cultivation of *Phytophthora fragariae.*

To 1 liter of distilled water, add 30 g of finely ground kidney beans (*Phaseolus vulgaris*) and 12 or 17 g of agar. Autoclave.

83. King's B medium agar (J. Lab. Med. 44:301)

A variation of NCSU medium 10.

20.0 g proteose peptone (Difco No. 3)	15 ml glycerol
	15 g agar
1.5 g $MgSO_4 \cdot 7H_2O$	1 liter water
1.5 g KH_2PO_4	

84. Klemmer and Lenny agar (Phytopathology 55:320)

For inducing oospore formation in *Phytophthora* spp.

680.0 mg KH_2PO_4	0.07 mg $MnCl_2 \cdot 4H_2O$
708.0 mg L-asparagine	0.05 mg $NaMoO_4 \cdot 2H_2O$
250.0 mg $MgSO_4 \cdot 7H_2O$	0.08 mg $CuSO_4 \cdot 5H_2O$
100.0 mg thiamine HCl	0.07 mg $MnCl_2 \cdot 4H_2O$
50.0 mg $CaCl_2$	15 g glucose
4.4 mg $ZnSO_4 \cdot 7H_2O$	15 g agar
1.0 mg $FeSO_4 \cdot 7H_2O$	1 liter water
1.34 mg $Na_2HPO_4 \cdot 7H_2O$	

Autoclave, and then add 60 ml of lipids from wheat germ oil.

85. Krigsvold and Griffin's agar (Plant Dis. Rep. 59:543)

For isolation of *Cylindrocladium* spp. from soil.

10 g sucrose	0.5 g $MgSO_4 \cdot 7H_2O$
1 g KH_2PO_4	20.0 g agar
1 g Bacto oxgall	1 liter water
15.0 g peptone	

Autoclave, cool the molten agar to 45–50°C, and amend it with 50 mg of streptomycin sulfate, 50 mg of chlortetracycline HCl, and 75 mg of quintozene (PCNB).

86. Lactose–casein hydrolysate agar (Bull. Torrey Bot. Club 89:240)

For inducing sporulation of *Bipolaris zeicola* (syn. *Helminthosporium carbonum*), *Drechslera graminea* (syn. *H. gramineum*), and *Exserohilum turcicum* (syn. *H. turcicum*).

37.5 g lactose	1 g KH_2PO_4
3.0 g casein hydrolysate	15 g agar
0.5 g $MgSO_4 \cdot 7H_2O$	1 liter water
2 ml microelement stock solution (see below)	

Adjust the pH to 6.0

MICROELEMENT STOCK SOLUTION

439.8 mg $ZnSO_4 \cdot 7H_2O$	723.5 mg $Fe(NO_3) \cdot 9H_2O$
203.0 mg $MnSO_4 \cdot 4H_2O$	

Dissolve each salt, one by one, in 1 liter of water. To get a clear solution, add H_2SO_4, drop by drop, while stirring the solution.

87. Leonian agar (Phytopathology 13:257)

An old medium for cultivating and inducing sporulation of members of the Ascomycotina and Coelomycetes (especially the Sphaeropsidales).

0.625 g peptone (or neopeptone)	1.250 g KH_2PO_4
	0.625 g $MgSO_4 \cdot 7H_2O$
6.250 g maltose (or glucose)	20.000 g agar
	1 liter distilled water
6.250 g malt extract	

88. Lima bean agar (Phytopathology 52:807)

For cultivation of *Phytophthora* spp.

Soak 50 g of lima beans (*Phaseolus lunatus*) for 10 h in 500 ml of water. Blend the beans with 500 ml of water containing 15 g of melted agar. Adjust the final volume to 1 liter. Autoclave.

89. Lima bean agar, modification I (Phytopathology 52:859)

For growing stock cultures of *Phytophthora sojae*.

5 g dried Fordhook 242 lima bean seed	20 g Bacto agar (Difco)
	1 liter distilled water

Autoclave the lima bean seed with agar for 20 min. If the medium is to be used in test tubes, reautoclave.

90. Lima bean agar, modification II (Plant Dis. Rep. 40:314)

For inducing sporulation of *Rhynchosporium secalis* and *R. orthosporum.*

18 g dried lima bean infusion and agar (Difco)	5 g agar
	1 liter water

91. Lima bean agar (frozen), modification III (Mycologia 57:85)

For cultivating and inducing production of sporangia and oospores of *Phytophthora* spp.

Steam a package of frozen green lima beans (about 250 g) in 1 liter of water. After 25–30 min, filter through cheesecloth, muslin, or cotton wool. Adjust the filtrate volume to 2 liters. Add 40 g of agar, and autoclave.

92. Lima bean decoction broth (Phytopathology 49:550)

For cultivation of some *Phytophthora* spp.

Boil 100 g of lima beans in 1 liter of water, strain through cheesecloth, adjust the volume to 1 liter, and autoclave.

93. Lutz medium

For cultivation of agarics and boletes and sporulation of *Thanatephorus cucumeris* (anamorph *Rhizoctonia solani*).

10.0 g malt extract	0.1 g ferric sulfate
1.0 g ammonium nitrate	0.025 g manganese sulfate
1.0 g ammonium phosphate	25 g agar
	1 liter water
0.1 g magnesium sulfate	

94. Malachite green–captan agar (Plant Dis. Rep. 49:114; Phytopathology 56:908)

For detection of mycelium of *Fusarium* spp. in soil and for selective isolation of *Pyrenochaeta terrestris*.

2.00 g $NaNO_3$	0.5 g KCl
0.50 g $MgSO_4 \cdot 7H_2O$	30.0 g sucrose
0.01 g $FeSO_4$	20.0 g agar
1.00 g K_2HPO_4	1 liter distilled water

Autoclave, cool the molten agar to 45–50°C, and amend it with the following:

50 mg malachite green	300,000 units Porcain penicillin G
50 mg streptomycin sulfate	
100 mg captan	100,000 units Na penicillin G
75 mg dicrystine	

95. Malt agar (Can. J. Bot. 43:1097)

A standard medium for identifying members of the Hymenomycetes.

12.5 g Bacto malt agar (Difco)	20.0 g Bacto agar (Difco)
	1 liter distilled water

96. Malt extract agar (MEA) (conservation)

For cultivation of fungi and bacteria.

20 g malt extract	1 liter water
15–20 g agar	

Heat the malt extract in water until it is dissolved. Then add the agar, and boil until the agar is well dissolved. Autoclave for 20 min. Do not oversterilize, or the agar will hydrolyze.

Commercial malt extract is suitable but must be without additions. The pH is about 3–4 and may be adjusted with NaOH to 6.5.

Useful modifications of MEA include media 96A–D.

96A. Malt extract agar for cultivation of species of Basidiomycotina

50 g malt extract
5 g malic acid

96B. Blakeslee's medium (Thom and Raper, 1945, p. 35)

20 g malt extract	20 g dextrose
1 g peptone	1 liter distilled water

96C. Malt extract agar plus 20% sucrose

For organisms requiring high osmotic pressure.

20 g malt extract
200 g sucrose

Add the sucrose after boiling, to avoid caramelization.

96D. Malt extract agar plus 40% sucrose

For hygrophobes.

20 g malt extract
400 g sucrose

Add the sucrose after boiling, to avoid caramelization.

97. Malt extract–dextrose–peptone agar (Phytopathology 24:1127)

For cultivation of pythiaceous fungi.

1 g KH_2PO_4	0.5 g $MgSO_4 \cdot 7H_2O$
1 g peptone	5.0 g malt extract
15 g dextrose	1 liter water
15 g agar	

Mix ingredients and autoclave.

98. Malt extract–yeast extract agar (Phytopathology 13:257)

Commonly used for growing fungi.

3.0 g malt extract	0.5 g $MgSO_4 \cdot 7H_2O$
2.0 g yeast extract	20.0 g agar
0.5 g KH_2PO_4	1 liter water

99. Maltose–casamino acid agar (Mycologia 45:450)

For cultivating and inducing sporulation of *Ceratocystis fagacearum* (anamorph *Chalara quercina*).

5.0 g maltose	0.2 mg Fe sulfate
1.0 g casamino acids (Difco)	0.2 mg Mn sulfate
	0.1 mg Zn sulfate
1.0 g KH_2PO_4	20.0 g agar
0.5 g $MgSO_4 \cdot 7H_2O$	1 liter water
0.5 µg biotin	

Adjust the pH to 6.0 after autoclaving. Yeast extract may be substituted for biotin in some cases.

100. Maltose-peptone broth (Phytopathology 49:406; Phytopathology 50:826)

For growing both fungi and bacteria and for inducing the production of zoospores of *Aphanomyces euteiches*.

To 1 liter of water, add 3 g of maltose and 1 g of peptone. Autoclave.

101. Mannitol-orthophosphate agar (Indian J. Mycol. Plant Pathol. 11:57)

For growing bacteria, including *Xanthomonas campestris* pv. *vignicola*, *Erwinia carotovora*, and *Clavibacter* (*Corynebacterium*) *tritici*.

10.642 g mannitol	200 mg $MgSO_4 \cdot 7H_2O$
4.300 g $NH_4H_2PO_4$	2 mg ferrous sulfate
2.000 g KH_2PO_4	1 liter water
430 mg L-histidine	

Adjust the pH to 7.0 after autoclaving.

102. Marmite-dextrose agar (Trans. Br. Mycol. Soc. 79:129)

Primary growth medium for inducing sporulation of *Thanatephorus cucumeris* (anamorph *Rhizoctonia solani*).

25.0 g Marmite yeast extract	20 g agar
	1 liter water
12.5 g dextrose	

Adjust the pH to 6.2 after autoclaving for more than 15 min. Remelted medium may not be suitable for sporulation.

103. Neopeptone-glucose agar (Phytopathology 40:104)

For growing and inducing sporulation of *Colletotrichum lindemuthianum*.

2.80 g glucose	1.23 g $MgSO_4 \cdot 7H_2O$
2.00 g neopeptone	20.00 g agar
2.72 g KH_2PO_4	1 liter water

Adjust the pH to 5.2–6.5. Galactose, sucrose, or xylose can replace glucose in the medium.

104. Nutrient agar

Widely used for cultivation of bacteria.

5.0 g bacteriological peptone	15 g agar
3.0 g beef extract	1 liter water, preferably distilled
2.5 g glucose	

For modifications of this medium, see NCSU media 8A–G. Commercially available agars (Oxoid CM3 or Difco) are preferable.

105. Nutrient broth–beef extract agar

For cultivation of bacteria.

5 g peptone	0.5 g KH_2PO_4
3 g beef extract	15.0 g agar
2 g K_2HPO_4	1 liter water

After autoclaving, add 50 ml of an autoclaved 10% solution of glucose and 1 ml of 1 M $MgSO_4 \cdot 7H_2O$. Adjust the pH to 7.2.

106. Oat leaf agar (Plant Dis. Rep. 39:25)

For inducing sporulation of *Stagonospora* (*Septoria*) *avenae*.

Grind 300 g of fresh, green oat leaves. Mix the pulp and juice with 1 liter of water containing 12–15 g of agar. Boil the mixture for 5 min, strain through cheesecloth, and autoclave the filtrate.

107. *Ophiostoma* (*Ceratocystis*) *ulmi* cultivation medium (Mycologia 74:376)

For inducing production of mycelium of the Dutch elm disease fungus and, in liquid form, for inducing production of conidia of the *Sporothrix* anamorph.

6.0 g glucose	100.00 mg pyridoxine HCl
0.5 g K_2HPO_4	0.20 mg Zn^{2+}
0.5 g $MgSO_4 \cdot 7H_2O$	($ZnSO_4 \cdot 7H_2O$)
0.4 g L-asparagine	0.02 mg Cu^{2+}
0.2 mg Fe^{3+}	($CuSO_4 \cdot 5H_2O$)
($Fe(NO_3)_3 \cdot 6H_2O$)	15 g agar
0.1 mg Mn^{2+}	1 liter water
($MnSO_4 \cdot 7H_2O$)	

Adjust the pH to 6.5.

108. Palm leaf decoction–palm wine agar (J. Niger. Inst. Oil Palm Res. 5:19)

For inducing sporulation of *Cercospora elaeagni*.

Prepare like carrot leaf decoction agar (medium 33), grinding 300 g of palm leaves instead of carrot leaves. To the autoclaved medium, add 60 ml of filter-sterilized palm wine.

109. Pea (frozen) agar (Plant Dis. 70:1100)

For growth of *Phytophthora fragariae*.

Put 150 g of frozen peas in 500 ml of deionized water, and blend at high speed for 3 min. Squeeze the slurry through 8–12 layers of cheesecloth. Then filter through coarse filter paper covered by a 2.5-cm-deep layer of Celite 545. Add 15 g of agar to the filtrate, and dissolve the agar in a steamer. Add water to bring the volume to 1 liter, and add 0.5 ml of an antifoam agent. Adjust the pH to 6.5, and autoclave for 30 min.

110. Pea seed broth (Mycologia 62:397)

For cultivation of *Phytophthora cinnamomi*.

Place 200 g of frozen peas in 500 ml of deionized water, and blend in a mixer. Centrifuge the mixture for 10 min at $4,000 \times g$. Pour off the supernatant and adjust its volume to 1 liter. Autoclave.

111. Pea seed medium (Phytopathology 47:186)

For growing and inducing production of sporangia and zoospores of *Phytophthora infestans*.

Soak dried yellow peas (*Pisum sativum*) in water overnight. Place a 2.5-cm-deep layer of soaked peas in 250-ml flasks, and add enough water to barely cover them. Autoclave for 40 min. Inoculate and incubate at 20°C for 7–10 days.

112. Peach agar

For growth of *Stigmina carpophila* (syn. *Coryneum carpophilum*).

Homogenize 200 g of peach (*Prunus persica*) fruit. Add the slurry to 1 liter of water, add 40 g of agar, and autoclave.

113. Peanut hull extract agar (Phytopathology 70:990)

For growing and inducing sporulation of numerous fungi, including *Cercospora arachidicola*, *Colletotrichum gloeosporioides*, and *Phaeoisariopsis personata* (*Cercosporidium personatum*).

Blend 180 g of peanut hulls in 1,700 ml of tap water for 1 min. Filter the slurry through a double layer of cheesecloth. Restore the volume to 1 liter, add 20 g of agar, and autoclave.

114. Peanut leaf–oatmeal agar (Phytopathology 61:1414)

For inducing sporulation of *Cercospora arachidicola*.

Blend 50 g of peanut (*Arachis hypogaea*) leaflets in 500 ml of water for 10–15 s and filter through cheesecloth. Boil 15 g of oatmeal in 500 ml of water for 15 min and filter through cheesecloth. Combine equal amounts of both filtrates, add 20 g of agar, and autoclave.

115. Peptone–malt extract–antibiotic agar (Trans. Br. Mycol. Soc. 82:720)

For inducing sporulation of *Gaeumannomyces graminis.*

10.0 g malt extract	15 g agar
2.5 g peptone	1 liter water

Adjust the pH to 5.5–6.0. Autoclave. Cool the molten agar to about 55–60°C, and add 40 mg of penicillin G, 50 g of streptomycin sulfate, and 60 mg of oxytetracycline (Terramycin).

116. *Phymatotrichopsis (Phymatotrichum) omnivorum* sporulation medium (Phytopathology 57:228)

0.00100 M	0.0087 M KCl
Ca(NO$_3$)$_2$·4H$_2$O	ferric tartrate, 1 µg/ml
0.00080 M KNO$_3$	20 g glucose
0.00015 M MgSO$_4$·7H$_2$O	25 g agar
0.00008 M KH$_2$PO$_4$	1 liter distilled water

Adjust the pH to 4.6. Grow with daily photoperiods of 16 h under blue or white light. Conidia form abundantly when pyridoxine (1 mg/ml) is added to the medium.

117. Phytone-dextrose agar (Phytopathology 53:1443)

For inducing copious sporulation of *Stemphylium* spp.

15 g phytone (BBL)	17 g agar
1 g yeast extract	1 liter water
15 g dextrose	

118. Pigeon pea seed meal agar (Indian Phytopathol. 36:152)

For growth of *Phytophthora drechsleri* f. sp. *cajani.*
Boil 40 g of seed meal ground from light-colored seeds of pigeon pea (*Cajanus cajan*) in 1 liter of distilled water, and filter through cheesecloth. Make up the volume of the filtrate to 1 liter, add 20 g of agar, and autoclave.

119. *Populus* leaf decoction agar (Mycologia 79:654)

For inducing production of ascocarps of *Mycosphaerella populorum* (anamorph *Septoria musiva*).
Place 25 g of naturally senescent *Populus* leaves in 250 ml of water, and steam them for 30 min. Pour off the water, add 5 g of malt extract and 15 g of agar, and make up the volume to 1 liter. Autoclave.

120. Potato-carrot agar

A very weak medium for inducing sporulation and for storage of fungi.

20 g potato, grated	20 g agar
20 g carrot, grated	1 liter tap water

Wash, peel, and slice the vegetables before grating them. Boil the grated vegetables for 1 h in tap water. Strain

through a fine sieve or cheesecloth (do not press or squeeze through), and then add agar to the filtrate. Restore the volume to 1 liter. Boil over a water bath until the agar completely dissolves. Autoclave at 15 lb/in^2 for 20 min.
Note. Do not use new potatoes. A strip of sterile filter paper placed on the agar surface is excellent for cultivating *Chaetomium* spp. and other cellulose-destroying fungi.

121. Potato-dextrose agar (PDA)

Numerous variations of this popular medium contain different amounts of potato and dextrose. See NCSU medium 5.
The "standard" procedure is to boil 200 g of peeled, thinly sliced potatoes in 1 liter of water until they are soft and cooked. Strain through cheesecloth, and restore the volume to 1 liter. Add 10–20 g of dextrose and 12–17 g of agar. Autoclave. As a substitute for dextrose, sucrose increases the sporulation of many fungi.
Note. PDA made from 500 g of peeled, sliced potatoes, 10 g of dextrose, and 1 g of NaCl per liter induces sporulation of a number of fungi, including *Alternaria solani, Botrytis* spp., *Glomerella cingulata* (anamorph *Colletotrichum gloeosporioides*), *Monilinia fructicola*, and *Rhizopus stolonifer.*

122. PDA, diluted

Place 40 g of peeled, sliced potato in 200 ml of water, and steam for 45 min. Strain through cheesecloth, restore the volume to 1 liter, and add 5 g of dextrose and 17 g of agar. Autoclave.

123. Potato-Marmite agar (PMA) (Trans. Br. Mycol. Soc. 39:343; Can. J. Bot. 41:879)

For inducing sporulation of *Thanatephorus cucumeris* (anamorph *Rhizoctonia solani*).

250 g potato	20 g agar
20 g dextrose	1 liter water
1 g Marmite yeast extract	

Steep peeled, sliced potato in water at 60°C for 1 h, strain through cheesecloth, and restore the volume to 1 liter. Add the other ingredients, and autoclave.
Transfer 2- to 3-day-old PMA blocks to the edge of 2% water agar plates, and keep them at room temperature in diffuse light.
See also medium 124.

124. Potato-Marmite agar, another version (Trans. Br. Mycol. Soc. 79:129)

For growing and inducing sporulation of *Thanatephorus cucumeris* (anamorph *Rhizoctonia solani*).

15 g dehydrated potato	20.0 g agar
40 g Marmite yeast extract	1 liter water
7.5 g dextrose	

Adjust the pH to 5.3.

125. Potato–peptone–yeast extract–dextrose agar (Ann. Phytopathol. Soc. Jpn. 52:169)

For inducing sporulation of *Thanatephorus cucumeris* (anamorph *Rhizoctonia solani*), in a soil-on-agar culture.

30 g potato	0.5 g dextrose
5 g peptone	20.0 g agar
5 g yeast extract	1 liter water

Boil slices of peeled potato in several hundred milliliters of water until they are soft. Discard the potato, and add the other ingredients to the extract. Autoclave.

126. Potato-sucrose agar

For cultivating fungi and inducing sporulation in *Fusarium* spp.

Boil 1,800 g of peeled and sliced or diced potatoes for 10 min in 4.5 liters of water. Strain through cheesecloth or muslin. The sterile filtrate can be stored in large glass containers in a refrigerator until use.

To prepare the medium, mix 500 ml of sterile potato filtrate with 500 ml of water. Add 20 g of sucrose and 20 g of agar. Heat the mixture over a water bath to completely dissolve the agar. Adjust the pH to 6.4–6.5 with $CaCO_3$. Autoclave at 15 lb/in² for 20 min.

127. Prune agar (Trans. Br. Mycol. Soc. 64:295)

For cultivation of *Nectria cosmariospora*.

Add 40 g of mashed prunes and 20 g of agar to 1 liter of water. Autoclave.

128. Prune–lactose–yeast extract agar (Plant Pathol. 9:57)

For identifying and inducing sporulation of *Verticillium albo-atrum* and *V. dahliae* and, with amendments, for inducing sporulation of *Botrytis* spp.

100 ml concentrated prune extract	1 g yeast extract (Difco)
	5 g lactose
30 g agar	1 liter distilled water

Prepare the prune extract by simmering 50 g of chopped dried prunes in 1 liter of water until they are soft. Strain through cheesecloth, and then filter. Restore the volume to 1 liter. Adjust the pH to 5.8–6.0 and autoclave.

Amending the medium with streptomycin (100 μg/ml) and erythromycin (100 μg/ml) (Plant Pathol. 12:15) induces sporulation of some *Botrytis* spp. (Trans. Br. Mycol. Soc. 85:621).

Note. Autoclaved prune extract, in batches, can be refrigerated until used.

Potassium thiocyanate (KCNS) may be added (5.0 g/liter) to suppress bacteria in isolation plates.

Pour at least 15 ml for each plate.

129. *Pyrenochaeta* isolation agar (Phytopathology 64:275)

For isolation of *Pyrenochaeta* spp.

3 g $NaNO_3$	1 g $MgSO_4 \cdot 7H_2O$
5 g sorbose	20 g agar
250 mg chloramphenicol	1 liter water

Autoclave.

130. Rabbit food agar (Plant Dis. 64:788)

For inducing sporulation of *Bipolaris* (*Helminthosporium*) *oryzae*.

Steep 50 g of commercial rabbit food for 20 min in 500 ml of distilled water. Filter through two layers of cheesecloth. Adjust the volume to 1 liter, add 15 g of agar, and adjust the pH to 6.5. Autoclave.

131. Radish agar (Phytopathology 54:1167)

For growing and inducing production of oospores of *Aphanomyces raphani*.

Soak 200 g of sliced radish (*Raphanus sativus*) root tissue in 1 liter of water at 60°C for 1 h, or soak 250 g of root tissue in 1 liter of water at 100°C for 45 min. Strain through cheesecloth, and add 20 g of agar to the filtrate. Autoclave.

Note. Variations of this medium for the cultivation of *A. raphani* include the addition of 10 g of glucose (Phytopathology 54:1167) and 0.5% peptone (Phytopathology 68:377).

132. Rape (canola) seed agar (Mycologia 59:161)

For inducing oospore production in *Phytophthora* spp.

Wash rape seed thoroughly in water, and then boil 100 g of seed in 1 liter of distilled water for 30 min. Filter the extract through cotton wool. Adjust the volume of the filtrate to 1 liter, adjust the pH to 7.0, and add 20 g of agar. Autoclave.

133. Rice node decoction agar (Indian J. Mycol. Plant Pathol. 16:329)

For inducing sporulation of *Pyricularia oryzae*.

Harvest mature culms of a susceptible rice cultivar. Cut the culms into pieces 2–3 cm long with a node in the center. Place enough of the pieces in 150-ml flasks to make a single closely packed layer. Add 50 ml of water to each flask. Plug and autoclave. After cooling, add 1% dextrose and 2% agar to the liquid. Autoclave.

134. Rice polish agar (Phytopathology 56:507)

For inducing abundant sporulation of *Bipolaris* spp., *Drechslera* spp., and *Pyricularia oryzae*.

Mix 20 g of rice polish with 500 ml of water, and steam without pressure for 15 min. Blend the suspension for several minutes, and mix with 500 ml of water containing 17 g of melted agar. Autoclave, cool, and pour into culture plates. Agitate the medium while pouring to prevent settling of the rice polish.

135. Richard's solution (Phytopathology 13:543)

An old general medium for cultivation of fungi.

10.0 g KNO_3	0.02 g $FeCl_3$
5.0 g KH_2PO_4	50.0 g sucrose
2.5 g $MgSO_4 \cdot 7H_2O$	1 liter distilled water

Autoclave.

136. Sachs's agar (Phytopathology 48:281; Can. J. Bot. 33:562)

For inducing sporulation of *Cochliobolus sativus*.

4.00 g $CaCO_3$	trace $FeCl_3$
1.00 g $CaNO_3$	20 g agar
0.25 g K_2HPO_4	1 liter water
0.25 g $MgSO_4 \cdot 7H_2O$	

137. Sanderson and Srb medium (Am. J. Bot. 52:72)

For growing and inducing sporulation of *Phoma medicaginis* (syn. *Ascochyta imperfecta*).

MINIMAL MEDIUM

10.0 g sucrose	0.1 g $CaCl_2$
2.4 g $NaNO_3$	1 ml trace element stock
2.0 g KH_2PO_4	solution (see below)
0.6 g $MgSO_4 \cdot 7H_2O$	20 g agar (Difco)
0.1 g NaCl	1 liter water

Note. For a complete medium, add 0.75% malt extract and 0.25% yeast extract.

TRACE ELEMENT STOCK SOLUTION

822.0 mg $ZnSO_4 \cdot 6H_2O$	9.00 mg H_3BO_3
139.8 mg $FeCl_3 \cdot 6H_2O$	7.60 mg $NaMoO_4$
58.5 mg $CuSO_4 \cdot 5H_2O$	1.95 mg KI
9.00 mg $MnSO_4 \cdot H_2O$	300 ml water

138. Schmitthenner's medium (Phytopathology 61:1149)

For inducing oospore formation in several *Pythium* spp.

2.50 g sucrose	270 mg asparagine
150.00 mg KH_2PO_4	150 mg K_2HPO_4
100.00 mg $MgSO_4 \cdot 7H_2O$	55 mg $CaCl_2$
2.00 mg thiamine HCl	10 mg cholesterol
4.40 mg $ZnSO_4 \cdot 7H_2O$	1 mg $FeSO_4 \cdot 7H_2O$
0.07 mg $MnCl_2 \cdot 4H_2O$	1 liter water

139. Soil medium for storage of fungi

Fill test tubes or bottles half full of fine-sieved, air-dried soil, and autoclave at 15 lb/in^2 for 20 min. When ready for use, inoculate and moisten at the same time with 2–3 ml of a suspension of spores or tissue in sterile distilled water. Incubate for 10 days and then store in a refrigerator or cool place. Retrieve cultures by sprinkling a few soil grains on a culture plate.

140. Soybean meal agar (Phytopathology 48:165)

For cultivating and inducing oospore formation in *Phytophthora* spp.

Grind 15 g of soybeans, and add the meal to 1 liter of distilled water containing 20 g of agar. Autoclave. Grow cultures in the dark at 20°C.

141. Soybean seed decoction–sucrose broth (Phytopathology 65:236)

For inducing production of sclerotia of *Macrophomina phaseolina*.

Boil 100 g of soybean (*Glycine max*) seeds in 1 liter of tap water for 15 min. Strain through cheesecloth. Adjust the volume of the filtrate to 1 liter, and add 20 g of sucrose. Autoclave.

142. *Stagonospora* (*Septoria*) *nodorum* sporulation agar (Phytopathology 41:571)

20.000 g carbon source	0.2 mg Zn^{2+}
0.425 g nitrogen source	0.1 mg Mn^{2+}
0.500 g $MgSO_4 \cdot 7H_2O$	20 g agar
0.200 mg Fe^{3+}	1 liter water

For a carbon source use galactose or maltose, equal amounts of glucose and sucrose, or sucrose alone. Glycine is a favorable nitrogen source. Good sporulation results from a combination of galactose, lactose, and maltose plus asparagine, glycine, or urea.

143. Sucrose–asparagine–mineral salt medium (Trans. Br. Mycol. Soc. 58:169)

For inducing oospore formation in *Phytophthora* spp.

10.00 g sucrose	1 mg thiamine HCl
1.00 g L-asparagine	1 ml trace element stock
0.50 g KH_2PO_4	solution (see below)
0.25 g $MgSO_4 \cdot 7H_2O$	1 liter water
0.10 g $CaCl_2$, anhydrous	

TRACE ELEMENT STOCK SOLUTION

4,403 mg $ZnSO_4 \cdot 7H_2O$	88 mg $Na_2B_4O_7 \cdot H_2O$
910 mg $Fe(SO_4)_3 \cdot 9H_2O$	72 mg $MnCl_2 \cdot 4H_2O$
393 μg $CuSO_4 \cdot 5H_2O$	50 mg $Na_2MoO_4 \cdot 2H_2O$
500 mg EDTA,	1 liter water
1% disodium salt	
(Difco)	

Oospores are formed on the Difco agar medium (10 g/liter) with added sterol (e.g., cholesterol, 10 mg/liter).

144. Sucrose-peptone agar (Plant Dis. 69:122)

For growth of the A- and B-cancrosis strains of *Xanthomonas campestris* pv. *citri*.

1% sucrose	0.03% $MgSO_4 \cdot 7H_2O$
0.5% peptone	1.50% Difco purified agar
0.05% K_2HPO_4	

Agars other than Difco may not support growth of the B-cancrosis strain of the bacterium.

145. Sucrose-proline agar (SPA) (Can. J. Bot. 40:809)

For inducing sporulation of *Bipolaris, Drechslera, Exserohilum,* and *Pyrenophora* spp.

6.0 g sucrose	10.0 mg $FeSO_4$
2.7 g proline	2.0 mg $ZnSO_4$
1.3 g K_2HPO_4	1.6 mg $MnCl_2$
1.0 g KH_2PO_4	20.0 g agar
0.5 g KCl	1 liter water
0.5 g $MgSO_4 \cdot 7H_2O$	

Autoclave. For sporulation, grow cultures in the dark at 20°C.

146. Sugar beet leaf decoction agar (Phytopathology 24:1101)

For inducing sporulation of *Cercospora* spp.

Finely grind 300 g of freshly picked young sugar beet (*Beta vulgaris*) leaves, and add them to 1 liter of water containing 12 g of melted agar. Boil the mixture for 5 min, strain through cheesecloth or muslin, and autoclave the filtrate at 110–115°C for 15 min.

147. Sugar beet molasses agar (Phytopathology 55:1370)

For inducing sporulation of *Cercospora beticola*.

Add 150 g of sugar beet molasses and 15 g of agar to 1 liter of water. Autoclave.

148. TCV agar

For detection of *Septoria* and *Stagonospora* spp. in seeds.

Amend 1 liter of autoclaved potato-dextrose agar (NCSU medium 5 or medium 121) at 60°C with 5 ml of a solution containing 0.02% thiophanate-methyl, 0.17% chloroneb, 0.02% vancomycin HCl, and four drops of Tween 20.

149. Tea agar (Natl. Acad. Sci. Lett. 5:87)

For isolation, growth, and storage of wood decay fungi.

Wash waste tea leaves in tap water, and dry them before storage. Add 150 g of the dried leaves and 15 g of agar to 1 liter of tap water, and autoclave. Cool the medium, and pour it into culture plates for isolation and growth of wood decay fungi. If storing the fungi in flasks or bottles, increase the amount of tea leaves to 250 g.

150. Tobacco leaf decoction agar (Phytopathology 31:97)

For sporulation of *Cercospora nicotianae*.

Harvest yellow green leaves from the lower quarter of greenhouse-grown tobacco plants at the blooming stage. Grind 600 g of leaves in 2 liters of water. Steam, without pressure, for 1 h. Strain through cheesecloth, adjust the volume to 2 liters, and add 60 g of agar. Autoclave.

151. Tomato paste–CaCO₃ agar (Trans. Br. Mycol. Soc. 89:402)

For inducing sporulation of *Drechslera teres*.

29 g tomato paste	13 g agar
(30% tomato)	1 liter water
3-6 g $CaCO_3$	

Higher concentrations of $CaCO_3$ will increase sporulation.

152. Tryptone–yeast extract agar (Plant Prot. Bull. [Taipei] 28:225)

For cultivation of *Xanthomonas campestris* pv. *citri*.

10 g tryptone	5 g glucose
10 g yeast extract	15 g agar
5 g K_2HPO_4	1 liter water

Adjust the pH to 7.0, and autoclave.

153. V8 juice agar (Phytopathology 45:141, 461)

An extremely useful medium for supporting the growth and inducing the sporulation of many fungi; often used instead of potato-dextrose agar.

A common formula is the following:

200.0 ml V8 juice	15–20 g agar
3–4.5 g $CaCO_3$	800 ml water

Dissolve the agar in water, add the V8 juice, and adjust the pH to 7–7.5 by changing the amount of juice or adding $CaCO_3$. Autoclave at 15 lb/in² for 20 min. Since the juice contains fibrous tissue, the medium is opaque. It can be partially clarified by centrifuging (at 3,000 rpm) or filtration through a single and then a double layer of filter paper.

154. V8 juice–benomyl agar (Trans. Br. Mycol. Soc. 77:218)

For inducing sporulation of *Drechslera graminea*.

Centrifuge 200 ml of V8 juice, and dilute it with 800 ml of water containing 10 mg of benomyl. Add 20 g of agar. Autoclave for no more than 10 min. The pH before autoclaving is about 4.5.

155. Vegetable plugs

For growth of numerous fungi.

Punch out cylinders of potato, carrot, turnip, etc., using an apple borer or a cork borer. Cut each cylinder diagonally into two pieces. Put a piece of water-saturated cotton wool at the bottom of each tube, and place a cylinder on top of the cotton. Plug the tubes with cotton. Autoclave at 20 lb/in^2 for 20 min.

156. Water agar (Phytopathology 50:644)

This medium is claimed to be superior to common media for isolating actinomycetes. Many sensitive fungi sporulate well on it. See NCSU medium 4.

157. Wheat straw agar

For growing and inducing sporulation of many fungi.

Autoclave 15 g of agar in tap water, and then add a few pieces of sterile wheat straw.

158. Yam-dextrose agar (Phytopathology 56:1336)

A substitute for potato-dextrose agar (PDA), especially in the tropics, for routine work; compare to cassava-dextrose agar (medium 36).

Prepare like PDA, but replace the potato with bitter or trifoliate yam (*Dioscorea dumentorum*) or white yam (*D. rotundata*), using 200 g of peeled, sliced tuber per liter. The medium prepared from the former is colored; the medium prepared from the latter looks like PDA.

159. Yang and Schoulties's medium (Mycopathol. Mycol. Appl. 46:5)

For inducing production of zoospores of *Aphanomyces euteiches*.

5.5 g D-glucose	0.40 g K_2HPO_4
4.0 g DL-asparagine	0.10 g glutathione
0.7 g KH_2PO_4	0.02 g $CaCl_2 \cdot 2H_2O$
0.4 g $MgCl_2 \cdot 6H_2O$	1 liter distilled water

Adjust the pH to 6.4 with 1 M KOH or HCl. Autoclave.

160. Yeast extract–sucrose agar (Physiol. Plantarum 59:249)

For cultivation of *Verticillium agaricinum*.

1 g yeast extract	15 g agar
20 g sucrose	1 liter water

Adjust the pH to 5.8. Autoclave.

161. Yeast extract–peptone–glucose agar (Mycology 80:77)

For growth of *Phytophthora syringae*.

1.25 g yeast extract	0.075 g $CaCl_2 \cdot H_2O$
1.25 g peptone	20.000 g agar
1.25 g glucose	1 liter water

Add 6 ml of wheat germ oil or 12 ml of maize (corn) oil and 6 ml of linseed oil. When pouring culture plates, agitate the medium to keep the oils in suspension.

162. Yeast extract–potato–peptone–dextrose agar (Trans. Br. Mycol. Soc. 77:660)

For growth of *Ustilaginoidea virens*.

Start with potato-dextrose agar (NCSU medium 5 or medium 121), and add 100 mg of yeast extract and 100 mg of peptone. Autoclave.

163. Yeast extract–glucose medium (YgG)

For inducing production of zoosporangia and zoospores of *Aphanomyces* spp.

4 g yeast extract (Difco)	1 g $MgSO_4 \cdot 7H_2O$
20 g glucose	1 liter water
1 g K_2HPO_4	

Autoclave.

III. Common Selective Media for Plant-Pathogenic Bacteria

Adding toxic chemicals, varying incubation temperatures, changing the physical environment (e.g., aerobic or anaerobic conditions), and withholding or adding certain substrates, such as vitamins and antimicrobial agents, are just a few of the techniques commonly employed to enrich media for desired bacteria or to eliminate undesired bacteria. Not all undesired bacteria, however, are eliminated, so a selective medium must be able to differentiate the desired from the undesired bacteria.

Listed below are a number of selective media developed for plant-pathogenic bacteria. When isolating an unknown plant-pathogenic species, use several different common agar media.

A. Standard Media for Culturing and Enumerating Plant-Pathogenic Bacteria from Plant Tissues and Soil

Genera of plant-pathogenic bacteria are relatively easy to differentiate by Gram stain, the presence of spores, flagellation, aerobic or anaerobic growth, and colony characteristics on certain agar media. The anaerobic test is important for identifying the genera *Erwinia* and *Clostridium*.

164. Asparagine-mannitol agar, or Thornton agar (Ann. Appl. Biol. 9:241)

For isolation and enumeration of bacteria from soil.

0.5 g asparagine	0.1 g $CaCl_2$
1.0 g mannitol	0.1 g NaCl
0.2 g $MgSO_4 \cdot 7H_2O$	0.5 g KNO_3
1.0 g K_2HPO_4	15 g agar
2.0 mg $FeCl_3$	1 liter water

Dissolve the K_2HPO_4, KNO_3, and asparagine, in that order, in distilled water, and add the $MgSO_4 \cdot 7H_2O$, $CaCl_2$,

NaCl, and FeCl$_3$ from a standard solution. Add the agar, melt, and filter at 100°C through ½ inch of absorbent cotton. Dissolve the mannitol in the filtrate, cool the mixture to 60°C, and check the reaction against bromothymol blue to pH 7.4. Autoclave.

165. Casein agar plus glucose (CAG)

For isolation of *Bacillus* and *Clostridium* spp.

10 g casein acid hydrolysate	4 g K$_2$HPO$_4$
	17 g agar
5 g yeast extract	1 liter water
5 g glucose	

After autoclaving the basal medium, aseptically add 10 ml of a previously autoclaved 50% glucose solution.

166. 523 medium (Phytopathology 60:969)

A good medium for cultivation of bacteria.

10 g sucrose	0.3 g MgSO$_4$·7H$_2$O
8 g casein hydrolysate	15.0 g agar
4 g yeast extract	1 liter water
2.0 g K$_2$HPO$_4$	

Dissolve the MgSO$_4$ separately in 50 ml of water, and mix with the rest of the medium before autoclaving. For fluorescent pseudomonads, add egg white to the autoclaved medium. The pH should be 6.9 after autoclaving.

167. Hugh-Leifson anaerobic growth medium (J. Bacteriol. 66:24)

2.0 g peptone	3.0 g agar
5.0 g NaCl	3.0 ml bromothymol blue,
0.3 g KH$_2$PO$_4$	1% aqueous solution

Dissolve the ingredients, and adjust the pH to 7.1. Transfer the basal medium to test tubes (13 mm in diameter), placing 5 ml of medium in each tube, and sterilize at 121°C for 20 min. Prepare a 10% aqueous solution of glucose, and sterilize by filtration. Add 0.5 ml of the sterile glucose aseptically to each tube of basal medium. Inoculate two tubes with the test bacterium. Cover one tube with a layer of sterile melted Vaseline or paraffin to a depth of about 5 mm. A color change from blue to yellow in both tubes is recorded as positive for anaerobic growth (fermentation).

King's medium B agar (KB) and its modifications

See NCSU medium 10.

Nutrient broth–yeast extract agar (NBY)

See NCSU medium 8F.

Nutrient-dextrose agar (NDA), nutrient-glucose agar (NGA), and nutrient-glucose broth (NGB)

See NCSU media 8A and B.

168. Soil extract agar (Can. J. Microbiol. 4:363)

For isolation and enumeration of bacteria from soil.

To 1 liter of soil extract, add 0.2 g of K$_2$HPO$_4$ and 15 g of agar. Adjust the pH to 6.8, and autoclave. To prepare the soil extract, autoclave 500 g of fertile soil in water to yield 1 liter of extract. After cooling, filter through filter paper. Refilter the first cloudy filtrate by passing it through the same filter.

169. Tyrosine agar (Phytopathology 49:457)

A common medium for isolation of bacteria.

25 g sodium caseinate	15 g agar
10 g sodium nitrate	1 liter water
1 g L-tyrosine	

Autoclave.

170. Tyrosine agar, modified (Mem. Osaka Univ. B Nat. Sci. 7:1; Int. J. Syst. Bacteriol. 16:313)

15.0 g glycerol	0.50 g NaCl
0.5 g L-tyrosine (Difco)	0.01 g FeSO$_4$·7H$_2$O
1.0 g L-asparagine (Difco)	20.00 g agar (Bacto)
0.5 g K$_2$HPO$_4$	1 liter water

Combine melted agar and dissolved dry ingredients. Then add the following solution of trace salts, usually at a concentration of 1 ml/liter (Int. J. Syst. Bacteriol. 16:313):

0.1 g FeSO$_4$·7H$_2$O	0.1 g ZnSO$_4$·7H$_2$O
0.1 g MnCl$_2$·4H$_2$O	100 ml water

Adjust the pH to 7.2–7.4 for characterization of *Streptomyces* spp. Autoclave.

171. Yeast extract–dextrose–calcium carbonate agar (YDC)

See NCSU medium 9 and its modifications. To obtain a milky white medium the CaCO$_3$ must be finely ground or it will precipitate to the bottom. Autoclave at 10 lb/in^2 for 1 h (Phytopathology 57:618), or autoclave the dextrose separately in 100 ml of water. Use a vortex mixer for tubes to thoroughly mix the CaCO$_3$ after removing the tubes from the water bath.

B. Diagnostic Media for Culturing and Enumerating Gram-Negative Bacteria

1. *Agrobacterium tumefaciens*

To isolate *Agrobacterium tumefaciens* from soil or maintain it in culture, use dehydrated PDA (Difco) supplemented with 0.5% CaCO$_3$. See NCSU medium 5A.

Since one or two biovars of *A. tumefaciens* may occur in the same gall, and since the biovars vary in different geographic regions, it may be necessary to use a selective medium for each of the three groups.

a) Selective Media for Biovar 1 Strains

172. Brisbane and Kerr's medium 1A (J. Appl. Bacteriol. 54:425)

For isolation of *Agrobacterium tumefaciens* biovar 1 from plant tissues and from soil. The selectivity is based on L(−)-arabitol.

3.04 g L(−)-arabitol	0.25 g $MgSO_4 \cdot 7H_2O$
0.16 g NH_4NO_3	15.00 g agar
0.54 g KH_2PO_4	2.00 ml crystal violet,
1.04 g K_2HPO_4	0.1% aqueous solution
0.29 g sodium taurocholate	1 liter water

Mix the ingredients, and adjust the pH to 7. Heat to dissolve, dispense in 100-ml volumes into bottles, and autoclave. When plates are needed, melt the medium, cool to 50–55°C, and then add the following filter-sterilized stock solution:

1.0 ml cycloheximide, 2% aqueous solution
6.6 ml sodium selenite ($Na_2SeO_3 \cdot 5H_2O$), 1% aqueous solution

173. Schroth's medium (Phytopathology 55:645)

For isolation of *Agrobacterium tumefaciens* biovar 1 from soil.

10.0 g mannitol	0.100 g $MgSO_4 \cdot 7H_2O$
4.0 g $NaNO_3$	0.075 g $NaHCO_3$
2.0 g $MgCl_2 \cdot 6H_2O$	0.075 g $MgCO_3$
1.2 g calcium propionate	15.000 g agar
0.2 g $MgHPO_4 \cdot 3H_2O$	1 liter distilled water

Mix and dissolve the ingredients, and adjust the pH to 7.1 with 1 N NaCl. Dispense in 95- to 100-ml volumes into bottles, and autoclave. When plates are needed, melt the medium, cool to 50–55°C, add the following stock solutions, and then mix and pour plates:

1.0 ml berberine* (0.275 g/10 ml) suspended in sterile distilled water
1.0 ml sodium selenite (0.1 g/10 ml)
0.1 ml penicillin G (950,000 units/10 ml)
0.1 ml streptomycin sulfate or nitrate (0.234 g of base per 10 ml)
1.0 ml cycloheximide (0.224 g of active ingredient per 10 ml)
0.1 ml tyrothricin (pure)
1.0 ml bacitracin (6,500 units/10 ml)

*Berberine is insoluble. Do not attempt to sterilize by filtration.

A. tumefaciens biovar 1 produces domed, smooth, entire, yellowish or tan-colored colonies on this medium.

b) Selective Media for Biovar 2 Strains

174. Brisbane and Kerr's medium 2E (J. Appl. Bacteriol. 54:425)

For isolation of *Agrobacterium tumefaciens* biovar 2 from plant tissues and soil. Selectivity is based on erythritol.

0.16 g NH_4NO	0.25 g $MgSO_4 \cdot 7H_2O$
3.05 g erythritol	5 ml malachite green,
1.04 g K_2HPO_4	0.1% aqueous solution
0.54 g KH_2PO_4	15.00 g agar
0.29 g sodium taurocholate	1 liter water
1 ml yeast extract,	
1% aqueous solution	

Mix the ingredients, and adjust the pH to 7.0. Heat to dissolve, dispense 100-ml volumes into bottles, and autoclave 10–15 min. Cool the medium to 50°C, and then add filter-sterilized cycloheximide (1.0 ml of 2% aqueous solution) and sodium selenite ($Na_2SeO_3 \cdot 5H_2O$) (6.6 ml of 1% aqueous solution).

175. New and Kerr's medium (J. Appl. Bacteriol. 34:233)

For isolation of *Agrobacterium tumefaciens* biovar 2 from soil. The medium is not wholly selective for biovar 2 but is a useful semiselective medium.

5.0 g erythritol	0.2 g NaCl
2.5 g $NaNO_3$	0.2 g $MgSO_4 \cdot 7H_2O$
0.1 g KH_2PO_4	2.0 µg biotin
0.2 g $CaCl_2$	18.0 g agar
2.0 ml Fe-EDTA,	1 liter water
65% aqueous solution	

Prepare Fe-EDTA by mixing 278 mg of $FeSO_4 \cdot 7H_2O$ and 372 mg of Na_2-EDTA in 100 ml of water.

Mix the ingredients, adjust the pH to about 7, and dispense in 95- to 100-ml volumes. Autoclave, cool to 50–55°C, and then add the following filter-sterilized antibiotic stock solutions to give the following final concentration:

1.00 ml bacitracin (650 units/ml) = 100 mg
2.50 ml cycloheximide (0.01 g/ml) = 250 mg
1.00 ml sodium selenite (0.1 g/ml) = 100 mg
0.10 ml tyrothricin (0.001 g/ml) = 1 mg

A. tumefaciens colonies on this medium are domed, entire, and pearly white, often becoming tan-colored after 4–5 days.

c) Selective Media for Biovar 3 Strains

176. Roy-Sasser medium (Phytopathology 73:810)

Very effective for isolation of *Agrobacterium tumefaciens* biovar 3 from plant tissues and from soil. Colony morphology is very distinct.

0.20 g $MgSO_4 \cdot 7H_2O$	1.0 g H_3BO_3
0.90 g K_2HPO_4	0.5 ml chlorothalonil,
0.70 g KH_2PO_4	(Bravo 500), 4%
4.00 g adonitol	aqueous solution
0.14 g yeast extract	15.0 g agar
0.20 g NaCl	1 liter water

Mix the ingredients, and adjust the pH to 7.2. Autoclave 10–15 min, cool to 50°C, and then add the following ingredients, dissolved separately in a few milliliters of distilled water and filter-sterilized:

80 mg triphenyltetrazolium chloride
20 mg D-cycloserine
20 mg trimethoprim (add 1 drop of HCl to the distilled
    water)

Colonies of biovar 3 on this medium can be counted after 4 days of incubation at 27°C. Colonies have dark red centers and white edges. Compare colonies with those of a known isolate plated at the same time, as some pseudomonads grow on this medium.

177. Brisbane and Kerr's medium 3DG (J. Appl. Bacteriol. 54:425)

For isolation of *Agrobacterium tumefaciens* biovar 3 from plant tissues and from soil. Selectivity is based on D-glutamic acid and (+)-sodium tartrate.

SOLUTION A

5.75 g (+)-sodium tartrate · $2H_2O$
15.00 ml D-glutamic acid, 4% aqueous solution neutralized to pH 7 with NaOH
0.29 g sodium taurocholate
6.24 g $NaH_2PO_4 \cdot 2H_2O$
1.00 ml yeast extract, 1% aqueous solution
2.50 ml Congo red, 1% aqueous solution
5.84 g NaCl
4.26 g Na_2HPO_4
0.25 g $MgSO_4 \cdot 7H_2O$
500 ml distilled water

SOLUTION B

1.12 g $MnSO_4 \cdot 4H_2O$
15.00 g agar
500 ml water

Mix the ingredients for each solution separately. Heat to dissolve. Dispense solutions A and B separately in 50-ml amounts, and autoclave. Before pouring plates, add the following (per 50 ml of solution B):

1.0 ml cycloheximide, 2% aqueous solution
0.5 ml sodium selenite, 1% aqueous solution

Mix solutions A and B at 50°C. A precipitate forms. Redistribute the precipitate by recapping the bottle and inverting it several times. D-Glumatic acid can be replaced by 0.16 NH_2NO_3, but some fluorescent pseudomonads can grow on this medium.

Agrobacterium spp. grow well on Brisbane and Kerr media as white, domed, mucoid colonies. Other bacteria may grow on these media, but colonies of them will be small.

d) Selective Medium for *Agrobacterium tumefaciens* and *A. radiobacter*

178. Mannitol-nitrate agar (Phytopathology 55:645)

For isolation of *Agrobacterium tumefaciens* and *A. radiobacter* from soil. This medium excludes 99% of other species growing on nonselective media. Omitting streptomycin helps to improve the recovery of some biovar 1 strains.

10.0 g mannitol	1.2 g Ca propionate
2.0 g $MgCl_2$	0.1 g $MgSO_4 \cdot 7H_2O$
0.2 g $Mg_3(PO_4)_2$	4.0 g $NaNO_3$
75.0 mg $NaHCO_3$	20.0 g agar
75.0 mg $MgCO_3$	1 liter water

After autoclaving, cool the medium to 50°C. Add the following:

275 mg berberine	250 mg cycloheximide
100 mg sodium selenite	1 mg tyrothricin
60 mg penicillin G	100 mg of bacitracin
30 mg streptomycin sulfate	

Adjust the final pH to 7.1 with 1 N NaOH.

2. *Erwinia* spp.

a) *Erwinia "amylovora"* (Non–Soft Rot) Group

King's medium B agar (KB)

See NCSU medium 10.

Yeast extract–dextrose–calcium carbonate medium (YDC)

See NCSU medium 9.

179. *Erwinia amylovora* differential medium
(Plant Dis. Rep. 62:167)

Commonly used for isolation of *E. amylovora* from plant tissue.

To 1 liter of nutrient agar (NCSU medium 9) add 5 g of glucose and 0.2 ml of Tergitol anion 7 (Union Carbide). Autoclave, cool to 50°C, and then add 3 ml of a 1% solution of thallium nitrate and 50 mg of cycloheximide.

180. High-sucrose (40% sucrose) medium
(Phytopathology 63:1425)

A selective medium for *Erwinia amylovora,* whose colonies develop characteristic craters.

160.0 g sucrose	20.0 ml cycloheximide,
12.0 g nutrient agar (BBL)	0.1%
0.8 ml crystal violet (0.1%	380.0 ml water
in absolute alcohol)	

Autoclave, disperse into gridded, plastic culture plates (petri plates), and dry in an incubator at 30°C for 2 h before use. Make an appropriate dilution of the bacterial suspension, and spread 0.1 ml uniformly over the agar surface with a bent glass rod. Incubate at 28°C for 60 h. After incubation, view each colony at 15 or 30× with an oblique light source.

b) *Erwinia "carotovora"* (Soft Rot) Group

If an *Erwinia* pathogen belonging to the soft-rotting group is suspected, isolate the bacteria on modified crystal violet–pectate medium (medium 181). Pitting on the medium is diagnostic for soft-rotting *Erwinia* spp.

181. Modified crystal violet–pectate medium (CVP)
(Phytopathology 64:468; Phytopathology 72:898; Kelman and Dickey, 1989)

For isolation of pectolytic bacteria from soil, plant debris, and insects; selective for gram-negative bacteria. Useful for differentiating *Erwinia* spp. from *Agrobacterium* and *Pseudomonas* spp.

Preheat a 5-cup Waring Blendor by rinsing with hot water. Add 500 ml of boiling distilled water to the blender, and then add the following ingredients at low blending speed, using a rheostat:

1.0 ml crystal violet, 0.075% (w/v) aqueous solution
   = final concentration of 1.5 µg/ml
4.5 ml 1 N NaOH (8 g NaOH in 200 ml of distilled
   water)
6.0 ml CaCl$_2$·2H$_2$O, 10% (w/v) aqueous solution,
   freshly prepared
2.0 g agar (Difco)
1.0 g NaNO$_3$
2.5 g trisodium citrate dihydrate
0.5 ml sodium lauryl sulfate, 10%

Blend at high speed for 15 s; then blend slowly, and add 9 g of sodium polypectate; then blend at high speed for another 15 s. Pour into a 2-liter flask, add 0.5 ml of 10% sodium lauryl sulfate (sodium dodecyl sulfate [SDS], laboratory grade), cap with aluminum foil, and autoclave for 25 min. The pressure should drop *slowly* to avoid bubbles. Pour plates *as soon as possible* and allow to dry for 48 h at room temperature (until *no surface is water present*) before use. The final pH should be 7.2.

Note. When trisodium citrate dihydrate is added, strains of *P. marginalis* and pectolytic strains of *P. fluorescens* do not form pits or form only very shallow pits, making it much easier to detect the deeper, cuplike pits made by pectolytic strains of *Erwinia* spp. in background populations. Soft-rot *Erwinia* colonies under oblique light are iridescent, translucent, and crisscrossed with internal markings.

182. Miller-Schroth medium (MS)
(Phytopathology 62:1175)

For differentiation of *Erwinia* and *Xanthomonas* spp. and isolation of *Erwinia* spp. from soil.

Prepare by pouring and storing in 200-ml bottles for not much longer than 2 weeks. Heat, while mixing, 15.0 g of agar in about 800 ml of distilled water. When the agar is dissolved, add the following ingredients, in the order listed, dissolving each before adding the next:

10.0 g mannitol or sorbitol
0.5 g nicotinic acid
3.0 g L-asparagine, anhydrous
2.0 g K$_2$HPO$_4$
0.2 g MgSO$_4$·7H$_2$O
2.5 g sodium taurocholate (Difco)

Put the following liquid ingredients together in a flask or beaker, and mix into the medium after the dry ingredients are dissolved:

0.10 ml Tergitol anion 7 (sodium heptadecyl sulfate)
10.00 ml nitrilotriacetic acid, 2% solution (20 g + 14.6 g
   of KOH per liter)
9.00 ml bromothymol blue, 0.5% aqueous solution
2.50 ml neutral red, 0.5% aqueous solution
5.00 ml 1 N NaOH (do not use if a precipitate forms in
   the medium)
15.00 g agar

Autoclave, and cool to 50°C. Then add the following:

1.75 ml thallium nitrate, 1.0% aqueous solution
50.00 ml cobalt chloride, 14 mM aqueous solution
5.00 ml cycloheximide, 1% aqueous solution

After all the liquid ingredients are added, bring the medium to a 1-liter volume with distilled water; the pH should be 7.2–7.3 before autoclaving and 7.4 afterwards.

Erwinia colonies are red to orange on this medium, while those of pseudomonads and other bacteria are green to blue.

183. Paton's medium, modified
(Nature [London] 181:61)

A general-purpose medium for isolation of pectolytic bacteria (strains of *Erwinia* and *Pseudomonas* spp.).

5.0 g peptone	12.0 g agar
5.0 g Lab-lemco (Oxoid)	(Oxoid No. 3)
5.0 g calcium lactate	1 liter water

Adjust the pH to 7.2, and autoclave. Pour plates when the medium has set. Then pour 5 ml of cold pectate overlayer (see below), and let the plates set for 4 h. Let the medium dry for several days at 25°C (or 12 h at 37°C) before using.

PECTATE OVERLAYER

2.00 g sodium polypectate
0.01 g Na_2-EDTA
trace bromothymol blue (to give color)
100 ml distilled water

Heat the water to 80°C on a hot plate, add the sodium polypectate a little at a time, with constant stirring, then add the Na_2-EDTA, and maintain at 80°C. Stir until the polypectate has dissolved. Add enough bromothymol blue to just color the medium. Adjust the pH to 7.2. Dispense 5-ml volumes into bijou bottles. Autoclave at 115°C for 4 min.

Use a sterile bent glass rod rather than a loop on this fragile medium.

Pectolytic pseudomonads often produce very shallow pits which cannot be detected after more than several days of incubation. Look at the plates by oblique light 1 or 2 days after inoculation. If in doubt, transfer inoculum to a potato slice.

A MODIFICATION OF THIS MEDIUM

Prepare a 2% pectate gel composed of 1 liter of tap water plus 5.0 g of $CaCl_2 \cdot 6H_2$ and 1 g of K_2HPO_4, 1 g of $(NH_4)_2SO_4$, or 1 g of NH_4Cl. Pour the calcium agar base into a culture plate, allow it to solidify, and pour an equal amount of pectate solution is poured over it.

c) *Erwinia rubifaciens*

184. *Erwinia rubifaciens* isolation medium (glycerol–eosin–methylene blue agar) (Phytopathology 60:557)

For isolation of *E. rubifaciens* from soil.

5.0 g $(NH_4)_2SO_4$	40 mg neomycin sulfate
2.0 g K_2HPO_4	40 mg novobiocin
0.4 g eosin Y	10 ml glycerol
250 mg cycloheximide	15 g agar
65 mg methylene blue	1 liter distilled water

Autoclave, and then add filter-sterilized cycloheximide.

3. *Pseudomonas* spp. (Pseudomonads)

a) Fluorescent, Oxidative-Positive, Pectolytic or Nonpectolytic Pseudomonads

185. *Pseudomonas cichorii* selective medium (PCSM) (Ann. Phytopathol. Soc. Jpn. 48:425)

For isolation of *P. cichorii* from soil and plant debris. Colonies can be identified with a stereomicroscope.

0.5 g KH_2PO_4	100 mg thiram-benomyl
3.0 g $Na_2HPO_4 \cdot 12H_2O$	wettable powder
8.0 g sodium tartrate	10 mg cetrimide
5.0 g $(NH_4)_2SO_4$	25 mg cycloheximide
25 mg $MgSO_4 \cdot 7H_2O$	10 mg Na ampicillin
24 mg $Na_2MoO_4 \cdot 2H_2O$	1 ml phenol red*
10 mg Fe-EDTA	1 ml methyl violet[†]
50 µg L-cystine	25 mg K tellurite
50 mg potassium	15 g agar
pheneticillin	1 liter water

*Dissolve 200 mg of phenol red in 10 ml of a 5% aqueous solution of NaOH. Add 1 ml to the medium.

[†] Dissolve 10 mg of methyl violet in 2 ml of ethanol, and bring the volume to 10 ml with distilled water. Add 1 ml to the medium.

Autoclave, and cool to 50°C. Dissolve the potassium tellurite in 5 ml of distilled water, sterilize it by passing the solution through a 45-µm membrane, and add it to the autoclaved, cooled medium.

186. FPA medium (Phytopathology 62:998)

For isolation of pectolytic, oxidase-positive, fluorescent pseudomonads from soil.

To 500 ml of King's medium B (NCSU medium 10), buffered to adjust the pH to 7.0, add the following:

20 g Bacto-Proteose Peptone No. 3 (Difco)
5 g citrus pectin (Sunkist Growers, Corona, CA)
1.5 g K_2HPO_4
1.5 g $MgSO_4 \cdot 7H_2O$

Mix these components, and adjust the pH to 7.0. Add 15 g of agar (Difco) to 500 ml of distilled water, and combine the two mixtures. Autoclave, and cool to about 45°C. Then add 10 ml of a freshly prepared antibiotic mixture containing 45 mg of novobiocin, 75,000 units of penicillin G, and 75 mg of cycloheximide. Mix the antibiotics, and dissolve them in 1 ml of 70% ethanol and then 9 ml of sterile distilled water, before adding them to the basal medium. Buffer to adjust the pH to 7.0. Pour about 12 ml of medium per culture plate.

Modified crystal violet–pectate medium (CVP)

See medium 181.

King's medium B agar (KB)

See NCSU medium 10.

Nutrient broth–yeast extract agar (NBY)

See NCSU medium 8F.

b) Fluorescent, Oxidative-Negative Pseudomonads

187. BCBRVB (modified KB medium) (Schaad, 1988, p. 61)

For detection of oxidase-negative fluorescent pseudo-monads in plant debris and insects.

Note. The medium may be toxic to certain oxidase-negative *Pseudomonas* spp.

Autoclave 1 liter of King's medium B (NCSU medium 10), and cool to 45°C. Then add a mixture of the following in 70% ethanol (all amounts are per milliliter of solution):

10.0 mg bacitracin	75.0 mg cycloheximide
6.0 mg vancomycin	250.0 mg benomyl
0.5 mg rifampicin	

188. Modified KB medium (KBBC) (Phytopathology 77:1390)

For isolation of *Pseudomonas syringae* pv. *syringae* from seeds or plants, plus some strains of *P. s.* pv. *pisi* and *P. s.* pv. *tomato.*

To 900 ml of King's medium B (NCSU medium 10), autoclaved and cooled to 45°C, add the following ingredients:

100 ml boric acid, 1.5% (w/v) aqueous solution,
    autoclaved and cooled
0.8 ml cephalexin, stock solution at 10 mg/ml,
    in distilled water
0.2 ml cycloheximide, stock solution at 100 mg/ml,
    in 75% methanol
16 g agar

189. M-71 agar (Phytopathology 62:674)

A semiselective medium for isolation of *Pseudomonas syringae* pv. *glycinea* from soil and plants.

Autoclave the tetrazolium HCl compound and cyclo-heximide separately in 5 ml of stock solutions and add to the cooled agar (45–50°C) before pouring plates. Selectivity depends on boric acid (H_3Bo_3).

5.00 g glucose	0.05 g cycloheximide
10.00 g peptone	1.00 g boric acid
1.00 g casein hydrolysate	20.00 g agar
0.05 g triphenyl- tetrazolium HCl	1 liter water

190. *Pseudomonas syringae* pv. *glycinea* isolation medium (SVCA) (Phytopathology 68:1196)

For isolation of *P. s.* pv. *glycinea* from leaves. Colonies appear domed and mucoid.

50.0 g sucrose	16 g agar
8.0 g nutrient broth	990 ml water
0.4 ml crystal violet, 1% aqueous solution	

Autoclave, and then cool to 45°C. Add 10 ml of cyclo-heximide (1% aqueous solution).

191. Modified sucrose-peptone agar (MSP) (Phytopathology 77:1390)

A useful selective medium for isolation of *Pseudomonas syringae* pv. *phaseolicola* from bean seed and plants; also for isolation of *P. avenae* from plant debris and soil.

20.0 g sucrose	0.25 g $MgSO_4 \cdot 7H_2O$
5.0 g peptone	20.00 g agar
0.5 g K_2HPO_4	1 liter water

Adjust the pH to 7.2–7.4 with 40% NaOH. Autoclave, and cool to 45°C. Then add a mixture of the following stock solutions, which have been millipore-filtered through a 0.22-μm membrane:

2.0 ml cycloheximide (100 mg/ml, in 75% methanol)
8.0 ml cephalexin (10 mg/ml, in distilled water)
1.0 ml vancomycin (10 mg/ml, in distilled water)
1.0 ml bromothymol blue (15 mg/ml, in 95% ethanol)

Colonies of *P. s.* pv. *phaseolicola* form a blue fluorescent pigment and levan.

c) Nonfluorescent, Oxidative-Positive Pseudomonads

192. Sorbitol–neutral red agar (SNR) (Phytopathology 67:1113)

For isolation of *Pseudomonas avenae* and other non-fluorescent, oxidase-positive pseudomonads from plant debris and soil.

3.0 g K_2HPO_4	1.0 g KNO_3
1.0 g NaH_2PO_4	0.3 g $MgSO_4 \cdot 7H_2O$

10 ml neutral red,
    0.2% aqueous solution
    (69% active)
	15 g agar
	1 liter water

Autoclave, and cool to 45°C. Then add 2 ml of an alcoholic solution of cycloheximide (100 mg/ml) and 50 ml of a 10% aqueous solution of D-sorbitol, which has been millipore-filtered through a 0.22-μm membrane.

193. CB medium (Appl. Environ. Microbiol. 48:743)

For isolation of strains of *Pseudomonas cepacia* from plant tissues, soil, and "environmental water and water-related samples."

    1 mg C-390 (9-chloro-9-(4-diethylaminophenyl)-
        10-phenylacridan) (Norwich Eaton)
    75 mg polymyxin B sulfate
    1 liter plate-count agar (PCA) (Difco)

Prepare aqueous stock solutions of 0.1% C-390 and 7.5% polymyxin B sulfate. Add 1 ml of each to 1 liter of the rehydrated PCA medium before autoclaving. The pH should be about 6.95.

194. TTC medium, modified (SM-1) (Plant Dis. 67:1084)

Excellent for isolation of *Pseudomonas solanacearum* from soil.

5.0 g glucose	17.0 g agar
10.0 g peptone	1 liter water
1.0 g casamino acids (Difco)	
(casein hydrolysate)	

Melt the agar in 600 ml of water. Add glucose, peptone, and casamino acids to 400 ml of water. Combine 200 ml of the two mixed solutions in each of five 250 ml pharmaceutical bottles. Autoclave. Add the following compounds to each 200-ml portion of the cooled melted agar:

    triphenyltetrazolium chloride, 50 mg/ml, as a sterile
        1% aqueous solution
    crystal violet, 50 µg/ml
    thimerosal, 5 µg/ml
    polymyxin B sulfate, 100 µg/ml
    tyrothricin, 20 µg/ml
    Chloromycetin, 5 µg/ml

Autoclave triphenyltetrazolium chloride and crystal violet for 7 min. Sterilize the remaining compounds by dissolving in 1 ml of 70% ethanol, hold for 30 min, and then dilute to the desired concentration with sterile distilled water. Store at 4°C until the medium is to be used. Add the mixture just before pouring the basal medium into culture plates.

Colonies of *P. solanacearum* after 4–5 days are similar to those on tetrazolium chloride medium (TTC or TZC) (Phytopathology 44:693; see also NCSU medium 12).

When dealing with strains sensitive to Chloromycetin, adjust the concentration according to their sensitivity.

If fungal growth appears, add 50 mg of cycloheximide and 80 mg of chlorothalonil to the medium.

195. *Pseudomonas solanacearum* isolation medium III (Phytopathology 71:220)

For isolation of *P. solanacearum* from plant tissues and soil.

To each liter of tetrazolium chloride medium (NCSU medium 12) add the following:

crystal violet, 50 ppm	tyrothricin, 20 ppm
merthiolate, 0.005%	Chloromycetin, 5 ppm
polymyxin B sulfate,	
100 ppm	

The addition of chlorothalonil (80 ppm) and cycloheximide (50 ppm) is optional (Can. J. Microbiol. 29:433) if fungal contaminants are a problem. After 2–3 days at 30°C colonies are round, pulvinate, fluidal, and tan in color.

196. PCAT medium (Phytopathology 72:706)

For isolation of *Pseudomonas cepacia* from plant tissues and soil.

2.00 g azelaic acid	0.04 g chlorothalonil
0.20 g tryptamine	(Bravo 500), 1:24
0.10 g MgSO$_4$·7H$_2$O	aqueous suspension
4.00 g K$_2$HPO$_4$	15 g agar
4.00 g KH$_2$PO$_4$	1 liter water
0.02 g yeast extract	

Dissolve the magnesium sulfate in the water (distilled), add the azelaic acid, and heat while stirring until it dissolves. Then add the remaining ingredients. Autoclave at 121°C for 10 min, and cool. Then adjust the pH to 5.7.

d) Other Media for Isolating Pseudomonads

197. *Pseudomonas* differential medium (Int. Rice Res. Inst. Newsl. 14:27)

For differentiating the four *Pseudomonas* spp. that attack rice.

1.0 g NH$_4$H$_2$PO$_4$	12 g bacteriological agar
0.2 g KCl	(Oxoid No. 1)
0.2 g MgSO$_4$·7H$_2$O	1 liter water
1 ml bromothymol blue,	
1.6% alcohol solution	

Adjust the pH to 7.2 after autoclaving. Add 1 g of filter-sterilized arginine. Streak the bacterial isolates on the agar surface. After 24 h observe for growth and fluorescence under black light. If no growth appears after 48 h, spread about 1 ml of a 1.5% sucrose solution over the medium, incubate for 24 h, and then observe again for growth and fluorescence.

Growth and fluorescence without sucrose indicates *P. fuscovaginae*. Growth without fluorescence indicates *P. glumae*. No growth after 48 h but growth after the addition of sucrose solution indicates *P. syringae* pv. *oryzicola*. No growth after the addition of sucrose indicates *P. avenae*. The isolates must be pure, and the origin and pathogenicity from rice must be certain.

198. *Pseudomonas fuscovaginae* semiselective medium (KBS) (Int. Rice Res. Inst. Newsl. 14:29)

For isolation of *P. fuscovaginae* from rice samples.

20.0 g casamino acids (Difco)	15 ml glycerol
1.5 g K_2HPO_4	15 g agar
1.5 g $MgSO_4 \cdot 7H_2O$	1 liter water

Adjust the pH to 7.0–7.2 with NaOH before autoclaving. After autoclaving, cool the medium to 45°C. Add the following ingredients:

5 ml Cetrimide, aqueous stock solution (20 µg/ml)	50 mg penicillin G
	20 mg bacitracin
	50 mg cycloheximide
20 mg trimethoprim	

Dissolve the last four of these compounds in 4 ml of 70% ethanol. Incubate at 28°C for 6 days.

Other fluorescent pseudomonads and a few other pathogenic bacterial species may grow on this medium, but the growth of most saprophytic bacteria is inhibited.

199. *Pseudomonas syringae* pv. *savastanoi* semiselective medium (PVF-1 agar) (Phytopathology 79:185)

For isolation of *P. s.* pv. *savastanoi* from American olive (*Osmanthus americanus*) and oleander (*Nerium oleander*) leaves, stems, and fruits.

30.00 g sucrose	0.4 g $MgSO_4 \cdot 7H_2O$
2.50 g casamino acids (Difco)	10 ml glycerol
1.96 g $K_2HPO_4 \cdot 3H_2O$	16 g agar
0.40 g sodium dodecyl sulfate (SDS)	1 liter water

A pale blue fluorescent pigment appears after 3–5 days under UV light. The medium can be refrigerated (at 4°C) for 15 days without losing its selectivity. Adjust the medium to pH 7.1 with HCl.

200. *Pseudomonas syringae* pv. *tomato* isolation medium (J. Appl. Biol. 61:163)

To King's medium B (NCSU medium 10) add 9 mg of basic fuchsin and 1.4 mg of triphenyltetrazolium chloride.

4. *Xanthomonas* spp. (Xanthomonads)

a) Differential Media

201. Yeast extract–nutrient agar (YNA)

A general medium for isolation of *Xanthomonas* spp. from plant tissues.

5 g yeast extract
23 g nutrient agar (Difco)

Autoclave.

202. Glucose-yeast-chalk agar (GYCA)

For examining pigmentation. After 1 month of storage some *Xanthomonas* spp. may lose their viability.

5 g yeast extract	15 g agar
5 g glucose	1 liter water
40 g $CaCO_3$, not refined	

Autoclave. Before the agar solidifies, mix the ingredients well in a vortex mixer, and set the medium quickly in cold water to keep the $CaCO_3$ from precipitating. Pour plates.

203. Yeast extract–sucrose–peptone agar (YSP) (N.Z. J. Sci. 5:393)

For isolation of *Xanthomonas campestris* pv. *albilineans*.

5 g yeast extract	15 g agar
20 g sucrose	1 liter water
10 g peptone	

Autoclave.

204. Yeast-salts broth (YSB) and yeast-salts agar (YSA) (N.Z. J. Sci. 5:393)

Very useful for storage and as a basal medium for some cultural tests.

0.5 g $NH_4H_2PO_4$	5.0 g NaCl
0.5 g K_2HPO_4	5.0 g yeast extract
0.2 g $MgSO_4 \cdot 7H_2O$	

Autoclave.
To make YSA, add 12.0 g of agar.

Yeast extract–dextrose–calcium carbonate agar (YDC)

See NCSU medium 9.
Slants are recommended for cultures stored for 3 months or less.

b) Semiselective Media

(1) For *Xanthomonas campestris* pv. *campestris*

205. SX agar (Phytopathology 64:876)

For isolation of *Xanthomonas campestris* pv. *campestris* and other pathovars of *X. campestris* (pvs. *incanae, nigromaculans, dieffenbachiae, vesicatoria,* and *maculifoliigardeniae*) and several other starch-positive species (e.g., *X. begoniae, X. corylina,* and *X. fragariae*) from plant tissue, soil, and contaminated seeds by direct plating.

10.0 g soluble potato starch	2.0 ml methyl green, 1% aqueous solution
1.0 g beef extract	
5.0 g NH_4Cl	15.0 g agar
2.0 g KH_2PO_4	1 liter water
1.0 ml methyl violet 2B, 1% solution in 20% ethanol	

After autoclaving, add 5.0 g of cycloheximide in 10.0 ml of methanol, add water to bring the volume to 100 ml, and filter-sterilize through a 0.22-μm membrane. Adjust the pH to 6.8–6.9.

Colonies of *X. campestris* after 5 days at 30°C are 3 mm in diameter, mucoid, glistening, convex, circular with entire margins, translucent with bluish purple centers, and surrounded by a clear starch digestion zone of 8–10 mm.

206. Starch-methionine medium (SM agar) (Plant Dis. 67:632)

For isolation of *Xanthomonas campestris* pv. *campestris* from leaves and plant debris in soil. Most strains of *X. c.* pv. *campestris* give a higher recovery on SM agar than on SX agar (medium 205). Other *X. campestris* pathovars (e.g., pvs. *begoniae* and *fragariae*) may also be isolated on SM agar, which differentiates some pathovars by color and colony morphology.

1.0 g KH_2PO_4	2.0 ml methyl green,
2.6 g Na_2HPO_4	1% aqueous solution
1.0 g NH_4Cl	1.0 ml triphenyl-
2.0 g NaCl	tetrazolium chloride,
0.2 g $MgSO_4 \cdot 7H_2O$	1% aqueous solution
67.0 mg $CaCl_2 \cdot 2H_2O$	1.0 ml trace element
1.0 g glucose	solution (see below)
0.2 g L-methionine	50 mg cycloheximide
10.0 g soluble potato starch	20 g agar
1.0 ml methyl violet 2B,	1 liter water
1% solution in 20% ethanol	

Dissolve all ingredients except starch, cycloheximide, and tetrazolium chloride in 800 ml of water. Before boiling, the pH should be 7.0. Suspend the starch in 200 ml of water, add to the boiling medium, and autoclave. Autoclave the tetrazolium compound separately at 121°C for 5 min. Cool the autoclaved medium to about 50°C before adding the tetrazolium and cycloheximide. The final pH should be 6.8.

TRACE ELEMENT STOCK SOLUTION

250 mg EDTA	25 mg $NaMoO_4 \cdot 2H_2O$
500 mg $FeSO_4 \cdot 7H_2O$	18 mg $Na_2B_4O_7 \cdot 10H_2O$
10 mg $ZnSO_4 \cdot 7H_2O$	10 mg $CoSO_4 \cdot 7H_2O$
10 mg $CuSO_4 \cdot 5H_2O$	100 ml water
10 mg $MnSO_4 \cdot H_2O$	

207. NSCAA agar (Phytopathology 74:268)

Very useful for isolation of *Xanthomonas campestris* pv. *campestris* from crucifer seeds.

23.0 g nutrient agar (Difco)	15 g agar
15.0 g soluble potato starch	1 liter water

Autoclave, and cool to 50°C. Add the following ingredients:

5.0 ml cycloheximide (add 5 g to 10 ml of methanol, bring to 100 ml with sterile distilled water, and filter-sterilize through a 0.22-μm membrane)

1.0 ml nitrofurantoin (add 50 mg to 5 ml of 50% dimethylformamide)

1.0 ml vancomycin (add 12.5 mg to 25 ml of water and filter-sterilize through a 0.22-μm membrane)

The prepared medium can be stored for about 60 days at low temperature (4°C).

208. BSCAA agar (Phytopathology 74:268)

A selective medium for isolation of *Xanthomonas campestris* pv. *campestris* from crucifer seeds.

10.0 g soluble potato starch	0.2 g glycine
1.0 g K_2HPO_4	15.0 g agar
1.0 g KH_2PO_4	1 liter water
0.2 g $MgSO_4 \cdot 7H_2O$	

Add the $MgSO_4$ first, and be sure it is fully dissolved. Autoclave, and cool to 50°C. Then add 0.2 ml of methyl green (1% aqueous solution) and 5 ml of cycloheximide (5 g in 10 ml of methanol), add sterile distilled water to bring the volume to 100 ml, filter-sterilize through a 0.22-μm membrane, and add to the basal medium.

209. Starch-cycloheximide agar (Phytopathology 74:268)

Selective medium for isolation of *Xanthomonas campestris* pv. *campestris* from crucifer seeds.

10 g soluble potato starch	0.2 g methyl green,
1 g K_2HPO_4	1% aqueous solution
1 g KH_2PO_4	15.0 g agar
0.2 g glycine	1 liter water
0.2 g $MgSO_4 \cdot 7H_2O$	

Autoclave, and cool to 50°C. Add the following stock solutions:

2.0 mg nitrofurantoin (5 mg/ml, in 50% dimethylformamide)

0.1 mg vancomycin (0.25 mg/ml, filter-sterilized through a 0.22-μm membrane)

200.0 mg cycloheximide (100 mg/ml, filter-sterilized as above)

The prepared medium can be stored for about 60 days at low temperature.

(2) For *Xanthomonas campestris* pv. *carotae*

210. Modified D-5 agar (Plant Dis. 69:758; Phytopathology 76:1109)

Very useful for isolation of *Xanthomonas campestris* pv. *carotae* from carrot (*Daucus carota*) seeds and soil.

10.0 g D-cellobiose*	0.3 g $MgSO_4 \cdot 7H_2O$
3.0 g K_2HPO_4	0.2 g cycloheximide*
1.0 g NaH_2PO_4	15.0 g agar (Bacto)
1.0 g NH_4Cl	800 ml water

*Autoclave, and cool to 55–60°C. Dissolve the cellobiose in 200 ml of water, and add it and the cycloheximide, after filter-sterilizing them through a 0.22-μm membrane, to the basal medium.

D-5A is a modification of this medium, with the addition of glutamic acid (5 mg/liter), L-methionine (1 mg/liter), cephalexin (10 mg/liter), and bacitracin (10 mg/liter) .

211. XCS agar (Int. J. Syst. Bacteriol. 33:409)

Excellent for isolation of *Xanthomonas campestris* pv. *carotae* from seeds; also good for isolation of other pathovars of *X. campestris,* including pvs. *alfalfae, dieffenbachiae, malvacearum, manihotis, nigromaculans, pelargonii,* and *vesicatoria.*

10.0 g lactose	1.0 g NH_4Cl
4.0 g D(+)-trehalose	0.5 g yeast extract
0.2 g thiobarbituric acid	15.0 g agar
0.8 g K_2HPO_4	1 liter water
0.8 g KH_2PO_4	

Combine all ingredients except the agar. Adjust the pH to 6.6. Autoclave, and cool to 50°C. After filter-sterilizing, add the following stock solutions:

1.0 ml tobramycin (80 mg in 10 ml of water)
0.5 ml Na salt of ampicillin (20 mg in 10 ml of water)
0.5 ml vancomycin (10 mg in 10 ml of water)
5.0 ml cycloheximide (5 g in 10 ml of methanol,
   with water added to bring the volume to 100 ml)

(3) For *Xanthomonas campestris* pv. *citri*

Tryptone–yeast extract agar

See medium 152.

(4) For *Xanthomonas campestris* pv. *fragariae*

212. *Xanthomonas campestris* pv. *fragariae* growth medium (Plant Dis. 64:178)

Good for growth of *X. c.* pv. *fragariae.*

0.8% nutrient broth	1.000% glucose
0.5% casein hydrolysate	0.001% $FeSO_4 \cdot H_2O$
0.1% yeast extract	0.001% $MnSO_4$

(5) For *Xanthomonas campestris* pv. *juglandis*

213. BS medium (Phytopathology 71:336)

A semiselective medium for isolation of *Xanthomonas campestris* pv. *juglandis* from English walnut (*Juglans regia*) buds and catkins. Other pathovars of *X. campestris* (e.g., pvs. *begoniae, campestris, incanae, malvacearum, phaseoli,* and *vesicatoria*) also grow on this medium.

3.00 g $K_2HPO_4 \cdot 3H_2O$	10.00 mg brilliant cresyl
1.50 g KH_2PO_4	blue (dye content 55%)
2.00 g $(NH_4)_2SO_4$	10.00 mg methylene green
0.25 g L-methionine	15.00 g agar
0.25 g nicotinic acid	1 liter water
10.00 g soluble potato	
starch (Difco)	

Before autoclaving, adjust the pH to 6.8–7.0, using 8 ml of 1 N NaOH per liter. The medium can be stored and re-melted when needed.

(6) For *Xanthomonas campestris* pv. *phaseoli*

214. MXP agar (Phytopathology 77:730)

A semiselective medium for isolation of *Xanthomonas campestris* pv. *phaseoli* from plant tissues, debris, seed, and soil.

8.0 g soluble potato starch	0.8 g K_2HPO_4
1.0 g glucose	0.6 g KH_2PO_4
0.7 g yeast extract	10.0 g KBr
30.0 μl methyl violet 2B,	15 g agar
1% solution in 10%	1 liter water
ethanol	
60.0 μl methyl green,	
1% aqueous solution	

Zones of starch inhibition are enhanced by the two dyes.

After autoclaving, cool to 40°C. Add the following filter-sterilized stock solutions to the basal medium:

1.0 ml chlorothalonil (1.2 ml/38.8 ml of water) = 15 mg
10.0 ml cephalexin (1 g/500 ml of water) = 20 mg
10.0 ml kasugamycin (1 g/500 ml of water) = 20 mg
5.0 ml gentamicin (100 mg/500 ml of water) = 2 mg

The stock solutions can be stored for about 3 months.

Some strains of *X. c.* pv. *phaseoli* are highly sensitive to gentamicin. Each purchase should be tested to determine the minimum and maximum damage rate for efficacy in plating efficiency and inhibition of contaminants.

(7) For *Xanthomonas campestris* pv. *pruni*

215. XPS medium (Plant Dis. 66:39)

A selective medium for isolation of *Xanthomonas campestris* pv. *pruni* from plant tissues and oil.

2.0 g alginic acid	0.8 g K_2HPO_4
0.2 g 8-azaguanine	0.1 g $MgSO_4$
2.0 mg nicotinic acid	15.0 g agar
3.0 mg cysteine	1 liter water
0.8 g KH_2PO_4	

After autoclaving, cool to 50°C. Then aseptically add 80 mg of chlorothalonil (75% wettable powder) and 16 mg of kasugamycin.

(8) For *Xanthomonas campestris* pv. *translucens*

216. XTS agar (Phytopathology 75:260)

For isolation of *Xanthomonas campestris* pv. *translucens* from wheat seeds; also good for isolating other pathovars of *X. campestris,* including pvs. *alfalfae, pelargonii,* and *vesicatoria.*

23 g nutrient agar (Difco)	1 liter water
5 g glucose	

Autoclave, and cool to 50°C. Add the following stock solutions, filter-sterilized through a 0.22-μm membrane, in 75% ethanol:

2.0 ml cycloheximide (1 g/10 ml; use only if assaying seeds) = 200 mg
0.8 ml gentamicin (80 mg/5 ml; omit if antagonism appears) = 8 mg
1.0 ml cephalexin (50 mg/5 ml) = 10 mg

(9) For *Xanthomonas campestris* pv. *vesicatoria*

217. Tween A, B, and C media (Plant Dis. 70:887; Phytopathology 74:882)

Three semiselective media for isolation of *Xanthomonas campestris* pv. *vesicatoria* from soil and plant material; also useful for isolation of *X. c.* pvs. *dieffenbachiae, nigromaculans, pelargonii,* and *phaseoli.*

BASAL MEDIUM

10.00 g peptone (Bacto)	10 ml Tween 80
10.00 g KBr	15 g agar (Difco)
0.25 g CaCl$_2$	1 liter water

Dissolve all of the ingredients, except Tween 80, agar, and (if necessary) boric acid, in water. Adjust the pH to 7.4 with NaOH. Dissolve the agar, and autoclave.

Autoclave Tween 80 separately, and add it immediately to the basal medium when it is removed from the autoclave. To keep the Tween from bubbling, keep a stirring rod in the medium during autoclaving.

TWEEN A MEDIUM

Add the following to the basal medium:

50 mg cycloheximide	12 mg 5-fluorouracil
35 mg cephalexin	0.4 mg of tobramycin

Add the antibiotics aseptically as aqueous solutions after autoclaving. Tween A is less selective for strains less tolerant of boric acid.

TWEEN B MEDIUM

For isolation of *X. c.* pv. *vesicatoria* from leaves.
Add the following to the basal medium:

300 mg boric acid	12 mg 5-fluorouracil
50 mg cycloheximide	0.4 mg tobramycin
65 mg cephalexin	

Add the antibiotics aseptically as aqueous solutions after autoclaving.

TWEEN C MEDIUM

For isolation of *X. c.* pv. *vesicatoria* from soil.
Add the following to the basal medium:

600 mg boric acid	12 mg 5-fluorouracil
50 mg cycloheximide	10 mg methyl green
65 mg cephalexin	

Add the antibiotics aseptically as aqueous solutions after autoclaving.

(10) For *Xanthomonas campestris* pv. *vignicola*

Mannitol-orthophosphate agar

See medium 101.

5. *Xylella fastidiosa* and Other Fastidious Bacteria

ELISA and other serological methods usually give better results than bacteriological media in identifying fastidious bacteria.

218. PW medium (Curr. Microbiol. 6:309)

For isolation of fastidious bacteria causing elm leaf scorch, periwinkle wilt, phony peach, Pierce's disease, plum leaf scald, ragweed stunt, and probably other strains from sap in petioles, stems, and roots.

4.00 g Phytone peptone	4 g L-glutamine
1.00 g trypticase peptone	6 g bovine serum albumin, fraction V
0.01 g hemin chloride	
1.20 g K$_2$HPO$_4$	12 g agar
1.00 g KH$_2$PO$_4$	1 liter deionized distilled water
0.40 g MgSO$_4$·7H$_2$O	
0.02 g phenol red	

Autoclave the basal medium. Filter-sterilize (through a 0.22-μm membrane) stock solutions of glutamine and the bovine serum albumin, and then add them to the basal medium after autoclaving.

219. BCYE medium (Appl. Environ. Microbiol. 42:357)

For isolation of the same strains of *Xylella fastidiosa* as those isolated on PW medium (medium 218).

10.0 g yeast extract	0.25 g ferric pyro-
2.0 g activated charcoal	phosphate, soluble
0.4 g L-cysteine HCl·H$_2$O	17.00 g agar
10.0 g ACES buffer	1 liter water
(Sigma)	

Autoclave the basal medium, and adjust the pH to 6.9 with KOH. Dissolve the L-cysteine HCl and ferric pyrophosphate in 20 ml of water, filter-sterilize through a 0.22-μm membrane, and add to the cooled basal medium.

220. CS-20 (Phytopathology 72:1006; Schaad, 1988, p. 96)

For isolation of the same isolates of *Xylella fastidiosa* as those isolated on PW medium (medium 218) and BCYE medium (medium 219).

2.00 g soytone	0.40 g MgSO$_4$·7H$_2$O
2.00 g tryptone	6.00 g L-glutamine
0.85 g (NH$_4$)$_2$HPO$_4$	1.0 g dextrose
15.00 ml hemin chloride,	2.0 g soluble potato starch
0.1% solution	5.0 ml phenol red,
in 0.05 N NaOH	0.2% solution*
1.00 g L-histidine HCl	12.00 g agar
1.00 g KH$_2$PO$_4$	1 liter water

*Dissolve 0.2 g of phenol red in 30 drops of 5 N NaOH (using a Pasteur pipette), and add water to bring the volume to 100 ml.

Add each ingredient separately, and allow to dissolve. Autoclave, and then shake gently to distribute the starch. The pH should be 6.6–6.7.

C. Diagnostic Media for Culturing and Enumerating Gram-Positive Bacteria

1. Coryneform Bacteria (*Arthrobacter, Clavibacter, Corynebacterium, Curtobacterium,* and *Rhodococcus*)

Note. Always compare unknowns with known coryneform bacteria.

Nutrient broth–yeast extract agar (NBY)

See NCSU medium 8F. The medium of choice for growth, pigmentation, and differentiation of colonies of coryneform bacteria (except for *Clavibacter xyli*).

221. CNS medium (J. Syst. Bacteriol. 32:315; Schaad, 1988, p. 106)

Allows better selection and faster growth of some species, especially *Clavibacter michiganense* subsp. *nebraskense*. Used for isolation from plant tissues and soil.

8.0 g nutrient broth	5.0 g LiCl*
2.0 g yeast extract	6.5 g agar
2.0 g K$_2$HPO$_4$	480 ml double-distilled
0.5 g KH$_2$PO$_4$	water

*LiCl is optional. It is temporarily toxic to freshly isolated *C. m.* subsp. *nebraskense*. It can be omitted from the medium, or the plating of plant suspensions (in a buffer containing Na) can be delayed for 1–2 h (Appl. Environ. Microbiol. 52:591).

Autoclave, and cool to about 50°C. The pH should be 6.9 after autoclaving. Add the following ingredients from fragile sterile stock solutions:

glucose, 10% (w/v) solution (50.0 ml/liter)

1 M MgSO$_4$·7H$_2$O (1 ml/liter)

4.0 ml cycloheximide (1 g/100 ml of double-distilled water); store refrigerated

2.5 ml nalidixic acid (0.1 g in 7 ml of double-distilled water and 3 ml of 1 N NaOH); store refrigerated

3.2 ml polymyxin B sulfate[†] (8,000 USP units/mg, or 1 g/100 ml of double-distilled water); store refrigerated

0.6 ml chlorothalonil (Bravo 500), 50% active ingredient (1.2 ml/38.8 ml of double-distilled water); store at room temperature

[†] Since polymyxin B sulfate preparations can vary, the medium should be pretested to determine the growth of single colonies (Plant Dis. 68:536).

Mix well by swirling the 1-liter flask; pour plates.

On this medium *C. m.* subsp. *michiganense* colonies are light yellow, convex, and glistening and have a butter-like odor.

222. SC agar for *Clavibacter xyli* subspecies (Science 210:1365)

Add the following ingredients, dissolved in deionized water, in the order given:

8.0 g phytone peptone or soytone

15.0 ml hemin chloride, 0.1% (w/v) solution in 0.05 N NaOH

1.5 g KH$_2$PO$_4$

0.5 g K$_2$HPO$_4$·3H$_2$O

0.2 g MgSO$_4$·7H$_2$O

17.0 g cornmeal agar

Autoclave. Filter-sterilize the following three stock solutions, and add to the cooled (50°C) medium:

1.0 ml glucose, 50% (w/v) aqueous solution

10.0 ml L-cysteine, free base, 10% (w/v) aqueous solution

10.0 ml bovine serum albumin, fraction V, 20% (w/v) aqueous solution

The final pH should be 6.6–6.7 without adjustment.

223. D2 medium (Phytopathology 60:969)

For isolation of *Corynebacterium* spp. from soil.

10.0 g glucose	2 g yeast extract
5.0 g LiCl	1 g NH$_4$Cl
4.0 g casein hydrolysate	15 g agar
0.3 g MgSO$_4$·7H$_2$O	1 liter water
1.2 g tris*	

*Tris(hydroxymethyl)aminomethane.

Mix all ingredients. Adjust the pH to 7.8 with HCl. Autoclave, and cool to about 50°C. The pH should be 6.9. Add 40 mg of polymyxin sulfate (300 USP units) and 2 mg of sodium azide. Prepare fresh for use, as sodium azide and polymyxin break down in time.

224. SCMS medium (Plant Dis. 68:1095)

For isolation of *Clavibacter xyli* subsp. *cynodontis* from plant tissues and soil.

To 1 liter of autoclaved and cooled SC agar (medium 222) add the following, dissolved in 10 ml of water and filter-sterilized through a 0.22-μm membrane:

50 mg cycloheximide
35 mg hymexazol
10 mg colistan methane sulfonate (1,190 units/mg)
50 mg polymyxin B sulfate (7,400 USP units/mg)

Colonies take 7–14 days to become readily visible (0.1–1 mm in diameter).

225. SMCMM selective medium (Ann. Phytopathol. Jpn. 54:540)

For isolation of coryneform species, especially *Clavibacter michiganensis* subsp. *michiganensis,* from plant tissues and soil.

20.00 g glycerol	40 mg cycloheximide
5.00 g peptone	20 mg nalidixic acid
5.00 g LiCl	2 mg NaNO$_3$
3.00 g yeast extract	3 mg tetrachloro-
2.00 g K$_2$HPO$_4$	isophthalonitrate
0.50 g KH$_2$PO$_4$	hydrate, 70%
0.25 g MgSO$_4$·7H$_2$O	15 g agar
80 mg K$_2$Cr$_3$O$_7$	1 liter water

Filter-sterilize stock solutions of cycloheximide and nalidixic acid through a 0.22-μm membrane, and add to the autoclaved and cooled agar.

226. *Clavibacter michiganensis* subsp. *michiganensis* medium (SCM) (Phytopathology 78:121)

Semiselective medium for detection of *C. m.* subsp. *michiganensis* from seeds.

10.0 g sucrose	1.5 g boric acid
2.0 g K$_2$HPO$_4$	0.1 g yeast extract
0.5 g KH$_2$PO$_4$	15.0 g agar
0.25 g MgSO$_4$·7H$_2$O	1 liter water

Autoclave, cool to 45°C, and add the following ingredients:

2 ml cycloheximide (100 mg/ml of 75% ethanol)
1 ml nalidixic acid salt (10 mg/ml of 0.1 N NaOH)
1 ml Chapman tellurite (1% solution)
10 ml nicotinic acid (10 mg of free acid per ml of water)

All stock solutions except tellurite should be filter-sterilized through a 0.22-μm membrane.

227. Tetrazolium chloride medium (TTC or TZC) (Phytopathology 44:693)

See also NCSU medium 12.

10 g glucose	10 g peptone
1 mg casamino acids	18 g agar
(casein hydrolysate)	1 liter water

Autoclave, and add 1 ml of a 1% (w/v) aqueous solution of 2,3,5-triphenyltetrazolium chloride, autoclaved separately, to each 200-ml portion.

228. RSD broth medium (Int. J. Syst. Bacteriol. 34:107)

Add and dissolve the following ingredients, in the order given:

1.0 g yeast extract
1.0 g (NH$_4$)$_2$HPO$_4$
15.0 ml bovine hemin chloride, 0.1% (w/v) solution
   in 0.05 N NaOH
1.0 g L-cysteine, free base
1.0 ml bromothymol blue, 0.8% (w/v) aqueous solution
0.2 g MgSO$_4$·7H$_2$O
0.2 g KCl
2.0 g glucose
10.0 ml bovine serum albumin, fraction V,
   20% (w/v) aqueous solution
1 liter water

Mix all ingredients but the bovine serum albumin. Adjust the pH to 7.2 with 0.1 N NaOH or HCl before autoclaving. Filter-sterilize the bovine serum albumin, and add it after the medium has cooled to room temperature. The final pH should be 6.7–6.8.

2. *Streptomyces*

229. Pridham yeast extract–malt extract–dextrose agar (YME) (Antibiotics Annu. 1956/1957, p. 947)

For identification of *Streptomyces.*

4 g yeast extract	20 g agar
10 g malt extract	1 liter distilled water
4 g dextrose	

Adjust the pH to 7.3 with NaOH before adding the agar. Dissolve the medium in water by boiling, and then autoclave (pH 6.8–7.0). The medium is also available as ISP

Medium 2 from Difco. Prepare according to label directions.

Most *Streptomyces* isolates produce abundant aerial mycelium.

230. Peptone–yeast extract–iron agar (PYI) (J. Bacteriol. 76:239).

36 g peptone iron agar   1 liter distilled water
1 g yeast extract

Combine the ingredients, and adjust the pH to 7.0–7.2 with NaOH or HCl when needed. Liquefy the agar by steaming or boiling at 100°C for 15–20 min. Streak strains of *Streptomyces scabies* on culture plates, and incubate them for 4 days at 28°C.

Melanoid pigments surround colonies of most strains of *S. scabies.*

Oatmeal agar for culturing *Streptomyces*

See NCSU medium 13D.

Streptomyces ipomoea growth medium (SPA)

See NCSU medium 11.

3. *Bacillus* (Mostly Gram-Positive)

231. Mundt-Hinkle medium (Appl. Environ. Microbiol. 32:694)

To 1 liter of nutrient broth (Difco) add the following ingredients, and autoclave:

3 g yeast extract   5 g agar (for slants)
5 g glucose   17 g agar (for plates)
2 g cycloheximide   1 liter water

Casein agar plus glucose (CAG)

See medium 165.

4. *Clostridium*

Casein agar plus glucose (CAG)

See medium 165.

232. Double-layer pectate medium (J. Appl. Bacteriol. 33:609)

For isolation of pectolytic clostridia.

BASAL LAYER MEDIUM

10.0 g tryptone
5.0 g yeast extract
5.2 g $CaCl_2 \cdot 6H_2O$
2.4 g Lab-lemco powder (meat extract)
0.4 g cysteine hydrochloride
15.0 g Davis agar or 18.75 g Bacto agar
1 liter distilled water
240,000 units polymyxin B sulfate

Dissolve the ingredients, except for the polymyxin, and adjust the pH to 6.8. Autoclave.

UPPER LAYER MEDIUM

20 g sodium polypectate of high molecular weight
60 ml ethanol, absolute
1 g Na_2-EDTA
1 liter water

Mix the sodium polypectate with ethanol. Add EDTA slowly, with constant stirring. Adjust the pH to 7.4. Autoclave.

PREPARATION OF PLATES

Cool the basal layer medium to 55–60°C. Add 10 ml of a sterile aqueous solution of polymyxin B sulfate (24,000 units/ml) to 1 liter of medium. Mix. Pour 20 ml of the basal medium into 9-cm culture plates. When the basal layer medium has solidified, overlay it with 8.5 ml of the sodium pectate solution. Thoroughly dry the medium, and then expose it to a stream of sterile air for 45–60 min on a laminar airflow clean bench. Just after drying and inoculating, place the plates in an H_2/CO_2 atmosphere, 90:10 by volume, in anaerobic jars with a room-temperature catalyst (Oxoid, Ltd.). Store the plates for 16 h or more before use.

233. Potato infusion medium (PIM) and potato infusion agar (PIA)

200.0 g potatoes   0.5 g Davis agar (PIM)
1.0 g $(NH_4)_2SO_4$ or
0.5 g cysteine HCl   15 g Davis agar (PIA)
5 g glucose   1 liter distilled water
3 g $CaCO_3$

These media are the same except that PIM has 0.5 g of agar and PIA has 15 g of agar.

PREPARATION OF PIM

Peel, cut into pieces, and immerse the potatoes with the cysteine HCl in some water. Pass the cut-up potatoes through a mincer, and then mix them in a blender for less than 30 s. Add enough water to make a pourable slurry.

Add the glucose, $(NH_4)_2SO_4$, $CaCO_3$, and 0.5 g of agar to the remaining water. Heat to dissolve the agar. To prepare aliquots, add appropriate volumes of the potato slurry and the agar solution, with the remaining ingredients, to a 500-ml flask. Mix well, and pour 20-ml volumes into small, narrow-necked bottles or test tubes with screw caps, to give a 25-mm-thick layer of potato solids. Autoclave. Immerse the bottles or tubes, with the caps loose, in a boiling water bath for 20 min before use, to expel the oxygen. Tighten the screw caps, and cool the medium.

PREPARATION OF PIA

Cut the potatoes into pieces, and cook them in about 500 ml of water. Strain through muslin or cheesecloth, and add all the other ingredients except the agar. Place the agar in the remaining water, and dissolve it at 100°C. Combine the potato filtrate and the other solution. Autoclave. Cool the medium to 50 to 55°C, and agitate to suspend the $CaCO_3$. Pour 18- to 20-ml volumes into 9-cm culture plates. When the agar has set, dry the plates by exposure to a stream of sterile air for 45 min in a laminar-airflow clean air bench. Then place the plates in an atmosphere of H_2/CO_2, 9:10 by volume, in anaerobic jars at room temperature for 16 h or more before use.

5. Spiroplasmas

The presence of spiroplasmas in plant tissues can be detected by spiroplasma cultivation in artificial media, serological procedures such as ELISA, the use of DNA probes or the polymerase chain reaction, or (in the case of very high titers of spiroplasmas) observation of phloem exudate via dark-field or phase optics in light microscopy. For details of these methods, except the use of DNA probes, see Schaad (1988). Identification of the species may be accomplished by serological approaches, such as the growth inhibition and metabolic inhibition tests (Schaad, 1988), by comparison of protein profiles following polyacrylamide gel electrophoresis, or by application of DNA methodologies.

ISOLATION OF SPIROPLASMAS FROM INFECTED TISSUES

1. Surface-sterilize the sample material by immersing it in 70% alcohol or 1% sodium hypochlorite solution for 1 min for succulent tissues to 5 min for tough stem and root tissues (surface sterilization ensures that surface mollicutes are not cultivated instead of, or in addition to, phloem-restricted mollicutes). Wash several times with sterile water.
2. Mince the specimen in a sterile dish containing 5 ml of liquid growth medium, and allow the pieces of plant tissue to remain in the liquid for 10–15 min at room temperature, so that the spiroplasmas are released from the cut phloem elements.
3. Leave the tissue pieces behind, collect the liquid, and filter it.
4. Make a dilution (1:50) of the filtrate in a fresh tube of broth to dilute out plant inhibitors (e.g., polyphenols) released from plant tissues.
5. Incubate both the original filtrate and the subculture at 30–32°C. *Spiroplasma citri* (the causal agent of citrus stubborn and horseradish brittle root) grows well aerobically, while *S. kunkelii* (the causal agent of corn stunt) may respond better to anaerobic conditions in a candle jar or anaerobic chamber.
6. Tubes will become only slightly turbid even with log phase growth, so it is wise to check daily for the presence of spiroplasmas by dark-field or phase microscopy.

Spiroplasmas will *not* grow in standard bacteriological media. Several media have been developed and used successfully. Three commonly used formulations are presented below.

234–236. C3G, LD8A3, and LD59 (Schaad, 1988, p. 142)

Three media for isolating and culturing plant-pathogenic spiroplasmas. C3G is a commonly used simple medium, LD8A3 gives superior growth, and serum-free LD59 is excellent for primary culture of the corn stunt spiroplasma. Always use freshly prepared media!

Prepare the basal medium (without serum or serum substitute) by dissolving all ingredients in deionized water. Adjust the pH to 7.5 with 2 N NaOH. Autoclave, allow the medium to cool to 50°C, and aseptically add serum or serum substitute. Quantities listed below are per liter of solution.

C3G

15.0 g "mycoplasma broth base"	200.0 ml horse serum
	720.0 ml deionized water
120.0 g sucrose	

LD8A3

15.0 g "mycoplasma broth base"	0.4 g pyruvate
	2.0 g lactalbumin hydrolysate
1.0 g fructose	
120.0 g sucrose	100.0 ml fetal bovine serum
0.6 g L-arginine	
0.6 g L-asparagine	7.0 g HEPES buffer
0.4 g L-methionine	820.0 ml deionized water
0.4 g α-ketoglutarate	

LD59

15.0 g "mycoplasma broth base"	0.4 g pyruvate
	2.0 g lactalbumin hydrolysate
1.0 g fructose	
120.0 g sucrose	7.0 g HEPES buffer
0.6 g L-arginine	200.0 ml serum substitute (BSA-lipids solution)*
0.6 g L-asparagine	
0.4 g L-methionine	720.0 ml deionized water
0.4 g α-ketoglutarate	

*A solution of 0.66 ml of lipid stock 1 (see below) and 3.5 ml of lipid stock 2 (see below) in bovine serum albumin (BSA) at 37°C (9 g of BSA in 200 ml of deionized water).

LIPID STOCK 1

Dissolve the following in 2 ml of 100% ethanol:

0.6 ml Tween 80	60 mg palmitic acid
20.0 mg phosphatidic acid	40 mg of cholesterol
0.075 ml linoleic acid	

LIPID STOCK 2

Dissolve the following in 10 ml of 100% ethanol:

12 mg lysophosphatidyl choline from egg yolk
12 mg lysophosphatidyl choline from soybean
20 mg phosphatidyl choline from egg yolk

References

Hundreds of other mycological and bacteriological media could be listed, and descriptions of new media are published almost every month. For media to isolate or enumerate specific species of fungi and bacteria in plant tissues, seed, or soil not listed here, see Booth, 1971; Dhingra and Sinclair, 1995; Lelliott and Stead, 1987; Schaad, 1988; and Tuite, 1969, 1988.

Complete references to these and other works cited in this appendix are given following Appendix J.

Fixatives

The mixtures listed below are useful for killing and maintaining organisms as they appear naturally.

1. Chromic acid–acetic acid

This fixative can be weak, medium, or strong, depending on the need.

WEAK CHROM-ACETIC

2.5 ml of 10% aqueous chromic acid
5.0 ml of 10% aqueous acetic acid

Add distilled water to a volume of 100 ml.

MEDIUM CHROM-ACETIC

7 ml of 10% aqueous chromic acid
10 ml of 10% aqueous acetic acid

Add distilled water to a volume of 100 ml. The addition of 2% maltose or urea or 0.3–0.5% saponin aids penetration.

STRONG CHROM-ACETIC

10 ml of 10% aqueous chromic acid
10 ml of 10% aqueous acetic acid

Add distilled water to a volume of 100 ml, and then add maltose, urea, or saponin as for medium chrom-acetic.

2. Flemming's solution

This fixative can be prepared weak or strong.

WEAK SOLUTION

25 ml of 1% chromic acid
10 ml of 1% acetic acid
60 ml of distilled water

Before use, add 5 ml of 2% osmic acid.

STRONG SOLUTION

75 ml of 1% chromic acid
5 ml of glacial acetic acid
20 ml of 2% osmic acid

If desired, this solution can be diluted with an equal volume of distilled water.

3. Formalin–acetic acid–ethanol (FAA)

13 ml of formalin
5 ml of glacial acetic acid
200 ml of 50% ethanol

Mounting and Staining Fungi for Microscopic Examination

I. Mounting and Staining Media

Before details of fungi are observed microscopically, it is common to make semipermanent mounts of plant tissue. Many mounting media incorporate a stain for more critical observation of features such as cell walls, nuclei, and spore details and to identify specific substances.

Water mounts usually give the truest colors and allow direct observation of living material, which is of special value when phase microscopy is used. Other mounting media have certain advantages over water: 1) they evaporate much more slowly; 2) they can be sealed with fingernail polish or a product such as Glyceel for long-term preservation; and 3) stains are more easily incorporated, allowing more critical observation and identification of specific substances.

Some mounting media remain liquid, while others harden. Media that do not harden require that mounts be sealed; those that harden are often ringed with a sealant.

Note. A preliminary mounting in 70% alcohol combined with separating out masses of mycelium or extraneous plant material often eliminates air bubbles when the permanent mounting medium is added later.

Host-parasite relationships may be studied by using non-fixed or fixed freehand or paraffin sections or whole mounts. Detailed information on preparing paraffin sections is given in several plant anatomy books and is not discussed here.

1. Glycerin jelly mounting medium (Johansen, 1940)

Dissolve pure gelatin in distilled water (1:6, w/w), soak for 2 h, and then add seven parts of glycerin (w/w) and 1 g of phenol crystals for each 100 g of mixture. Warm with constant stirring (about 15 min) until all the flakes dissolve. While the mixture is still warm, filter it through two or three layers of cheesecloth into small bottles that can later be warmed in a hot water bath. When the jelly is used, it is best to cut out small portions and melt them, because the mixture deteriorates from continued remelting.

To mount a fungus, place enough warm jelly on a slide, put the specimen on it, and apply a coverslip. Remove any excess glycerin with a piece of blotting paper or paper towel at the edge of the coverslip. Immediately seal the mount (see Appendix F) for long-term storage, or store for several days.

2. Glycerin jelly–methyl green (GJMG) stain and mounting medium (Phytopathology 55:246)

GJMG is a combination nonspecific stain and mounting medium. Hyphal cell walls and protoplasts are stained. Parts of the same fungus are sometimes stained differently. Prepare glycerin jelly as outlined in 1 above. Saturate the cool glycerin jelly by adding, dropwise, 3% methyl green in 50% ethanol.

To mount fungal mycelium, spores, freehand sections of fruiting bodies, etc., place a drop of Patterson's or Shear's fluid (see 3 below) on a clean glass slide and heat *gently* over a flame until the water and ethanol have evaporated. Then apply a small block of GJMG, and melt it over a flame. Stir the fungal material with a needle into the melting medium, apply a coverslip, and seal (see Appendix F).

3. Patterson's or Shear's fluid mounting medium

There are several variations of this standard combination mounting medium–stain fluid for fungi. Two are given here.

MYCOLOGIA 32:269

300 ml potassium acetate (2%) in water	180 ml ethanol (95%)
	120 ml glycerin

Place the specimen in a drop of the fluid on a 12-mm coverslip, and heat gently over a flame until most of the liquid has evaporated. Place a drop of pure glycerin on the mount, and heat slightly again. Turn the coverslip upside down onto a 22-mm coverslip. Press down firmly, and wipe away all excess glycerin from the margin of the smaller coverslip. Place a large drop of balsam on a microscope slide, heat gently, and then place the coverslips, with the 12-mm one underneath, onto it. When the balsam has spread enough to seal the glycerin, press the whole mount down until the balsam exudes slightly from the edges of the larger coverslip.

MYCOLOGIA 32:570

200 ml glycerin	300 ml ethanol (95%)
500 ml water	10 g potassium acetate

A small amount of certain compatible stains (e.g., erythrosin, eosin, or light green) may be added for staining and preserving fungal structures in one operation.

4. Hoyer's mounting medium (Alexopoulos and Beneke, 1952)

30 g gum arabic, lump	20 g glycerin
200 g chloral hydrate	50 ml distilled water

Soak the gum arabic in distilled water for 24 h. Add the chloral hydrate, and let stand for several days or until all the material has dissolved. Add the glycerin, and the mounting solution is ready to use.

Certain specimens (e.g., those of Myxomycetes and Ascomycotina) are wetted with absolute ethanol for 1 min, treated with 2% KOH for 1 min, and then washed in 70% ethanol before mounting in Patterson's or Shear's fluid (see 3 above).

5. Isaac's staining mountant (Stain Technol. 33:261)

A combined fixing, staining, and mounting medium that becomes permanently hardened.

50.0 ml distilled water	1.50 g ferric alum
10.0 ml formic acid, 85–90%	0.50 g chrome alum
	0.15 g Bismarck brown
20.0 g gum arabic, finely ground	30.0 ml glycerin
	20.0 ml chloral hydrate
0.5 g hematoxylin	

Dissolve the gum arabic in the formic acid and water. With a mortar and pestle, grind the hematoxylin, ferric alum, and chrome alum to a fine powder. Dissolve the powder in the gum arabic solution, and keep in an embedding oven at 60°C for 24 h. Cool to room temperature, add the Bismarck brown, and mix. Centrifuge at 20,000 rpm for 45 min at room temperature or 25°C to remove precipitates. Decant the liquid from the centrifuge tube, and add the glycerin and chloral hydrate while stirring continuously. Do *not* shake. The solution should be a deep chestnut brown. At use, to each 10 ml of solution add enough methyl green, while mixing slowly, to give the stain solution a neutral tint (check with a neutral density filter). The mountant will keep in well-stoppered bottles for several months. Prepare small quantities as needed. Nuclei are clearly differentiated in all but apical cells. Stained material is reddish purple at first but changes to a dense blue black within several months. Thin slides can be kept for 2 years or longer.

6. Lactic acid mounting medium

Use by itself as a mounting medium, or add cotton blue or trypan blue to give it some color.

7. Lactophenol mixtures (Science 70:430; Stain Technol. 11:9)

Lactophenol is a widely used mounting medium because its refractive index is close to that of the hyphae and spores of hyaline fungi. The medium remains liquid.

20 g phenol crystals	20 g lactic acid
40 g glycerin (sp. gr. 1.25)	(sp. gr. 1.21)
	20 ml distilled water

Gently warm the phenol crystals in water until dissolved, and then add the lactic acid (16 ml) and glycerin (31 ml). Add 0.01–0.1% dye (cotton blue, trypan blue [aniline blue], or acid fuchsin), if desired. Warming the preparation before adding the coverslip speeds the staining. Lactophenol is a good, general mountant for permanent preparations. It can be sealed with several coats of colorless fingernail polish if there is no seepage under the coverslip.

8. Necol mounting medium

Mix four parts of acetone with one part of diacetone alcohol (containing 1% benzyl abietate and 1% triacetin). Add cellulose acetate to this solution until it flows like glycerin. A small drop of the fluid is placed on a fungal colony growing on a substrate and allowed to dry for about 15–30 min. It forms a thin, colorless film in which the fungus is embedded. When dry, carefully remove the specimen with a scalpel and mount in pure glycerin.

9. Polyvinyl alcohol mounting medium (Bull. Br. Mycol. Soc. 13:31)

1.66 g polyvinyl alcohol	10 ml lactic acid
10.0 ml distilled water	1 ml glycerin

Dissolve the polyvinyl alcohol in distilled water. While stirring vigorously, add lactic acid and then glycerin. If necessary, filter and let stand for 24 h. The mountant can be used directly for specimens stained in lactophenol–cotton blue. After examining the preparation, warm for 10 min at 40°C to harden for observation in oil immersion. For complete hardening, hold for 24–36 h at 40°C.

10. Wallerite mounting medium (Dade and Waller, 1949)

10 g clear gelatin	28 g phenol crystals
28 g glacial acetic acid	

Dissolve the phenol crystals in the acetic acid *without* heat. Add gelatin. Let stand for several days until crystals are completely dissolved, and then mix in 10 drops of glycerin and incorporate thoroughly by stirring. Store in dark bottles. When the mountant is used, it may be thinned with glacial acetic acid. The medium hardens permanently about 24 h after application. Very wet preparations should be dehydrated in acetic acid for a few minutes and then drained.

11. Wittmann's mounting medium
(Pflanzenschutzberichte 41:91)

For direct mounting, fixing, and staining of fungal preparations.

30 g chloral hydrate	0.03 g aniline blue
20 ml lactic acid (90%)	0.02 g chlorazole black E
5 ml absolute ethanol	

12. Erythrosin staining mountant

Add 1 g of erythrosin to 100 ml of 10% ammonia. Mix equal parts of this solution with Gurr's water mounting medium (see 14 below). Erythrosin can be used for temporary mounts. It gives a very clear outline of the septation and structure of the sporogenous cells in hyaline fungi.

13. Erythrosin and ammonium hydroxide mountant

Add 10% NH$_4$OH to 3% erythrosin. This is a good temporary mountant for staining fungal spore appendages and mucilaginous sheaths of members of Coelomycetes and gives good differentiation between cell walls and cytoplasm.

14. Lacto-fuchsin mounting medium and stain
(Mycologia 47:611)

A mounting medium and stain superior to cotton blue. Staining is rapid, with cell walls standing out clearly. Dissolve 0.1 g of acid fuchsin in 100 ml of water-free lactic acid. It gives brilliantly stained specimens. When mixed half and half with Gurr's water mounting medium, it will solidify. The mixture keeps for months in a tightly stoppered dropper bottle. Seal with fingernail polish or Glyceel.

Note. Glyceel is available from Hopkins & Williams, P.O. Box 1, Romford, Essex, United Kingdom. Gurr's water mounting medium is available from Edward Gurr, Ltd., 42 Upper Richmond Road West, London, S.W. 14, U.K.

15. Lactophenol gum mounting medium
(Bot. Gaz. 77:343)

This mixture, used cold, is a good sealant for lactophenol-dye mounts but is not suitable for lactic acid–dye mounts. It dries quickly but tends to crack with age.

Dissolve 38 g of pure gum arabic in 50 ml of freshly distilled water, and then add 5 g of glucose and 6 ml of lactophenol. Filter through glass wool.

16. Maneval's acid fuchsin–aniline blue mounting medium and stain (Stain Technol. 11:9)

30 ml phenol, 5% aqueous	8 ml glacial acetic acid, 20%
4 ml ferric chloride (FeCl$_3$), 3% aqueous	

To the phenol, add 1–2 ml of 1% aqueous acid fuchsin or aniline blue and the glacial acetic acid. After 30 s to 3 min,

drain off the excess liquid, and then flood with tap water for 3–10 s. Add FeCl$_3$ to the mixture, if desired. Specimens can be mounted, stained, and immediately examined without fixing in lactophenol. If overstaining occurs, replace the phenol-dye mixture with lactophenol by carefully adding it to one edge of the coverslip and withdrawing the overstain at the opposite edge with a piece of blotter paper or paper towel.

17. Orseillin BB and crystal violet stain
(Can. J. Bot. 53:1338)

This technique is used for fungi in paraffin sections and water mounts. Treat unfixed fresh or dried fungal material, soaked in water, in a few drops of 50% acetic acid on a microscope slide. Allow the acid to evaporate, causing the fungus to flatten and adhere to the slide. Add a few drops of 3% acetic acid, saturated with filtered orseillin BB, to the specimen. After 10–30 min, gently rinse the slide with distilled water, and add a few drops of 1% aqueous crystal violet. After 1–2 s, rinse again with water. Do *not* overstain with orseillin BB because it is difficult to destain. Fungal structures and host cells stained include cell walls, cytoplasm, and to some extent, nuclei.

Orseillin BB is also a bacterial stain. The slides are made permanent by using Gurr's water mounting medium.

18. Safranin O stain (Mycologia 71:873)

A useful, rapid, nucleolar and condensed chromatin stain for higher fungi. It stains unfixed specimens and gives information on nuclear numbers and their positions in assimilative cells.

At the time of use, mix 79 ml of distilled water, 6 ml of 0.5% (w/v) aqueous safranin O, 10 ml of 3% (w/v) aqueous KOH, and 5 ml of glycerin.

Place one drop each of safranin O and 3% KOH on a slide and mix. Add the living fungal specimen, and place a coverslip over it. The central area of the fungus commonly remains unstained. Between the central part and the margin, however, the nucleoli and condensed chromatin commonly are stained. Usually the nucleoli of interphase nuclei are brightly stained, while the remainder of the nuclei around this structure appear as a hyaline zone. The cell walls and cytoplasm generally are lightly stained. The solutions keep more than 6 months, but stained preparations are useful for only a few weeks.

19. Thionin–orange G stain (Ann. Appl. Biol. 17:162)

Stain in 0.1 g of thionin and 100 ml of 5% phenol in distilled water for 1 h. Differentiate in a saturated solution of orange G in absolute alcohol, usually for 30–60 s. Mount in balsam. This is a useful stain for both observing fungal structures and staining bacteria that are large slime producers. Both freehand and paraffin sections can be stained (see below).

The orange stain can be removed by regrading the series to water and then running up again to xylol. The parasite is then very conspicuous against unstained walls.

20. Trypan blue stain (Stain Technol. 31:115)

Dissolve 0.1–0.5 g of trypan blue (aniline blue) in 100 ml of 45% acetic acid. When the desired hue is attained, replace the stain with pure lactophenol. Thin-walled specimens are rapidly stained an indigo blue. Heating usually hastens the process for thick-walled fungi. Permanent preparations may be made in Wallerite (see 10 above) or Gurr's water mounting medium.

To stain nuclei of unfixed *Rhizoctonia solani* and related fungi, place a drop of 0.5% lactophenol–trypan blue on the mycelium between the center and margin of the colony. Place a coverslip over the stained area, and examine the specimen by placing the culture plate on the microscope stage or by cutting out the agar around the coverslip and placing it on a slide. The nuclei in cells at the outer edge of the stained area appear dark blue while the protoplasm is light blue or pinkish purple. The septa also stain blue.

21. Malachite green–acid fuchsin stain (Stain Technol. 55:13)

For staining fungi (including yeasts), bacteria, and pollens. The prepared stain is stored in dark bottles for 8–10 days before use.

Add the following ingredients in the order given, shaking after each addition:

> 20 ml 95% ethanol
> 20 mg malachite green (2 ml of a 1% solution
>   in 95% ethanol)
> 50 ml distilled water
> 40 ml glycerol
> 100 mg acid fuchsin (10 ml of 1% aqueous solution)
> 5 g phenol
> lactic acid
>   6 ml for bacteria
>   2–3 ml for most fungi
>   1 ml for yeasts

To destain, prepare a solution containing 50 ml of distilled water, 35 ml of glycerol, and 15 ml of lactic acid. Place fungal material (spores and hyphae) on a microscope slide in one or two drops of the stain. Stir with a needle, and warm the slide *gently* over a flame until it fumes. Cool. Hyphae and spores stain red. Use the same procedure to stain fungi in host tissues. After the preparation is stored for 24 h, fungi stain bluish purple and host tissues stain green.

Stain bacteria (and yeasts) by smearing a sample on a slide and heating it gently to evaporate the water. Then follow the same procedures as outlined above. Bacteria stain purple with this stain and yeasts red.

22. Vital staining of fungi with fluorescein diacetate (Soil Biol. Biochem. 9:59 and 18:315; Mycologia 76:34)

This preparation stains and fluoresces only living parts of fungal mycelium and viable spores. Store a stock solution of fluorescein diacetate (FD), 2–5 mg/10 ml of acetone, at −20°C until use. Then dilute the solution to 2–10 µg/ml of phosphate buffer (pH 7.5) or distilled water. Mix equal volumes of the solution and the fungus (as mycelium or spores) to be stained; leave 3 min for thin-walled cells to several hours for thick-walled oogonia, oospores, or resting spores.

Double staining with FD and ethidium bromide (EB) is sometimes used for certain fungi. Prepare the stain solution by mixing equal volumes of diluted stock solutions of FD and EB, 50 µg/ml, in phosphate buffer. After staining for a few minutes, examine the specimen under a fluorescent microscope.

Fluorescent microscopy (Phytopathology 74:1353; Mycologia 76:1049) with an epi-illumination fluorescent microscope and ultraviolet light (390–400 nm) distinguishes live from dead cells without staining. Dead cells fluoresce naturally. Only the cytoplasm of spores and mycelial cells fluoresce. The technique is accurate, fast, and efficient. It is not changed by dormancy or conditions during incubation. The method can also be used for studying host-parasite interactions because only the dead host cells (containing cytoplasm) fluoresce.

II. Stains for Freehand Sections

1. Ruthenium red stain

Place nonfixed specimen sections in a culture plate containing a 0.01% aqueous solution of ruthenium red. View under a microscope to observe the staining progress. When the desired staining has occurred, remove the sections, wash 1–2 min in distilled water, and mount in glycerin.

2. Sudan IV stain (Phytopathology 51:382)

Wash fixed tissue specimen sections with water to remove fixative, and then immerse them in a saturated solution of Sudan IV in 70% ethanol. Wait 30 min, and wash the sections in 50% ethanol to get the desired intensity. Finally, mount in glycerin.

3. Acid fuchsin–cotton blue stain

Wash fixed tissue specimen sections with water, and then cover the sections on a microscope slide with 0.5% lacto-acid fuchsin. Heat *gently* over a flame for several minutes until the liquid steams, but do *not* boil. Drain off excess stain, and place a drop of 0.25% cotton blue in lactophenol over each section. Heat until the liquid is steaming but not boiling. Wash off excess stain with water, and mount sections in glycerin.

4. Basic fuchsin–tannic acid stain
(Stain Technol. 40:253)

To 100 ml of distilled water, add 10 g of aqueous tannic acid and 1 g of basic fuchsin. Mix well, and centrifuge at 2,000 rpm or more for 15–45 min to remove the precipitates. Pour off and use the supernatant. Stain fungi in a drop of the solution. Add a drop of 50% glycerin, apply a coverslip, and blot dry. Fingernail polish may be used as a sealant.

5. Giemsa stain (Can. J. Bot. 34:321 and 39:1497;
Proc. Linn. Soc. N.S.W. 78:122)

This is a good stain for fungal nuclei and is often used for differentiating bacteria in host tissue. Different brands are available as powders or in prepared stock solutions that vary in the proportion of dyes present. In general, all give good results.

To prepare a stock staining solution from powder, triturate 3.8 g of Giemsa powder with 250 ml of pure glycerin and 250 ml of absolute methanol. Dilute before use by adding 15 ml of phosphate buffer (pH 6.9) for every 1 ml of stain.

STAINING PROCEDURE

1. Fix spores and mycelia for 10–12 min in a 3:1 mixture of absolute ethanol and glacial acetic acid or a 3:1:1 mixture of absolute ethanol, glacial acetic acid, and lactic acid.
2. Rinse in 95% ethanol.
3. Transfer to 70% ethanol, where it can be stored for several days before staining.
4. Rinse in distilled water.
5. Immerse in 1 M HCl for 5 min at 25°C.
6. Remove the material, and immerse in 1 M HCl at 60°C for 7 min. To be successful, the temperature and duration of the HCl treatment must be varied depending on the material.
7. Rinse in distilled water and then in phosphate buffer (pH 6.9), five changes of each.
8. Place in the staining solution for 2 h (dilute the stock solution by adding 15 ml of phosphate buffer to every 1 ml of stain).
9. After staining, rinse in phosphate buffer and then in distilled water.
10. Air dry, and mount in euparal.

The nuclei stain rose to pink while the cytoplasm generally remains clear or light blue.

For special fungal nuclei stains, see Dhingra and Sinclair, 1995, and Tuite, 1969.

III. Stains for Paraffin Sections

Dewax paraffin sections affixed to microscope slides by immersing the slides in xylene for 5 min. Repeat by immersion in fresh xylene. Hydration is done in a graded series of ethanol (EtOH). Staining of sections is usually done in staining jars (Coplin jars are preferred because they hold the slides in a vertical position) filled with the reagent in a series. The slides are transferred from one jar to the next.

Various stain schedules are available for differentiating pathogens in host tissue. Only a few are described here. For a more complete listing, see Dhingra and Sinclair, 1995.

1. Pianeze IIIb stain (Ann. Mo. Bot. Gard. 1:241)

Fungal mycelium and host tissue can be differentiated in both lignified and nonlignified tissue. Host tissue stains green and fungal mycelium deep pink.

0.50 g malachite green	150 ml water
0.10 g acid fuchsin	50 ml ethanol (95%)
0.01 g martius gelb	

Hydrate sections and stain for 15–45 min. Remove excess stain with water, and decolorize in acidified 95% ethanol (EtOH). For permanent mounts, clear the sections with carbol-turpentine. Remove the clearing agent with xylene, and mount in balsam. For better differentiation, stain the sections for 45 min, place a few drops of 100% EtOH on each section, and restain for 1–2 min. Decolorize in acidified EtOH for 30–60 s, clear the sections in clove oil, wash with xylene to remove excess clove oil, and mount in balsam.

2. Magdala red–licht grün (light green) stain (Science 52:63)

Mycelium, fungal spores, and bacterial cells are stained brilliant red while host tissues remain green. If tissues are resistant to staining, first place sections for 2–5 min in a freshly prepared 1% KMnO₄ solution and then wash with water.

Bring the sections to 85% ethanol, and stain for 5–10 min in magdala red (20 g in 100 ml of 85% EtOH). Remove excess stain by washing with 95% EtOH. Stain for 1–3 min in light green (2% in clove oil plus a few drops of EtOH), wash with 100% ethanol or carbol-turpentine, and mount in balsam. Determine the staining time by microscopically examining the sections during staining. Light green staining is rapid. If overstaining occurs, red becomes tinged with purple.

3. Margolena stain (Stain Technol. 7:25)

Margolena stain turns cutin and spore walls yellow, cellulose yellowish pink, lignified elements green, middle lamellae reddish, and hyphae and spores purple.

Dewax and hydrate sections. Stain for 1 h in thionin (0.1 g in 100 ml of 5% aqueous phenol). Rinse with water, and stain with light green (0.5 g in 100 ml of 95% EtOH) until the sections are green. Wash with water to remove all the green color except from the xylem. Pass the sections to 100% EtOH, and then stain with an orange G–erythrosin mixture (prepare separately a saturated solution of orange G in 100% EtOH and erythrosin in clove oil: mix one part of orange G to two parts of erythrosin). Clear the sections in xylene, and mount in balsam.

4. Methylene blue–erythrosin stain (Phytopathology 7:389)

Fungal cell protoplasm is stained deep blue and host cell protoplasm light blue. Cellulose cell walls and hyphae stain deep pink to red. Fungal cell walls in lignified elements stain pink while lignified cell walls are blue.

Dewax sections, and bring to 100% EtOH. Stain for 3–30 min in methylene blue (a filtered, saturated solution in 100% EtOH). Rinse with water to remove excess stain. Then rapidly dehydrate in 100% EtOH, and quickly remove the sections. Flood the sections with erythrosin or eosin (a filtered, saturated solution in clove oil) and observe under a microscope for 1 to about 10 min for loss of the blue stain and uptake of red stain until the desired contrast is obtained. Incline the slides, and wash with a 1:1 mixture of 100% EtOH and xylene, which removes excess clove oil and prevents the formation of red precipitates. Clear sections in xylene and mount. Methylene blue sometimes washes out of fungal protoplasm. If this happens, fix the sections for 30 min in a 10% aqueous solution of tannic acid *before* staining with methylene blue. Wash with water, dehydrate to 100% EtOH, and stain.

5. Periodic acid–Schiff reagent (New Phytol. 54:277; Can. J. Bot. 44:1106)

For staining and differentiating a number of fungi (e.g., rusts and wood-decay fungi) in tissues of higher plants. Fungal and cellulosic plant materials stain magenta, and lignified material is unaffected. *Peronospora* spp. stain with difficulty.

STAIN SOLUTIONS

1. 1% aqueous periodic acid solution.
2. Schiff's reagent. Pour 100 ml of boiling water over 0.5 g of basic fuchsin or pararosaniline to dissolve it. Cool to 50°C, filter, and add 10 ml of 1 N HCl plus 0.5 g of anhydrous potassium metabisulfite ($K_2S_2O_5$). Store overnight or longer to clear. The solution should be colorless or straw colored and will be usable for 6 months if kept in the dark in a tightly stoppered, full bottle in a refrigerator.
3. Potassium metabisulfite solution. Add 5 ml of 10% aqueous $K_2S_2O_5$ and 5 ml of 1 N HCl to 90 ml of distilled water. Store in tightly stoppered bottles. The solution is usable as long as it smells fairly strongly of sulfur dioxide.
4. Light green SF. 0.3% light green in 95% EtOH.

PROCEDURE

Embed, section, dewax, and hydrate the sections. Immerse in the periodic acid solution for 2–5 min (if plant tissues are to be counterstained, do not exceed 3 min). Wash the sections in running tap water for 10 min, and immerse in Schiff's reagent for 5–15 min. (The sections remain colorless or light pink). Then, without washing, transfer the slides to the $K_2S_2O_5$ solution. Make at least two changes within 10 min. Wash the sections in running tap water for 10 min, dehydrate as usual, clear, and mount. If a counterstain is required, stain for a few seconds in the light green SF solution. Clear the sections in clove oil, rinse with xylene, and mount in balsam.

6. Safranin O and fast green stain (Sass, 1958; Proc. Okla. Acad. Sci. 17:69)

A common stain used on many types of plant material.

1. Dissolve 2 g of safranin O in 100 ml of methyl Cellosolve, and then, in order, add 50 ml of 95% EtOH, 50 ml of distilled water, 2 g of sodium acetate, and 4 ml of formalin.
2. Dissolve 0.5 g of picric acid crystals in 100 ml of methanol.
3. Add four or five drops of NH_4OH to 100 ml of 95% EtOH.
4. Fast green FCF. Mix equal amounts of methyl Cellosolve and 100% EtOH, and then saturate the mixture with fast green FCF. Add enough of this solution to a 1:3 mixture of 100% EtOH and clove oil to give the desired stain. Keep in dropper bottles because it can be used again. Aniline blue may be substituted for fast green FCF. Prepare in a solution of 1:1 methyl Cellosolve and 100% EtOH before diluting with clove oil.

PROCEDURE

Dewax sections. Bring freehand sections to 35% EtOH, others to 70% EtOH. Stain in safranin O solution for 2–24 h. Remove excess stain by washing in running tap water for a few seconds. Dehydrate for 10 s or longer in the picric acid–EtOH solution. Stop the action of the picric acid by immersion in the ammoniated EtOH for 2 min or less. Wash with 95% EtOH, and counterstain in fast green FCF for no more than 15 s. Pour the stain back into the bottle, and rinse the sections with used clove oil. Clear in a mixture of two parts of clean clove oil, one part of 100% EtOH, and one part of xylene. Wash in xylene for a few seconds, and then mount in balsam. If aniline blue is used, increase the staining time to about 60 s and clear in methyl salicylate.

7. Flemming's triple stain (Sass, 1958)

Fix tissues in 25 ml of 1% chromic acid, 10 ml of 1% acetic acid, and 60 ml of distilled water. Before use on tissues, add 5 ml of 2% osmic acid.

Dewax sections and bring them to 70% EtOH. If sections are fixed, bring to 35% EtOH. Stain in safranin O solution (see safranin–fast green schedule above) for 2–6 h, depending on the specimen. Rinse three times in tap water, and stain in 1% aqueous crystal (gentian) violet solution for 10–60 min. Rinse again in water; and immerse in 50%

EtOH, twice in 95% EtOH, and then three or four times in 100% EtOH. Stain with orange G (saturated solution in clove oil) for no more than 10 s. Wash off excess stain with used clove oil, and then immerse the slides in new clove oil. After a few seconds, examine the slides under a microscope. When satisfied with the violet color, rinse the sections with clove oil–xylene, wash in three changes of xylene, and mount them in balsam.

8. Harris's hematoxylin stain (Sass, 1958)

Stains fungal structures and differentiates bacteria in host tissues. First, dewax and hydrate paraffin sections; wash freehand sections with water to remove all fixing fluid.

In 1 liter of 50% EtOH, dissolve 2 g of hematoxylin crystals, 1 g of aluminum chloride, and 6 g of red mercuric oxide powder. Boil the mixture for 30 min. Filter and complete the volume to 1 liter with 50% EtOH. Add 10 drops of HCl to acidify the solution.

Dewax and hydrate sections, stain for 30 min in the stain solution, and then rinse with distilled water to remove all excess stain. Destain for 5 s in water acidified by adding about five drops of concentrated HCl in 100 ml. Stop the action of the acid by rinsing sections in water made alkaline (if needed) by adding a few drops of NH_4OH. The sections will stain blue. Again rinse with water, and dehydrate as usual. If desired, counterstain the sections with orange G or erythrosin.

9. Iron-hematoxylin stain (Sass, 1958)

For staining fungal structures, including nuclear phenomena, and differentiating bacteria in host tissues.

Dewax sections in xylene, and pass through a 1:1 mixture of xylene and 100% EtOH for at least 10 min. Place sections in a celloidin solution (1 g of celloidin, 50 ml of 100% ethanol, and 50 ml of ether). Remove the slides, and suspend them in air until the sections are opaque but not dry. Soak them for 5 min in 70% EtOH and then for 2 min in 35% EtOH. Wash thoroughly in water. Place the slides in fixing solution (30% aqueous solution of ferric ammonium sulfate or iron alum prepared fresh each time; the crystals of iron alum are violet—do *not* use crystals that have changed color). Thin sections should be in the fixing solution for 1 h, thick sections for up to 2 h. Wash in running tap water for 5 min, and then rinse in distilled water. Stain for 1–24 h in hematoxylin (0.5% aqueous solution; prepare a 10% stock solution in 100% EtOH, and then dilute with distilled water). Wash off excess stain with water. Destain with large volumes of 2% ferric ammonium sulfate or ferric chloride while viewing in a microscope. Make observations quickly. Immediately return the sections to the destaining solution. When destaining is as desired, quickly put the sections under water. Wash at least 1 h in running tap water, and then dehydrate 5 min each in 50, 70, and 95% EtOH. Counterstain, if desired, in orange G or fast green FCF for 1 min. Wash sections in clove oil diluted 1:1 with 100% EtOH. Clear in a 1:1:1 mixture of clove oil, 100% EtOH, and xylene. Finally, wash in pure xylene for 10 min, and mount in balsam.

If not counterstaining, place sections, after dehydration, in a 1:1 mixture of EtOH and xylene. Make two changes, 5 min each, in pure xylene before mounting in balsam.

For more stains useful in staining paraffin sections, see Dhingra and Sinclair, 1995, and Sass, 1958.

Staining Bacteria for Microscopic Examination

I. Gram's Stain
(Annette et al., 1985; Dhingra and Sinclair, 1995; Gerhardt, 1981; Johnston and Booth, 1983)

This basic and essential stain, with its several modifications, divides most species of bacteria into two broad groups: gram positive and gram negative. The technique involves making a fairly turbid suspension of bacteria, from an 18- to 24-h agar culture, in sterile water. Smear a loopful of the suspension on a clean glass slide (the glass slide should first be washed in acid and be grease free). After the bacterial smear has dried in air, fix it by passing the underside of the slide rapidly several times through a flame. Then flood the film of bacterial cells with a basic pararosaniline dye and then with iodine. Gram-positive bacteria retain the primary dye and stain blue to dark violet when treated with a decolorizing agent. Gram-negative bacteria take up the safranin O counterstain and are pink to red.

Only a few bacteria cannot be identified as gram positive or gram negative. Red bacterial cells within a violet-stained smear are usually dead. Bacterial spores do not stain.

HELPFUL SUGGESTIONS

1. Use reagents that are less than 12 months old. The iodine solution must be kept in a brown bottle in the dark until use.
2. The bacteria must be from a young (18- to 24-h-old) colony.
3. For controls, use known gram-positive and gram-negative species.
4. Avoid clumps of bacteria when preparing smears.

A. Solutions

1. Hucker's crystal violet–ammonium oxalate

1. Dissolve 2 g of crystal violet (Difco, certified) in 20 ml of 95% ethanol.
2. Dissolve 0.8 g of ammonium oxalate (Fisher Scientific) in 80 ml of distilled water.
3. Mix solutions 1 and 2, and store for 24 h before use. (The dye solution needs to be filtered through filter paper into the storage bottle.)

2. Lugol's iodine

Dissolve 2 g of potassium iodide (KI) in about 25 ml of distilled water, and then add 1 g of iodine. Stir in a closed container until dissolved or leave overnight in the dark, and then add the remaining water (275 ml).

Note. Grinding the solids together first with a mortar and pestle and adding water slowly will speed the dissolution. The solution will keep for a month or two in a dark bottle. If bacterial cells known to be gram positive do not retain the violet color, the iodine is usually the problem.

3. Decolorizing agents

Use 95% ethanol (slowest), acetone (fastest), or a 50:50 mixture of acetone and alcohol (intermediate).

4. Safranin O counterstain

Prepare a stock solution by dissolving 2.5 g of safranin O (Fisher, certified) in 100 ml of 95% ethanol. Dilute this solution 1:10 with distilled water for use.

B. Staining Procedure

1. Flood the fixed bacterial film with 2% aqueous crystal violet (or methyl violet 6B) and 1% ammonium oxalate in distilled water for 1 min.
2. Pour off the stain, and wash with a gentle stream of tap water for a few seconds. Drain off the excess water, and blot dry with a paper towel.
3. Flood the stained bacterial film with Lugol's iodine solution for 30–60 s.
4. Wash in tap water again for a few seconds, and blot dry.
5. Decolorize by washing in a gentle stream of 95% ethanol for 20–30 s or until no more stain comes away; then blot dry.
6. Rinse in tap water for about 2 s.
7. Counterstain by flooding with safranin O solution for 10–60 s.
8. Wash briefly in tap water, gently blot dry, and examine under oil immersion.

C. Potassium Hydroxide (KOH) Solubility Test

A few otherwise gram-positive bacteria (mainly *Bacillus* spp.) may give a gram-negative or variable reaction when stained. The gram reaction can be determined by a 3% aqueous KOH solubility test.

Make a loopful suspension from an older bacterial colony in a drop of KOH on a glass slide. If strands of viscid material are seen when the loop is raised 2–3 cm, then the bacterium is gram negative. If such strands are not evident on repeated raises of the loop, the bacterium is considered gram positive.

II. Staining Flagella

Determining the number and arrangement of flagella on bacterial cells is important taxonomically. Bacterial species with known flagellar arrangement (e.g., *Erwinia carotovora* with peritrichous flagella or *Pseudomonas marginalis* with polar flagella) should be included for comparison in every determination. To be successful, the cultures should be young (24–48 h old) and ideally in the log or exponential phase, usually grown at 27°C. The slides need to be extremely clean (use alcohol). The microscope condenser must be in focus and centered.

A. Silver Impregnation Stain (J. Bacteriol. 89:899)

Bacteria and flagella stain a dark brown to black when viewed on a light or golden background.

PREPARATION AND PRECAUTIONS

1. Prepare a faintly turbid suspension of bacteria grown on yeast extract–dextrose–calcium carbonate (YDC) agar or other standard medium at 27°C for 24–48 h. Carefully add 1 ml of sterile distilled water to the base of the slant, being careful not to shake the slant.
2. For a control, use known species that produce flagella.
3. Place a loopful of distilled water on an alcohol-cleaned slide.
4. Place a loopful of the bacterial suspension just touching the loopful of distilled water on a slide so that the two diffuse together.
5. Air dry the slides.

REAGENT A. FERRIC TANNATE MORDANT

5.0 g tannic acid	2 ml formalin, 15%
1.5 g ferric chloride	1 ml NaOH, 1%

Mix the ingredients, and bring the volume to 100 ml with distilled water.

REAGENT B

Note. Ammoniated silver nitrate ($AgNO_3$) must be used within 4 h after preparation and kept in dark bottles.

Dissolve 2% $AgNO_3$ in 100 ml of distilled water. Remove and save 10 ml of the $AgNO_3$ solution, and then add ammonium hydroxide (NH_4OH) dropwise to the remaining 90 ml until a heavy, brown precipitate is formed. Keep adding NH_4OH until the precipitate dissolves. From 10 ml of the saved 2% $AgNO_3$ solution, back titrate dropwise until the solution is faintly cloudy and persists even after shaking. Adjust the pH to 10.0 with NH_4OH and $AgNO_3$.

PROCEDURE

1. Cover the bacterial film with reagent A for 2–4 min, and then gently rinse with glass-distilled water.
2. Add a few drops of reagent B (pH 10.0) for about 30 s. Immediately wash the smear with distilled water.
3. Air dry the slides, and examine under oil immersion directly or under a coverslip sealed with Canada balsam. Silver-plated preparations break down when exposed to air after about a week.

B. Rhodes Silver-Plating Stain (J. Gen. Microbiol. 18:639)

Prepare a ferric tannate mordant by adding, in order, 10 ml of 10% tannic acid (w/v), 5 ml of a saturated solution of potassium alum, and 1 ml of a saturated solution of aniline blue. Shake to dissolve the curd, and then add 1 ml of 5% ferric chloride. Let the black solution stand for 10 min before use.

Prepare an ammoniated $AgNO_3$ solution as described above. Flood the bacterial film on a glass slide with the ferric tannate mordant for 3–5 min, and then wash thoroughly with distilled water. Add very hot (almost boiling) ammoniated $AgNO_3$ solution to the bacterial smear. After 3–5 min, wash with a gentle stream of water and examine under oil immersion.

For additional flagella stains, see Schaad, 1988, Dhingra and Sinclair, 1995, and Gerhardt, 1981.

III. Staining Cell Walls (J. Bacteriol. 67:252)

1. Flood an unfixed bacterial smear with 5% tannic acid for 1.5–2 min, and wash with tap water. Stain in 0.5% aqueous crystal violet for an additional 1.5–2 min. Wash again with water, and stain with 0.5% Congo red for 2–3 min or longer, depending on the specimen. Finally, wash, dry, and examine under oil immersion.
2. To identify an organism as a mycoplasma or spiroplas-

ma, use the method developed by J. Worley as outlined by Schaad, 1988. The procedure should be performed by an experienced electron microscopist.

IV. Staining Spores
(J. Bacteriol. 68:389)

A. *Streptomyces* spp. and Related Genera

Stain the bacterial preparation on a glass slide for 2 min with a 2:2:1 mixture of 0.1% Bismarck brown, 0.1% toluidine blue, and a saturated solution of $(NH_4)_2SO_4$. Wash with water, and mount under a microscope. The hyphae stain bright yellow, while the spores are blue. Red brown granules can be seen in the hyphae. A blue stain may be picked up by some nonsporulating aerial hyphae.

B. Malachite Green–Safranin Stain (Gerhardt, 1981)

Bacterial cells stain red, and spores are green.

Suspend a colony of bacteria growing on a standard agar medium in a drop of water on a glass slide, and let stand until air dry. Flood the smear with 5.0% (w/v) aqueous malachite green for 10 min. Wash thoroughly under a gentle stream of tap water, and dry briefly. Counterstain by flooding with 0.5% (w/v) aqueous safranin for 15 s. Rinse thoroughly with water, and blot dry with a paper towel. Observe the cells at 40×.

V. Staining Bacteria in Host Tissues
(Phytopathology 18:803, 52:103, and 56:154; Proc. Am. Soc. Hortic. Sci. 84:661; J. Agric. Res. 44:605; Ann. Appl. Biol. 17:162)

Many of the stain combinations used to differentiate various fungal structures are also useful for differentiating bacteria in host tissues. Several of the better combinations include Harris's hematoxylin in combination with orange G and iron hematoxylin–safranin–aniline blue (Phytopathology 56:154) and thionin–orange G (Ann. Appl. Biol. 17:162). Gram's stain is also commonly used to differentiate bacteria in host tissue.

Both freehand and paraffin sections can be stained with thionin–orange G. There are several modifications of the recipes outlined below.

1. Freehand sections. Wash in water to remove the fixing solution. Stain in thionin for 5 min, and wash again in water. Dehydrate by passing the sections through 35, 70, and 95% EtOH. Stain for several minutes in orange G, wash in 100% EtOH, clear in xylene, and mount in balsam.

   Bacterial cells stain a purple violet, cellulose walls are green to yellow, lignified elements are blue, host nuclei are pale blue, and fungal nuclei turn deep purple.

2. Paraffin sections. Dewax and hydrate sections, and then stain them for 1 h in 0.5% thionin (in 5% aqueous phenol solution) before dehydrating them by passage through a graded series (35, 70, and 95%) of EtOH ending in absolute EtOH. Stain for 30–60 s in a saturated solution of orange G in 100% EtOH. Wash in absolute EtOH and then in a 50:50 mixture of xylene and absolute EtOH. Rinse with pure xylene, and mount in balsam.

Cements for Sealing Mounts

1. Fingernail polish (Dade and Waller, 1949)

For best results, apply to a dry and clean slide. Remove moisture from around the edges of the coverslip because it prevents a good seal. Keep the brush flexible by suspending it in acetone vapors. Two coats of nail polish, well dried, are suggested followed by a thick layer of PVA (see below) diluted 1:1 with water to provide a watertight seal.

2. PVA (Trans. Br. Mycol. Soc. 56:317)

Commercial, water-based PVA (polyvinyl acetate) is a durable reinforcement over two coats of fingernail polish. The sealant is milky white at first but dries to a tough, transparent film.

3. Glyceel (Trans. Br. Mycol. Soc. 58:341)

This cellulose-based sealant is excellent for permanent and semipermanent slides. It is superior because two coats of well-dried fingernail polish become brittle with time and lack mechanical strength. Glyceel can be obtained from Hopkins & Williams, P.O. Box 1, Romford, Essex, U.K.

Cleaning Solutions

I. Glassware

New glassware may give off free alkali. Place it in 1% HCl overnight, wash in tap water, and rinse twice in distilled water.

Place used petri dishes, other culture plates, and test tubes in 3% Lysol or in boiling soapy water for 10–15 min; wash with soap in hot water; and then rinse in hot water, in cold water, and finally in distilled water. Glassware for physiologic studies should be put in a sulfuric acid–potassium dichromate mixture (see 2 below) for 12–24 h and rinsed five or more times in hot running tap water and at least twice in distilled water. Glassware that will not be used for accurate volumetric work can be oven dried.

1. Acid-ethanol

Add 30 ml of 37% HCl to 970 ml of 95% ethanol.

2. Sulfuric acid–potassium dichromate

In a large beaker, dissolve 75 g of $K_2Cr_2O_7$ in 500 ml of warm water. Place the beaker in cold water, and add *very slowly*, while stirring with a glass rod, 500 ml of concentrated H_2SO_4.

Note. Never use this cleaning agent on sintered glass filters.

Glassware should be left in mixture 1 or 2 at least overnight. The mixture is then poured off for reuse, and the glassware is rinsed repeatedly in running tap water and twice in distilled water.

II. ELISA Plates

The 96-well ELISA plates, used mainly in serological assays, are usually discarded after use. They can be reused, however, if cleaned in a mixture of sodium hydroxide (NaOH) and ethanol (EtOH). Dissolve 500–600 g of NaOH in 1 liter of water. Mix equal parts of NaOH and 100% EtOH. Stir the mixture for about 2 h. While stirring, add deionized water until the mixture appears uniform. Discard the normally yellow liquid when it turns brown after repeated use. If precipitates form, remove them by filtration (Plant Dis. 67:18).

Temperature and Humidity Control

I. Cabinets and Growth Chambers

A. Temperature

Growing microorganisms or plants at temperatures above ambient (about 25°C) requires ordinary laboratory incubators (low-temperature ovens) or growth chambers fitted with thermostats that switch on heaters. Operation below ambient requires some sort of refrigeration. Cabinets may be cooled by circulation of refrigerant 1) through a heat-exchange coil in each cabinet or 2) in coils through a reservoir of coolant (brine or water), which is then pumped to the cabinet and circulated through either a heat-exchange coil within it or a surrounding water jacket.

The first method leads to excessive icing and interference with humidity control, making the second, or indirect, method necessary. Indirect cooling is advantageous because the secondary coolant can be circulated through each cabinet at any desired temperature within the operating limits of the cabinets and is usually above 0°C. This favors accurate temperature control plus avoids the formation of ice on the coils.

Temperature control in cooled incubators or growth chambers, often set up in a series with a single coolant flow line, usually depends on running the coolant continuously, balanced by heaters that are switched on and off (within a tolerance of ±0.5–1°C) by using contact thermometers, capillary bulb sensors, or good-quality, bimetallic strip thermostats. Humidity control requires very accurate temperature control, normally with electronic temperature sensors.

Cabinets and growth chambers must be designed to avoid temperature gradients or layering effects by using internal forced-air circulation.

B. Humidity

Accurate humidity in incubators and growth chambers depends on indirect cooling with coil temperatures above freezing and very accurate temperature control. Water vapor is introduced as steam or through an atomizer. An effective sensor system is needed to control heat input to the element producing steam or to switch the pump operating the atomizer. A number of control devices are available for controlling relative humidity (RH), which depend on 1) dew point control operating through a closed air circuit, thus avoiding use of a wet bulb or RH element, or 2) control through differential setting of wet and dry bulb contact thermometers. The latter system is simpler, less accurate, and more suitable for small cabinets. When very precise control of RH is required, the best choice is the first method.

There is a wide variety of incubator cabinets and growth chambers available that include recording equipment and are designed for programmable cycling operation so that temperature and/or humidity levels can be accurately controlled and preset at timed intervals. It is important that the cabinets or chambers have adequate heating and cooling devices that can produce changes in temperature or humidity within a relatively short period of time (temperature rise and fall rates should not be less than 5°C per hour). An instrument with thermostat setting arms, temperature indicator, and trace recorder (to keep track of trends in temperature and humidity) is essential for all but the simplest incubator cabinets. When operating conditions are especially critical or the plant or microorganism material is very valuable, installation of a visible or preferably an audible device to warn of electrical failure is critical.

II. Small, Airtight Containers

A constant RH must be maintained to determine its effect on fungal growth, sporulation, spore germination, spore longevity, and many other physiological processes. Elaborate equipment is available (see above), but salt or acid solutions work well in maintaining a constant RH in the atmosphere above small, clear, moisture-proof containers. Saturated salt solutions, sulfuric acid, or glycerin solutions in different concentrations can maintain the RH in the air space above them (Dhingra and Sinclair, 1995; Winston and Bates, 1960; Johnston and Booth, 1983). The most serious disadvantage in using sulfuric acid is the possible presence of SO_3 in the overlying atmosphere.

PRECAUTIONS

1. The container must be airtight.
2. Containers with more than a liter of air space above the solution require an air-circulating device.
3. For maximum differentiation, the surface of the solution should be in contact with as much air as possible.

Sulfuric Acid Mixtures

H_2SO_4 (%)	RH at 25°C (%)
0	100.0
5	98.5
10	96.1
15	92.9
20	88.5
25	82.9
30	75.6
35	66.8
40	56.8
45	46.8
50	36.8
55	26.8
60	17.2
65	9.8
70	5.2
75	2.3
80	0.8

Aqueous Glycerin Mixtures[a]

Specific gravity	RH at 25°C (%)
1.261	0
1.260	3.7
1.255	17
1.243	27
1.227	39
1.216	47
1.207	51
1.192	60
1.186	62
1.168	70
1.165	71
1.156	74
1.135	80
1.130	81
1.107	86
1.082	90
1.049	95
1.000	100

[a] Use a good industrial grade of glycerin with distilled water. To prevent mold growth, add four drops of a saturated solution of copper sulfate in 100 ml of aqueous glycerin.

Saturated Salt Solutions at 20°C

RH (%)	Salt	RH (%)	Salt
0.0	P_2O_5	70.0	$NaCl + KCl$
5.5	$NaOH$	71.0	$NaNO_3 + KNO_3$
9.0	H_3PO_4	73.0	$NH_4Cl + KNO_3$ (1:1)
10.0	$ZnCl_2 \cdot H_2O$	75.0	K tartrate or $NaClO_3$
15.0	$LiCl \cdot H_2O$	76.0	$NaCl$ or $NaNO_3$
20.0	K acetate ($KC_2H_3O_2$)	78.0	$Na_2S_2O_3 \cdot 5H_2O$
32.5	$CaCl_2 \cdot 6H_2O$	79.5	NH_4Cl
35.5	$NaCNS$	80.5	$(NH_4)_2SO_4$
38.0	NaI	84.0	KBr or $AgNO_3$
39.0	Cr_2O_3	85.0	KCl
42.0	$Zn(NO_3)_2 \cdot 6H_2O$	86.0	$KHSO_4$
44.0	$K_2CO_3 \cdot 2H_2O$	87.0	Resorcinol
44.5	K_2HPO_4	88.0	K_2CrO_4
47.0	$KCNS$	90.0	$MgSO_4 \cdot 7H_2O$
48.5	KNO_2	92.0	Na tartrate or $NaBrO_3$
52.0	$NaHSO_4 \cdot H_2O$	93.0	$NH_4H_2PO_4$
54.5	$Na_2Cr_2O_7 \cdot 2H_2O$	93.5	KNO_3
55.0	Glucose	94.0	$CaH_4(PO_4)_2 \cdot H_2O$
55.5	$Ca(NO_3) \cdot 4H_2O$	95.0	$CaHPO_4 \cdot 2H_2O$
57.5	$MnCl_2 \cdot 4H_2O$	95.5	Pyrocatechol
58.0	$Pb(NO_3)_2 + NH_4NO_3$	96.5	KH_2PO_4
59.0	$NaBr \cdot 2H_2O$	97.0	$Pb(NO_3)_2$
65.0	Ag acetate	98.0	$CaSO_4 \cdot 5H_2O$
65.5	NH_4NO_3	98.0	K_2SO_4
67.0	$CoCl_2$	98.0	$K_2Cr_2O_7$

Saturated Salt Solutions at 25°C

RH (%)	Salt	RH (%)	Salt
0.0	P_2O_5	55.0	Glucose
7.0	$NaOH$ or $LiBr \cdot 2H_2O$	56.0	$MnCl_2 \cdot 4H_2O$
8.0	KOH	57.0	Na acetate + sucrose
8.5	$ZnBr_2$	57.5	$NaBr_2 \cdot 2H_2O$
9.0	H_3PO_4	58.5	$SrBr_2 \cdot 6H_2O$
11.5	$CaI_2 \cdot 6H_2O$	60.0	$FeCl_2 \cdot 4H_2O$
12.0	$LiCl \cdot H_2O$	61.5	$NH_4NO_3 + AgNO_3$
12.5	$LiCNS$	62.5	NH_4NO_3 or $CuBr_2$
13.0	Ca acetate + sucrose	63.0	$NaCl$ + sucrose
16.5	$CaBr_2$	64.0	$NaNO_2$
17.0	Ca acetate	66.0	$K_2S_2O_3$
17.5	$Ca(CNS) \cdot 3H_2O$	67.0	$CuCl_2 \cdot 2H_2O$
18.0	$LiI \cdot 3H_2O$	69.0	$NaCl + Na_2SO_4 \cdot 7H_2O$
20.0	$CaZnCl_2$ or ZnI_2	70.0	Na methane sulfonate
21.5	$KHCO_2$ (formate)	71.0	$SrCl_2 \cdot 6H_2O$
22.5	K acetate	71.5	$NaCl + KCl$
27.0	$NiBr_2 \cdot 3H_2O$ or MgI_2	72.5	Ca methane sulfonate
29.5	$CaCl_2 \cdot 6H_2O$	73.0	Na acetate
30.5	$KF \cdot 2H_2O$	74.0	$NaNO_3$
31.5	$MgBr_2 \cdot 6H_2O$	74.5	$BaBr_2$
32.5	$MgCl_2 \cdot 6H_2O$	75.0	K tartrate or NH_4Br_2
33.0	$SrI_2 \cdot 6H_2O$	75.5	$NaCl$
34.5	$MnBr_2 \cdot 6H_2O$	76.0	Urea
35.0	$Cu(NO_3)_2 \cdot 6H_2O$	78.0	NH_4Cl
37.5	$Ca(MnO_4)_2 \cdot 4H_2O$	80.0	$(NH_4)_2SO_4$ or KBr
38.0	NaI	80.5	$Zn(CNS)_2$
39.0	$FeBr_2 \cdot 6H_2O$	81.0	NaH_2PO_4
41.0	$Mg(ClO_4)_2 \cdot 6H_2O$	82.0	$AgNO_3$
41.5	$CoBr_2$	85.0	Sucrose
42.5	$CrCl_3$	87.0	$Na_2CO_3 \cdot 10H_2O$
43.0	$K_2CO_3 \cdot 2H_2O$	88.5	$ZnSO_4 \cdot 7H_2O$
45.5	$CeCl_3$	89.0	$MgSO_4 \cdot 7H_2O$
46.5	$KCNS$	90.0	$BaCl_2$
47.0	$BaCl_3 \cdot 7H_2O$	91.5	Mg silicofluoride
47.5	$Mg(CNS)_2$	92.0	Na acetate
49.0	$Co(NO_3) \cdot 6H_2O$	92.5	KNO_3
49.5	$K_4P_2O_7 \cdot 3H_2O$	93.0	NH_4HPO_4
50.0	$NH_4NO_3 + NaNO_3$	93.5	Pyrocatechol
50.5	$Ca(NO_3)_2 \cdot 4H_2O$	95.5	$Pb(NO_3)_2$
51.0	KBr + urea	96.0	KH_2PO_4
52.0	Sucrose + urea	97.0	$CaHPO_4 \cdot 2H_2O$
53.0	$Mg(NO_3)_2 \cdot 6H_2O$	97.5	K_2SO_4
54.5	$Ba(CNS)_2 \cdot 2H_2O$	98.0	$K_2Cr_2O_7$ or $KClO_3$

4. Only pure chemicals should be used.
5. The solutions are usable only when free of contaminants.
6. Sudden fluctuations in temperature should be avoided.

Prepare saturated salt solutions by dissolving the salt to saturation in boiling water. Partially cool the solution, and add more salt. After cooling, add more salt to the solution.

Let stand for a few days to 3 weeks to ensure saturation. A substantial excess of the chemical is desirable.

The temperature in the container must be kept constant. Use hygroscopic salts as flowable pastes. Mixtures of salts should be used with caution, and only those with common cations or anions should be used. Prepare the solutions separately before combining them. The ratio of two salts in a solution mixture is not critical as long as both solutions are saturated. A 1:1 ratio is preferred.

For the RHs of saturated salt solutions at 5, 10, 15, 30, 35, 40, and 45°C, see Dhingra and Sinclair, 1995. For further details on salt solutions to maintain a constant RH and on methods of measuring, see Winston and Bates, 1960.

pH Indicators

pH indicator compounds change color with changes in the substances in which they are dissolved. They vary widely in the pH ranges at which the color changes occur (see below). For this reason, care is necessary in choosing indicators for various tasks. The most accurate way of using indicator dyes is by comparing them with 1) standard solutions of known pH, e.g., buffer solutions containing the same concentration of indicator, or 2) standard colored glasses in a color comparator. These methods, when used carefully, are often accurate within 0.1–0.2 pH units. The simplest and most rapid method is to use indicator papers that can be dipped quickly into a test solution. The color obtained is then compared with the color chart that comes with the papers. More accurate measurement may then be made by using the paper with the most narrow range or paper impregnated with an individual indicator.

The color changes of indicators are gradual within a specific pH range. For example, bromothymol blue (see chart) is yellow in acid and blue in alkaline solutions. When an alkali is added gradually to an acid solution, a color change (yellow) begins at 6.0, and the color becomes bluish and darker as the pH increases. At pH 7.6, a dark blue color is reached. The different colors between pH 6.0 and 7.6 can be used to determine the pH. Outside its specific pH range, bromothymol blue or other indicator shows only whether the pH of the solution is below or above the range.

A list of indicator dyes covering a range of pH from 1.2 to 12 is given here. For additional information and other indicators (below 3.0 and above 10.0), see Cruickshank, 1960, and Johnston and Booth, 1983.

Indicator	pH range	Color change
Thymol blue (acid range)	1.2–2.8	Red to yellow
Methyl yellow	2.8–4.0	Red to yellow
Methyl orange	3.0–4.5	Red to yellow-orange
Bromophenol blue	2.8–4.6	Yellow to violet
Bromocresol green	3.6–5.4	Yellow to blue
Methyl red	4.4–6.2	Red to yellow
Chlorophenol red	4.5–7.0	Yellow to red
Litmus	5.0–8.0	Red to blue
Bromocresol purple	5.2–6.8	Yellow to violet
Bromothymol blue	6.0–7.6	Yellow to blue
Phenol red	6.6–8.4	Yellow to red
Neutral red	6.8–8.0	Red to yellow
Cresol red	7.2–8.8	Yellow to violet
Cresol purple	7.6–9.2	Yellow to violet
Thymol blue (alkaline range)	8.0–9.6	Yellow to blue
Phenolphthalein	8.2–10.0	Colorless to red
Thymolphthalein	9.3–10.5	Colorless to blue
Alizarine yellow GG	10.3–12.0	Colorless to yellow
BDH Universal	3.0–11.0	Red, orange, yellow, green, violet, red violet

PRECAUTIONS AND POSSIBLE ERRORS

1. In unbuffered solutions, the acidic or basic character of the indicator dye may cause a shift in the pH up to as much as 1 unit.

2. Protein solutions are difficult at best to measure colorimetrically. Amphoteric proteins combine with acidic and basic indicator dyes and interfere with the color changes.

3. Alkaloids may also interfere with color changes; the amount of interference is usually detected by first experimenting with blanks.

4. The composition of the solvent commonly affects the acid-base equilibrium, with similar changes sometimes occurring in the indicator dye. At room temperature, an error of up to about 0.5 unit may be induced by the presence of alcohol.

5. The presence of salts and other ions can affect the color of the indicator. Usually the effect is small. This can be ignored at salt concentrations below about 0.2 N.

6. Most indicators are designed for use at a room temperature of 20–25°C.

Buffers

The pH may change rapidly in in vitro experiments, when growing microorganisms in culture, or when plant tissues are disturbed. Unless the pH is checked, the experiment may be spoiled or the organism killed. A number of buffers, usually mixtures of a weak acid and a weak base and their salts, that resist changes in pH by controlling the concentration of hydrogen ions are described below. By varying the proportions of the components, the desired pH of a buffer is changed, usually within 1–2 pH units. The buffering power is greatest at a particular proportion, usually equimolar, but drops off quickly as the properties change.

If there is no chemical reaction other than acid-base equilibria, mixing two selected buffers is effective over the pH range of each individual buffer. Details of the preparation of the buffers given below, buffer theory and action, and their use in mycology and bacteriology are given in Cruickshank, 1965, and Munro, 1970. The formulas are based on those given in these publications. Variations occur in the ionic strength of different buffers and the same buffer at different pH levels. The salt concentration needs to be adjusted if an isotonic solution is required. The pH of certain buffers also changes as the temperature changes (Munro, 1970).

Common names of the most useful buffers are acetate (pH 3.6–5.6), citrate (3.0–6.2), citrate-phosphate (2.6–7.0), and phosphate (5.8–8.0). For additional buffers, e.g., boric acid–borax (7.6–9.2), Tris-HCl (7.2–9.0), barbital or veronal (6.8–9.2), succinate (3.8–6.0), glycine-NaOH (8.6–10.6), KOH-maleate (5.2–7.0), and carbonate-bicarbonate (9.2–10.6), see Cruickshank, 1965, Dhingra and Sinclair, 1985, and Munro, 1970.

Acetate buffer

Stock solution: 0.2 M acetic acid (11.55 g/liter) and 0.2 M sodium acetate (16.4 g/liter). Mix x ml of acetic acid to y ml of sodium acetate, and dilute to 1 liter.

pH	Acetic acid (ml)	Sodium acetate (ml)
3.6	463	37
3.8	440	60
4.0	410	90
4.2	368	132
4.4	305	195
4.6	255	245
4.8	200	300
5.0	148	352
5.2	105	395
5.4	88	412
5.6	48	452

Citrate buffer

Stock solution: 0.1 M citric acid (21.01 g/liter) and 0.1 M sodium citrate (29.41 g/liter). Mix x ml of citric acid to y ml of sodium citrate, and dilute to 1 liter.

pH	Citric acid (ml)	Sodium citrate (ml)
3.0	465	35
3.2	437	63
3.4	400	100
3.6	370	130
3.8	350	150
4.0	330	170
4.2	315	185
4.4	280	220
4.6	255	245
4.8	230	270
5.0	205	295
5.2	180	320
5.4	160	340
5.6	137	363
5.8	118	382
6.0	95	415
6.2	72	428

Citrate-phosphate buffer

Stock solution: 0.1 M citric acid (19.21 g/liter) and 0.2 M $Na_2HPO_4 \cdot 7H_2O$ (53.65 g/liter). Mix x ml of citric acid to y ml of phosphate solution, and dilute the mixture to 1 liter.

pH	Citric acid (ml)	Na_2HPO_4 (ml)
2.6	446	54
2.8	422	78
3.0	398	102
3.2	377	123
3.4	359	141
3.6	339	161
3.8	323	177
4.0	307	193
4.2	294	206
4.4	278	222
4.6	267	233
4.8	252	248
5.0	243	257
5.2	233	267
5.4	222	278
5.6	210	290
5.8	197	303
6.0	179	321
6.2	169	331
6.4	154	346
6.6	136	364
6.8	91	409
7.0	65	436

Phosphate buffer

Stock solution: 0.2 M NaH_2PO_4 (27.8 g/liter) and 0.2 M $Na_2HPO_4 \cdot 7H_2O$ (53.65 g/liter). Mix x ml of NaH_2PO_4 solution to y ml of $Na_2HPO_4 \cdot 7H_2O$, and dilute to 1 liter.

pH	NaH_2PO_4 (ml)	$Na_2HPO_4 \cdot 7H_2O$ (ml)
5.8	460.0	40.0
6.0	438.5	61.5
6.2	407.5	92.5
6.4	367.5	132.5
6.6	312.5	187.5
6.8	255.0	245.0
7.0	195.0	305.0
7.2	140.0	360.0
7.4	95.0	405.0
7.6	65.0	435.0
7.8	42.5	457.5
8.0	26.5	473.5

Collection, Preparation, and Mailing of Cultures and Specimens

The following information was adapted from Johnston and Booth, 1983, and Dhingra and Sinclair, 1995.

Occasionally, cultures and specimens of fungi and bacteria need to be sent to, and identified by, an expert or kept for reference. The following points should be kept in mind:

1. Adequate collection material (e.g., leaves, stems, and roots) of the specimens should be provided for proper examination.
2. Substrates with delicate fungal fruiting bodies or other structures should be properly secured inside a suitable container before mailing.
3. A representative part of all collections sent for identification should be retained because specimens cannot normally be returned to the sender.
4. For identification, fungi should be in good condition and sporulating well.
5. Specimens sent for identification should be dried as if they were herbarium specimens of flowering plants. Gently press leaf material flat while drying in a herbarium press. Wash stems and roots free of soil, and remove excess moisture before mailing.
6. Whenever possible, fresh isolates of microorganisms should be sent when growth is evident in the subculture. Avoid sending cultural mutants or older, nonsporulating strains. Some fungal cultures can be induced to sporulate with black light (Can. J. Bot. 39:706, 40:151, and 40:1577; Mycologia 55:151).
7. Reference fungal cultures should be grown on any firm, suitable agar substrate (e.g., potato-dextrose, cornmeal, V8 juice, or oatmeal agars) containing 2.5–3% agar in plastic culture (petri) plates, 1-oz universal containers, small McCartney bottles, or small test tubes (100 × 15 mm). (Do *not* send culture plates, which can be easily damaged or contaminated.) The cultures must arrive unbroken, free from mites, and not overgrown with contaminants. Contaminated cultures are usually destroyed immediately by specialists upon receipt. Always retain for your own use subcultures of the organisms sent for identification because cultures may not be kept alive after identification.
8. When a culture in a culture plate is sufficiently mature, it can be killed by placing formalin-soaked filter paper in the lid of the plate overnight or longer. Dry the culture so the entire colony can be carefully removed from the plate. Place the killed culture over 1.5% melted water agar that has been poured on the smooth side of 12-cm squares of ordinary commercial hardboard. Place the hardboard squares and cultures on raised pieces of cardboard in a dustproof container, and let dry for 2–5 days. Then loosen the culture with a razor blade or scalpel, remove it, and trim. (If the cultures are too dry, place them in a damp chamber for an hour or so.) Fix the dried agar disks to the underside of removable cardboard rings, which fit into flat, cardboard boxes of special design. Both the reverse and front sides of such cultures can be examined at any time.

For tube cultures, place formalin-soaked filter paper in the tube overnight, and carefully place the culture on the melted agar. Slice off the thick part of slope cultures to make them thinner and flatter. Otherwise treat as for culture plates. Dried slope cultures are usually stuck down with gum and placed in slide boxes.

There are several useful modifications of the above technique for preserving dried reference cultures.

a. Agar cultures from culture plates are laid on polyvinyl acetate (epiglass) uncovered for about 18 h and dried at 20–25°C. Cultures are then placed on Perspex acrylic plastic sheets lightly smeared with petroleum jelly. These cultures are more durable than those prepared on melted agar (Trans. Br. Mycol. Soc. 51:603).

b. Colonies in plastic culture plates are killed with formalin vapors, left uncovered for several days to partially dry, and then peeled off. Fifteen milliliters of hot 2.5% glycerin agar is poured off into each plate lid, and the partially dried agar culture is floated on the hot agar. The edges of the culture disk are smoothed as the lower layer of glycerin agar solidifies. The cultures are left uncovered to dry completely. Finally, the film of dried agar is peeled off and mounted in a specimen holder (Mycologia 59:541).

c. A sector of a colony is cut from an agar culture plate, killed with formalin vapors, placed on a glass slide, cut to fit into a 35-mm slide holder, and dried for 1–2 days in a dustproof container. When suitably dry, the slide is inserted into the slide holder. A coverslip, with a thick mounting material such as balsam placed at the four corners to raise the glass slightly, can be applied over the colony. Uncovered mounts can be conveniently placed in 35-mm photographic slide files (Mycologia 45:309).

Freeze-dried ampules are also satisfactory.

9. Dried specimens should be placed in ungummed, quality paper envelopes (about 15 × 10 cm). Larger portions of partially dried stems or roots can be wrapped separately in newspaper.

   Information concerning each specimen or culture should include, at a minimum, a serial or accession number, the scientific name of the host (if known) or substratum, the locality and date of collection, and the collector's name and address.

   Fresh leaf material showing virus symptoms should be placed loosely between cotton-wool pads in a plastic envelope left open at one end for aeration and then into a wooden or strong cardboard box.

   Avoid sending damp material in airtight containers or polyethylene bags. Such specimens are very likely to arrive decayed or overgrown with contaminants.

10. Before mailing specimens or cultures, all tubes, bottles, ampules, etc. should be adequately labeled by using an adhesive label or waterproof ink. Wrap each specimen separately, and pack it with cotton-wool or polystyrene granules in a wooden, metal, or strong cardboard box.

    The wrapped parcel, if it contains a plant-pathogenic organism, will need a special customs declaration and should have an international "Perishable Biological Substance" label. Strict federal and international rules cover the mailing of pathogenic material. Senders are advised to consult a copy of the International Postal Regulations and contact appropriate state and/or federal plant-quarantine officials well before mailing. Except when essential, avoid international air flights, which may cause delay and complicate clearing by customs and delivery.

11. Bacterial cultures and specimens are handled much like fungi, except that infected plant material should be sent by air mail as soon after collection as possible. Cultures should be grown on an agar medium containing 2% precipitated calcium carbonate in suspension. This will maintain the pH at an acceptable level so the bacteria remain alive and are in good condition when received.

APPENDIX L

Preserving Specimens

The following information was adapted from Johnston and Booth, 1983, and Dhingra and Sinclair, 1995.

I. Wet Specimens

A. Class, Plant Clinic, and Museum Specimens

1. Formaldehyde (40%) in water.
2. A mixture of 25 ml of 40% formaldehyde, 150 ml of 95% ethanol, and 1 liter of water.

B. Green Plant Parts

1. Immerse fresh plant specimens in a boiling mixture of one part of glacial acetic acid, saturated with normal copper acetate crystals, and four parts of water, and keep specimens in the solution until the green color returns and then in 5% formalin; or
2. immerse fresh plant material in a 5% copper sulfate solution for 6–24 h, and then wash in running tap water for 2–3 h. Preserve by immersing in 1 liter of distilled water to which 15 ml of sulfurous acid solution (5–6% SO_2) has been added. Place in sealed containers or mount on a plaque covered with a convex glass sealed with a mixture of tar (300 g), virgin wax (60 g), and one part of anhydrous lanolin.

C. Colored Fruits (Hesler's Preservative)

Place fruits in a solution containing 50 g of $ZnCl_2$, 25 ml of 40% formaldehyde, 25 g of glycerin, and 1 liter of water in sealed containers.

D. Colored Fungi

Keep the specimens in solution A1 or 2 described above.

1. Place fungi whose color is *not* soluble in water in a mixture of 10 g of mercuric acetate, 5 ml of glacial acetic acid, and 1 liter of water or in 25 g of $ZnSO_4$, 10 ml of 40% formaldehyde, and 1 liter of water.
2. Place fungi whose color *is* soluble in water in a mixture of 1 g of mercuric acetate, 10 g of neutral lead arsenate, 10 ml of glacial acetic acid, and 1 liter of 90% alcohol.

II. Dry Specimens

An old general procedure is to press leaves gently in a plant press between sheets of blotter paper, occasionally exposing the specimens to high temperatures. The normal green color fades in a short time. Problems with mite infestations and contaminants are common.

A better method for preserving the green color and checking mite infestations is to place infected leaves in a boiling mixture of one part glacial acetic acid, saturated with copper acetate, and four parts of water until the green color returns. Then wash specimens thoroughly in gently running tap water, stretch carefully over blotter paper, remove excess water, and place between dry blotter papers in a plant press. Change the blotter papers after 24 h, and return the leaves to the plant press. Specimens are ready for display in about a week. They can be placed in paper folders or Riker mounts or between two plastic or glass plates with sealed edges.

A third procedure is to laminate fresh, dry specimens between two clear plastic sheets. Leaves retain their green color indefinitely and can take rough handling.

Selected References for Appendixes

Alexopoulos, C. T., and Beneke, E. S. 1952. Laboratory Manual for Introductory Mycology. Burgess, Minneapolis.

Annette, E., Bazows, H. A., Hausller, W. J., and Shadomy, H. J. 1985. Manual of Clinical Microbiology. 4th ed. American Society for Microbiology, Washington, DC.

Booth, C., ed. 1971. Methods in Microbiology, vol. 4. Academic Press, New York.

Cruickshank, R. 1960. Handbook of Bacteriology: A Guide to the Laboratory Diagnosis and Control of Infection. E. & S. Livingstone, Edinburgh.

Cruickshank, R., ed. 1965. Medical Microbiology: A Guide to the Laboratory Diagnosis and Control of Infection. 11th ed. E. & S. Livingstone, Edinburgh.

Dade, H. A., and Waller, S. 1949. A new technique for mounting fungi. Commonw. Mycol. Inst. Mycol. Pap. 27.

Dhingra, O. D., and Sinclair, J. B. 1985. Basic Plant Pathology Methods. CRC Press, Boca Raton, FL.

Dhingra, O. D., and Sinclair, J. B. 1995. Basic Plant Pathology Methods. 2nd ed. CRC Press, Boca Raton, FL.

Fahy, P. C., and Persley, G. J., eds. 1983. Plant Bacterial Diseases: A Diagnostic Guide. Academic Press, San Diego, CA.

Gerhardt, P., ed. 1981. Manual of Methods for General Bacteriology. American Society for Microbiology, Washington, DC.

Johansen, D. A. 1940. Plant Microtechnique. McGraw-Hill, New York.

Johnson, L. F., and Curl, E. A. 1972. Methods for Research on the Ecology of Soil-borne Plant Pathogens. Burgess, Minneapolis.

Johnston, A., and Booth, C., eds. 1983. Plant Pathologist's Pocketbook. 2nd ed. C.A.B. International, Wallingford, United Kingdom.

Kiraly, Z., Klement, Z., Solymosy, F., and Voros, J. 1970. Methods in Plant Pathology with Special Reference to Breeding for Disease Resistance. Academiae, Budapest.

Lelliott, R. A., and Stead, D. E. 1987. Methods for the Diagnosis of Bacterial Diseases of Plants. Blackwell Scientific, Boston.

Meynell, G. G., and Meynell, E. 1970. Theory and Practice in Experimental Bacteriology. 2nd ed. Cambridge University Press, London.

Munro, A. L. S. 1970. Measurement and control of pH values. In: Methods in Mibrobiology, vol. 2. J. R. Norris and D. W. Ribbons, eds. Academic Press, New York.

Onions, A. H. S., Allsopp, D., and Eggins, H. O. W. 1981. Smith's Introduction to Industrial Mycology. 7th ed. Edward Arnold, London.

Razin, S., and Tully, J., eds. 1983. Methods in Mycoplasmology. Academic Press, San Diego, CA.

Saettler, A. W., Schaad, N. W., and Roth, D. A., eds. 1989. Detection of Bacteria in Seed. American Phytopathological Society, St. Paul, MN.

Sass, J. E. 1958. Botanical Microtechniques. 3rd ed. Iowa State University Press, Ames.

Schaad, N. W., ed. 1988. Laboratory Guide for Identification of Plant Pathogenic Bacteria. 2nd ed. American Phytopathological Society, St. Paul, MN.

Singleton, L. L., Mihail, J. D., and Rush, C. M., eds. 1992. Methods for Research on Soilborne Phytopathogenic Fungi. American Phytopathological Society, St. Paul, MN.

Streets, R. B. 1978. The Diagnosis of Plant Diseases. University of Arizona Press, Tucson.

Sykes, G. 1965. Disinfection and Sterilization. 2nd ed. D. Van Nostrand, New York.

Sykes, G. 1969. Methods and equipment for sterilizaton of laboratory apparatus and media. In: Methods in Microbiology, vol. 1. J. R. Norris and D. W. Ribbons, eds. Academic Press, New York.

Thom, C., and Baker, K. B. 1945. A Manual of the Aspergilli. Williams & Wilkins, Baltimore.

Tuite, J. 1969. Plant Pathology Methods: Fungi and Bacteria. Burgess, Minneapolis.

Tuite, J. 1988. Plant Pathology Methods: Laboratory Exercises. Purdue University, West Lafayette, IN.

Tuite, J., and Cantone, F. 1990. Plant Pathology Methods: Laboratory Exercises. Teacher's Supplement. Purdue University, West Lafayette, IN.

Winston, P. W., and Bates, D. H. 1960. Saturated solutions for control of humidity in biological research. Ecology 41:232.